O, Wow

a novel

Howard Jones

SPOTops BOOKS

Published by **SPOTops BOOKS**
Johnson City, TN
37601-1327 USA

ISBN: 978-0-578-04319-7

Visit **www.spotops.net**
for more information
about SPOT media.

To
the imaginal compass
of
Carl Dockery

~ 1 ~

"Run, Ants! Run! Scatter! Shatter! Splatter! Latter tatter!"

A lithe and slight-framed young man bellows across converging streams of men, women and children. Most are afoot. Many push or pull slapdash conveyances weighed down by haphazard remnants of possessions tossed together on short notice. A few bicycles dart like sparrows among pedestrians indifferently avoiding collision or hindrance. Occasionally someone flashes a fingered rebuke or yells back into his hoarse cant: "Run. Run. Run. You built it . . . now it's falling all around you! Fall! All! Appall! Fol-der-ol! Babel fable treble table, Mabel's able devil shovels revel . . . Ants pant slant rants . . .paint . . . faint . . . taint . . . saint . . . ain't . . . ." Burbled out of both rhyme and breath, he resorts to mutely aping those who cross his baleful eye.

His companion, Kristin Banner, looks wearily back to where he has taken his stand to berate this anonymous throng trudging warily by him into the gaping maw of the Holland Tunnel. On a day such as this, among these distressed evacuees, Peter Dooley is just another oddity to be ignored or escaped.

They are leaving Manhattan Island. Along with countless others they have joined a swelling exodus of Gotham Seaboard: Boston, New York, Philadelphia, Baltimore, DC. Decades of fiscal woe and bureaucratic neglect have deteriorated vital services. Recently hurricane Moira devastated urban infrastructure along the coast, especially on and around Long Island and lower Manhattan. Now teleview suddenly is rampant in threats and speculation of smuggled nuclear devices, of plagues unleashed, of booby-trapped bridges, of numbered days.

Brushing aside a stray lock of copper red hair from green and canny eyes, Kristin waits for exhaustion to end his impromptu exhortations. She slips into refuge of a battered and abandoned security check booth. Scarred, metal-sheathed concrete pylons shield against the onrushing horde of pilgrims weighed down by all they might carry or haul along in impromptu freight. Swarming physics of bouncing fleshly particles, they advance into the darkened gape of the tunnel. Occasional bicyclists dismount before the entrance and prepare to push their wheels along into its straited netherworld.

"We should have gotten bicycles while they were still available," she chides no one in particular.

Dooley does not hear, or pretends not to hear. Or does not care. He stares edgily about. Eyes float as two bright blue corks, rimmed in wary white. They bob across the river of traffic. Each glance tosses blue darts from beneath his haloing wreath of exuberantly silver hair. Physically unimposing, in his mid-twenties, Dooley does not present any daunting obstacle.

Like flotsam he is nudged and buffeted toward where she waits. He stares vaguely at her for a moment, then looks away, as if he cannot quite recall how they met. He is making spitting sounds with tongue inserted against teeth. Nervous tics flicker at his eyes.

Kristin leans back against the booth. Her backpack, legacy of rather distant student days, settles onto a concrete ledge, relieving her shoulders of its burden: a few belongings and provisions she managed to hurriedly toss together as they anguished whether to get out now.

How many protein bars to . . . to wherever? Maybe Harrisburg? What then?

Ancient slatherings of graffiti have been flung across a far wall. *Perps Stalk Twerps That Talk* vies with *New Yuck New Yuck* and *BERSERKERS RULE!* She glances across, but does not really read, an elaborately sprayed ontological credo: *Teleological Hereafter-Enraptured International Sectarian Movements give GO°D-HEAD!*

An unlikely froth of human guises sway and play like bath bubbles dancing at a drain. They pause, turn, then commit to the plunge. Beyond, from tunnel depths quaver echoes of shuffling footsteps, distraught children, shouts and curses of irate adults. Leery young crews, mostly male, saunter along. Gang signs flicker nervous fingers to troll tribal reassurances among reckless crowdings. They weigh favors and follies of diverse neighborhoods and boroughs. Caution watches for hints of companionable relief and blusters against yet unseen jeopardy.

How can this be happening?

A shimmering wash of luminous images, hardly visible in full daylight, plays spook games about Dooley's torso. Real bodies bump against his but the actual bodies ignore, or do not notice, his encircling wraiths of The Three Stooges, Buster Keaton, and Harold Lloyd who taunt and rebut each intrusion into his spectrally squired body space. Flailing weakly, washed dim in the brilliant flare of early autumn sunlight, the capering specters go largely unseen.

Kristin grimaces impatience. She reaches across to curtly switch off his iLink limner. Its glowing holofield of swirling buffoonery dies.

"Will you stop jerking off that crap? We don't have many power paks left."

He ignores her scold to resume his tirade against milling antagonism, shouting toward the shuffling parade, "You think you can escape Al Quidproquo?"

"'Quid pro quo' . . . do you mean, 'Qaeda'?" Kristin puzzles aloud for clarification.

Dooley ignores her question to mount into raging glossolalia, "Jive hive wives strive live knives . . . bust rust just us lust . . . double rubble trouble bubble . . . ."

"Pete, I don't like that. Stop! . . . look at me . . . Pete! Peter! Look at me!" Her voice lunges toward panic to lift above the din. She pounds his back with a fist, as if to save him from some choking obstruction. A lightly fetid breeze swirls smoke and the powdery smell of charred concrete as crowds amble on by, uncaring, unaffected.

Dooley snaps his head back and forth, bewildered to so suddenly return from frenetic reverie. She sees him fight an urge to switch on the limner. Looking back toward roiling columns of smoke rising in the distance he seems to be reconsidering or wondering or simply lost at sea. His head pivots to gaze distractedly at the tunnel maw . . . pondering options?

He watches her for a moment, puzzling through some secret conundrum until his mystified child voice asks, "Banner Ant, where can he be?" and vagrant gaze drifts haplessly back across the shuffling throng to dismiss its clamor, "These Ants are all useless, stupid . . . nasty."

"Banner Ant?" she glances sharp appraisal but finds no suggestion of malice. Her gut coils to counter a threatening flood of emotion. She looks away, thinks for a moment, then forces her voice flat, "He went missing . . . when Bill died." Looking back, she watches worry tease Dooley's expression, then surmises, "Your iLink is trolling for his signal . . . that's what's bugging you, isn't it? It's constantly polling, where's Ant?" Dooley does not answer but looks back into the tunnel adit as if reckoning its gaping prospect. She reminds him, "If you'd leave it switched off, it wouldn't nag you that he's gone. It's up to you."

From a passing euPhone a gravelly voice delivers a snatch of some song recently gone viral,

People in the alley way watch empty streets go by
Thinking of another day when margins melted from the sky.
Hydrants overflow into scattered flower beds,
Diminishing the water given to the dead
In their all-consuming fire
While Lazarus in the choir of Angels
Doesn't really want to watch a rich man in his thirst.

After a long while he removes his belt and reaches for her left hand, "Put this on your arm."

"But what . . .?"

"It'll be dark in there." His voice is mechanical, "This will help keep us from getting separated."

He places the belt, looped through its buckle, over her left wrist. The other end of the impromptu leash he wraps around his right wrist to hold tightly in his fist. Kristin winces uncertainly as the loop tightens, but accepts the bond as together they gauge the best path toward the tunnel adit.

A man and woman are passing by, pushing a supermarket shopping cart loaded with clothing and assorted gear. Bruised boxes and fraying plastic bags strain to retain these few remaining possessions. In the toddler seat a young girl perches happily; her legs clack a persistent rhythm, mimicking cart wheel chatter against the pavement. Recognizing him from teleview, the girl laughs and waves delight across her parents' shoulders, back toward the object of her discovery, "Look, Mama, it's Grasshopper!"

Neither parent pays any attention; they press on into shadows fallen just before the entrance. To Kristin the child's bright joy seems especially damning of their unfolding plight, but Dooley pays no attention to his young fan. Then, again, he has not much noticed fans since the untimely death of his partner, her brother, Bill Banner.

They step out from the toll booth to rejoin the hapless scramble.

"Can't we just hold hands?" Kristin tugs fretfully at the belt leash.

"No. This is better."

No hint why it is better. She peers closely at him as they transition from bright sun into cavern dimness.

He seems to wish she were not with him. Some old hints of resentment have lately resurfaced, percolating out of residues and stresses that once nearly foiled their raucous and sudden celebrity. Peter Dooley and Bill Banner had become comic mascots of a *Federal Emergency Response Administration, Llc. (FERAL)* public relations effort meant to soften public reaction to sudden and even more austere government policies imposed amid worsening economic tribulations occasioned in ongoing wars and besieged financial markets.

Serialized propaganda campaigns aimed at going viral across maxnet video sites sought to redress decades of profligacy. Preaching an ethic of hard work, frugality and stoic forbearance, Ant and Grasshopper recently sprang through teleview spots and occasional, carefully managed public appearances to cheer lead public awareness in revised invocations: America has grown too lavishly cavalier. Save more. Waste less. Shop, always shop, but be smart about it. Be happy, but careful about tomorrow.

Now Kristin is led by Peter Dooley—perhaps acting out some latter antic of Ant and Grasshopper—into turmoil lurching through a dark and cavernous unknown. Their wrists joined by his belt, they enter the tunnel. She tries to look back at faint veils of gray-green smoke rising above the pinnacled skyline but an urgent press of bodies shoves her forward into what seems an eternally receding hole. Pavement just before her falls nearly invisible as light quickly dies. Her feet have to find their own way through a clutter of dropped objects, disowned clothing, painfully encountered elbows and errantly tripping feet.

Stumbling over what may have been a corpse shutters her resolve. Slowly her eyes adjust to such scant illumination as is availed in a few fixtures still working overhead. She tries to suppress a new gasp of fear by softly intoning a song fragment learned among long ago campfire rounds of her girlhood,

Sing with me, I'll sing with you,
And so we shall sing together,
So we shall sing together.

Dooley leads along the tunnel wall to her right. They are shoved violently against it and occasionally rebuffed just as roughly by cursing figures loitering in the darkness. She steps on someone's foot and a powerful blow sends her reeling against Dooley. Someone in the crowd pushes back and both are buffeted back against concrete wall. For a moment they are trapped, unable to find an opening in the crush of bodies coursing by. She recalls once being caught by a rip tide while swimming in the ocean, of being sucked into a swirling blackness of airless desperation. Kristin feels a whimper rise spontaneously in her throat. "Peter, let's go back. Please."

He does not answer.

Except for occasional yells and angry grunts, amid plaintive cries of children, the darkness is oddly quiet. An abrading rhythm of feet striding pavement seems more like a pulsing fog than sound. Noises echo precariously in underground twilight to disperse clues of origins and tease any grasp of dimension or orientation. Inexorable press lends but one sensible direction, onward: its rhythm and pace of irresistible movement squelches any resolve to go back. Even if he gives in to her plea, they are unable to travel any way but forward.

At times, stumbling among foul encounters and aggravated complaints, she thinks she can see figures moving across an immense stage of secret illumination. At other times she is trapped in a packet smallness of a closet peopled with fellow aliens. Her eyes open and close involuntarily. They care not whether such darkness is inward or without. The same fear permeates all.

From somewhere in the distance a thin pencil of red light erupts to sweep the dimness. It searches briefly, hovers to paint a tiny dot of red on her breast bone, just below her chin, and then disappears. She remembers a director once using a tiny laser pointer as baton to conduct her movements in a commercial she was shooting. Who was pointing at her now? Why?

Again, the sliver of red light tags her chest.

From above, on an elevated maintenance catwalk, a hand brushes her hair, takes an ample handful and pulls up hard, snatching her gasping painfully onto her tiptoes. Another groping pair grabs at her shoulders, then reaches under her armpits to lift her from the pavement. Male voices are above, grunting and hurrying the effort. More grunts, as more hands join the abduction, moving across her body in a nightmare of rampant violation. She feels herself abruptly lifted upward and flails ineffectually against them. In panic she rises unwilling toward coarsely whispered complaints of male voices now shouting.

"Peter!" she shrieks. A hand chokes at her throat while bruising fingers dig into her armpits. Others take hold of wads of her clothing. Another hand covers her mouth, crushing her lips tightly to keep her from biting. The belt leash scrapes and painfully binds her wrist. One of the hands slides along her arm to discover the strap. It probes and yanks angrily at the restraint. Excruciating spasms lance from her wrist up her arm. Her shoulder threatens dislocation. Someone above mutters, hissing exasperation, "Pull! She's tied to somebody—give me your damn knife!"

"Ants!" Dooley bellows accusation, " Ants! I told you! I knew it would . . . get away, you goddamned bastards! Get away!"

A bright orange flash erupts about them as his iLink limner gathers energy from emotional eruption of his rage, translates it via internalized media archives to slam a stupendously rendered phantom fireball into the now sharply illumined void. It boils seething and fiery plumes up onto the walkway to envelope now startled marauders, accompanied by a roaring iLink surround sound that explodes cinematically across ruptured darkness to reverberate within the tunnel echo chamber. Uproar and cacophony build violently from harnessed energies of a suddenly screaming crowd bolting in seizures of terror that culminate their worst fears of the times.

In a suddenly spuming pleroma of light, accompanied by thunderously cascading soundtrack of choral voices singing *Carmina Burana, O' Fortuna*, she sees five young men frozen as jacklighted deer, straining to free her from the hindering belt leash and panicked by media chaos unleashed. Another man wears military-style light amplifying goggles. Standing off to the side, he has been coordinating and directing their efforts with a laser pointer.

Also vivid in the still blossoming fireball image, Dooley's eyes flare panicked damnation. Boiling up from his fervid psychic depths, a hideously imagined raptor stalks amid fiery wraiths of hell's threat unleashed. The stunned abductors, now cowering as prey, in unison forget abduction and let her go.

Kristin falls awkwardly back onto the pavement, pulling Dooley down on top of her. Again the limner erupts, streaking plumes of 3D-imaged fire from which rocket-driven darts explode to drive the gang in frantic retreat back toward the tunnel opening. Out of the subsiding fireball pursuers emerge to chase after now-comic miscreants suddenly scrambling for hasty exit: Buster Keaton, Harold Lloyd, Laurel and Hardy join the Keystone Cops in amazingly agile headlong pursuit down the catwalk. And they seem really pissed!

"A hand just reached down . . . and grabbed me," Kristin sobs, "then another one picked me up. And another."

A wide swath of vacant pavement now surrounds them. They manage to roll over onto their feet and stagger to huddle up against the wall. For the first time, no one shoves or pummels them. A surround of saving, empty twilight remains open, yet still is edged in alarm and cries of protest that incite Dooley's ongoing rant against abominable Ants.

Her ears still ringing from the explosive uproar, the men's startled outcries, and Dooley's brawling screech, Kristin cannot make out what he is saying. It has something to do with "chance." He shouts it over and over again in a spuming tirade that abuses her still belted wrist as he gesticulates wildly in the dark.

"What?" Kristin pleads through competing floods of anger and fear.

"Ants . . . chance against Ants!" His Grasshopper voice is sharp and bitter. Harold Lloyd and Buster Keaton swagger back toward them ahead of a strutting troupe of Keystone cops, returned from avid chase. The slapstick chimeras all transmogrify into large cartoon Ants bearing stolen picnic booty. Grasshopper drags Kristin from the wall and begins to trot forward along the tunnel wall. Somehow she stays on her feet and seems to gain hope in their sudden sprint.

A wide margin still separates them from the crowd, now warily fleeing threats of further explosions while large cartoon Ants swarm past and surge even more insistently to push on ahead, parting the mass of human evacuees trying to get out of the way of this graphic nightmare rampant among them.

The iLink ultracapacitor, depleted to its threshold minimum power level, automatically shuts down the limner. Platooned Ant specters fade suddenly and Grasshopper's diatribe eases into rambling exhortations against any insult, real or imagined. Against what complaints he rails Kristin is uncertain, but it seems that Ants are to blame for misfortunes lately visited upon them all. Still yanked about by the leash during his wild gesticulations, she topples along to keep up as best she can. They move forward, it seems, only for lack of something better to do.

Slowly, rhythmic pulsing builds again, ushering them along as hysteria subsides and the crowd gradually closes ranks in surly, buzzing aftermath.

Eventually a dimness ahead rouses into dusty radiance. She can just make out the profile of his face fringed in faint glimmering. Beyond, a thickly massed throng of heads undulates slowly up and down like lily pads on a pond disturbed in the wake of a passing boat. Then she reckons that it is not the surrounding crowd swaying up and down. It's him! His head bobs and rocks slowly as he surges through an exaggerated tidal rising, falling, rising, falling. She feels his arm, still strapped to hers, pull her along in pacing undulation.

"Pete, what are you doing?" she snaps. Annoyance strikes fear to spark exasperation. They move quickly toward gathering declarations of external light.

"Marching!" Grasshopper announces proudly, "just like in your song." He glances down at her feet. His frown lightly rebukes her poor showing of cadence.

"Please don't do that," she begs.

Her right hand tugs at the buckle and manages to loosen the leather loop still binding her wrist. Rolling it into a tight spiral she tries to return the belt to his jacket pocket. But he takes it and begins threading it casually back through the loops of his trousers.

"I'll walk like you, then," Grasshopper teases. He glances again at her feet, gives a brisk little hop to synchronize their paces and strives up onto tiptoes where he minces along in caricature of her once-signature stride down haute-couture runways.

"For we are marching to Pretoria, Pretoria, Pretori," he screeches in a superciliously continental, quavering countertenor falsetto. His torso swishes through turns calculated to usher buyers into a new season of fashion-mongering.

"My God, he's crazy!" a woman gasps from behind.

"Shut up, Hazel," a sour male voice clips.

"Yeah, Hazel, crazy ways'll faze a maze of hazy days," Grasshopper chides, then resumes his lusty serenade,

For we are marching to Pretoria, Pretoria, Pretoria.
We are marching to Pretoria. Pretoria, hooray!
Haze with me, I'll craze with you
So we all plays together,
And we all lays together,
So let's embraze together.

Grasshopper jumps and capers about. He berates a few reluctant hikers nearby to release their dismay into the rhythms of his marching song. Other voices are recruited, finally even those of Hazel and her companion.

In the void of Dooley's departed intent, Grasshopper is intrigued by a suddenly profound notion. It promises to transform this heretofore senseless jumble of noise and movement into terms he can grasp. It renders scattered intimations of hope, pleasure, joy, rage, fear and sadness into an intuitively evident maxim.

He proclaims it loudly for all to hear, "For lack of anything better to do—just do anything!"

He cackles inspired delight, "Its meaning is the doing. It's so maaahvelous! Simply maaaaahvelous! Explains everything . . . in its own good time. Miserable Ants are wishing—but Grasshopper's swishing. They're hoping; I'm hoppening! Hoppening everywhere—all around . . . always NOW." he exclaims in a trilling squeal of exultation to celebrate new dawn manifesting at the far tunnel exit. "Here and now, hoppening!"

As the laggardly dispersed chorus launches through another round of singing, Grasshopper exults in the elegant simplicity of it all. He cackles praise of a new tenor he hears joining his choral glee. He has no idea from whence they have come, nor where they may be headed, nor even why Ant's Sister refuses to sing. Or why she vacillates between staring coldly ahead, looking dolefully at him and crying as they trudge out toward the old New Jersey Turnpike and on to access to Interstate-78 to Harrisburg.

Whatever her reasons for not joining in, he relishes their outing. As they emerge from the tunnel Grasshopper gratefully breathes deeply, warm in the radiance of a lovely Indian Summer Day, sighing happily, "always now."

Gradually the throng eases into a more leisurely pace as the avenue widens out beyond the ramp way bottlenecks.

Grasshopper is only mildly curious at the grim wariness seen among his fellow travelers. So many pragmatic details and concrete features of industrial landscape to drink in, to opine loudly about. He is sorry the singing has ceased. Occasionally he tries to kick it up again by whistling a cheery little melody. But the crowd seems to have fallen back into their earlier surliness. Oh, well . . . Ants.

He overhears another remark about someone being insane. Looking intently about, he peers closely into several faces, but is unable to identify the poor unfortunate.

While he has not found the lunatic traveling among them, he does like the sound of the accusation. "Looooo-naaaa-tick!" He rolls it about on his tongue, feeling the extruding puissance of its phonemes flattening up against the roof of his mouth. He squeezes out hamburger patties of squeaky glee into sibilant whispers, "Innnnnn-saaaaannnne."

Ant's Sister grows even more distant, falling stubbornly silent as they depart the tunnel. Finally she sighs, rubs her chafed wrist pensively and fastens her ripped blouse as best she may by its remaining buttons. But her silence is ruthless. She totally ignores him.

Grasshopper suspects she is playing a little game with him, with them all. She is like that, a melodrama of shock and trauma because she doesn't want to whistle. Or doesn't know how. He clucks absently at the transparency of her mute ploy. She only glances at him dumbly and picks up her pace.

Skipping along to keep up, Grasshopper works through the jive classic *Three Little Fishes* gleaned from the iLink's local memory cache of stored resources built in producing Ant and Grasshopper vignettes but is unable to decide what the little fishes did once they "famm and famm all over the dam" since the episode never got that far. The iLink does not detect a maxnet signal and so cannot retrieve song lyrics online. The little fishes fimm off into rescinded limbos of flash memory as he hustles to remain abreast of Ant's Sister. But she still refuses to acknowledge him.

Off to the right, on a gentle incline of an overgrown embankment that sweeps out and downward to meet another plunging slope amid ponderous remains of former industrial prowess, Grasshopper sees a number of soldiers sprawled in a shallow hollow, languishing in warm sunlight. Most of them carry, or have lying nearby, shepherd staves: long sturdy, but flexible, poles made of synthetic fiber polymers, at one end of which is a variable high-voltage prod capable of stunning a horse, if need be. A well-trained platoon of so-equipped crowd handlers can dissuade and direct even the most unruly mob.

Hovering, then darting suddenly to assume new vantage above and along the evacuation route, are several remotely controlled usher drones bearing teleview cameras that permit monitoring the progress of the city's evacuation. Usher drones also are equipped with crowd control measures. High-voltage ion-discharge pellets can ionize a directed channel of air to permit unleashing into any unruly target a high-voltage bolt equally as vicious as delivered by the manually wielded shepherd staves. Usher drones also may carry a range of behavioral gases to invoke a spectrum variety of nausea, fear, or good cheer in a malingering crowd. Pepper guns can embed urgently burning pellets of capsicum to distract and deter mischief-makers or blind and distract entire groups with red clouds of capsicum spray.

Nearby another pair of guardsmen tinker with a large automatic rifle. Occasionally one glances up to disinterestedly survey the scattered passing of multitudes. Grasshopper waves cheerily and speculates about a picnic inviting armed Ants. "Maybe they're Soldier Ants," he speculates aloud, "with enough heavy artillery to take over the picnic . . . r e v o l u u u u t i o n!"

Kristin pretends not to hear him. She neither glances toward the soldiers nor acknowledges Dooley's enthusiasm. She hates his berserk manner. She hates that she has no idea where they are going nor how they will get there. She hates that Ira Crown can organize soldiers to keep them on the freeway, but does not provide sufficient transport to move them en masse to Relocation Centers. She hates that Dooley keeps calling her Sister. She hates that she is certain of very little but what is in her backpack and already it seems both too much to carry and too little to sustain. She hates that her feet are aching and that they still have a very long way to go.

Up ahead, a narrow valley between two ramps opens into a plain of their separation. Some of the passing crowd drift toward the inviting green, tempted by dilapidated signs advertising a nearby shopping mall.

A minicopter arises abruptly from behind a hillock. It hovers head-high a hundred paces beyond to block the wayward group trying to leave the highway. A swarm of usher drones are dispatched from behind the same rise. The smaller machines dart and flit about, dispersing quickly to cut off intended paths of escape. They waft about like oversized, cluster-laden dragonflies inspired by souls of Border Collies. The crowd recoils from threats of airborne high-voltage catapults and pepper sprays. An acrid sample of nausea gas is pushed across the group by the rumbling wash of distant minicopter rotors.

A public address speaker pummels the air with an amplified midwestern accent that orders the crowd back onto the pavement and shoulders of the highway. Several soldiers farther off in the valley nonchalantly watch the aerial to ground confrontation. They do not move, but stretch to display staves, heavy truncheons or plastic shields in case they may be ordered to engage and repel a vagrant few back onto the official evacuation route.

Grasshopper is entranced. He pauses as the vulturously hovering black machine beats the air with its rotors. It seems to dance and skim lightly on choreographed waftings that are answered in bouncing weeds and soldiers deploying beneath its sidling drift.

The usher drones dart and play to befuddle and intimidate vagrant defectors who look defiantly up, waving fists and middle-fingered salutes amid shouted protests of indignation. They are slow to give up reaching the advertised mall. Farther back along the highway a gathering audience pauses to watch the outcome of the confrontation. A collective sigh hopes for delivery from pavement already grown dreary. A shopping center!

The faint staccato of a machine gun on board the minicopter burps puffs of powder smoke that quickly dissipate in the wash of rotor blades. A line is drawn through tracing puffs of dust in the turf beyond the defiant band, which pauses to reconsider. Slowly they pull back from the dotted boundary and disconsolately return to the freeway.

The distant soldiers relax back into their casual postures and scatter along the grassy slope. Among the chastened evacuees more than a few loudly damn the President, Ira Crown, as a son of a bitch who ought to be shot.

Grasshopper muses aloud that the cannon fire sounded too faint. Probably they are using rubber bullets and puffer cartridges. "Use the real thing!," he urges the departing minicopter, "Don't let a bunch of spoilsports ruin the parade!"

Ant's Sister trudges on without breaking pace. She has ignored the entire confrontation and now refuses to talk about it or anything else. She is so like her brother, Grasshopper reminds himself as he again catches up to her. Narrow and stubborn, nothing is of interest unless it gets you somewhere. You'd think she knows where she is going, so intent is she getting there. For her it's the goal, not the journey. But maybe she does know! Sometimes Ants carry a lot of astonishing things. Maybe she does, too.

"Where're we going, Sis?" he asks lightly, trying to seem casually disinterested. Her lips flinch slightly, still refusing to chat with him.

"Hey, Sis. You're in charge. Where're we going?" he repeats a little louder. As he reaches for her arm he is startled by the vehemence of her jerking away to pull back beyond his reach.

"Leave me alone, Pete!" she snarls, "For God's sake, just shut up and fucking leave me alone!" Clenched jaws harden her face. A glaring threat of jade retribution glints in her eyes. "I don't know what you're doing but I don't like it! Cut it out and get away from me!" She reaches arms widely out and about, dismissing puzzled expressions on people still wary of the peculiar duo, avoiding both of them, "Everybody is going nowhere," she hisses, "Maybe you haven't noticed."

Grasshopper beams back at her, bolstered in intuition that cheery disposition is the most apt antidote to such impatience. "Don't worry," he chuckles as he holds up the frozen display of his iLink for her inspection, "they'll make it okay. Time's awaiting."

Kristin looks back angrily at him, without tears but as if there should be some. The flush of anger slowly relents. She does not seem to mind when finally he reaches out and puts his arm around her shoulder as if they are schoolyard chums. They stand for a while longer, over by the side of the roadbed. Then she moves abruptly from his casual embrace to rejoin the surging tide of refugees. Grasshopper watches her take a few steps. Then she pauses and turns to look searchingly back at his impossibly happy face.

"Come on," she relents, resigned to this dawning age of abject senselessness, "we're all we have left."

~ 2 ~

Kristin and Dooley cross the Hackensack and Passaic rivers and hike on over into Newark. They have replaced the depleted iLink powerpak with a fresh one. For a while it picks up a maxnet signal that enables lurching access to some servers, others seemingly crippled by outages and overloads. Kristin tries phoning the Ant and Grasshopper project's main production office near Cambridge, Massachusetts. A recorded message informs that the facilities are closed until further notice. She tries text messages, but no one responds. Finally she sends email to a number of contacts, one directed to Eric Cornrath's personal server, to explain briefly that they have been unexpectedly separated from car and driver and have joined foot traffic out of Manhattan, heading toward Harrisburg.

Web news concerning Amtrak is discouraging; Penn Station in Newark is swamped. There is a backlog of overbooked spaces. Entire rosters of trains have been canceled and there are only ambiguous intimations of future service. At Kristin's urging they continue on foot along I-78 toward Harrisburg.

They are not alone. Scattered clusters, many families with children, trudge generally outward from Gotham Seaboard's core, seeking hints of what to do, where to go, and how to keep on keeping on.

Overhead, helicopters occasionally rumble across the sky, oblivious to vicissitudes down upon the earth. Farther above them occasional military heavy-lift skypods drift lazily across azure skies, transporting supplies and personnel, their drably camouflaged tortoise-like profile seeming somewhat incongruous in their grace of buoyant progress. Smaller gray and silver-accent skypods drift closer to earth, monitoring suspicious activities on terra firma among vagrant features of depleted industry and vacant commerce. From time to time an usher drone, dispatched from its mother skypod, appears just overhead and off to the side to gain closer view, scans a sweep of the highway and its occupants by airborne camera, then zips away. Sometimes people yell questions that seem mere rhetorical exercises toward retreating drones, which ignore pleas and curses alike.

Once in a while bicyclists whiz by, sometimes in clusters, but usually no more than one or two. After a while pedestrians tire of complaining and yelling at them. Once or twice the crowd moves to the shoulders to let pass Army vehicles or FERAL resource vans. One gaudily painted van advertises *Uncle Don's Power Protein Bar* and *Uncle Don's WonderWet Water* as it bounces along shabby pavements of the aged Interstate system, long since fallen into desultory repair. Kristin watches the van bang away into the distance as she steps across a broken slab and resigns herself to a long and arduous trek.

Where Pulaski Skyway joins I-78 at last they find some indication of official presence: a posted sign declares I-78 to be an Authorized Evacuation Route. Sitting back approximately two hundred yards from the highway channel they spy a spacious tent set up near the access ramp. A large Red Cross logo beckons. A larger sign, pegged above it, advertises *GloBanc Services: Credit <=> Worth*. Above the tent entrance a banner identifies *FERAL Civil Guard Recruiting Post #37*.

They descend in hopes of finding water, food, and relief from pervasive dystopia. Inside the tent, Kristin wends her way through scattered clumps of fellow exiles who obviously are avoiding contact with her. In their eyes she has become suspiciously alien, tainted in the sullen aftermath of Grasshopper's tirades.

At a booth situated near stacked pallets of bottled water, she engages a brightly smiling young woman who wears a FERAL name tag: Francine. As Francine monitors the restless queue of people getting bottles of water she chit-chats about her job, how she came to be here and responsibilities she has shouldered "in this time of national peril". Mostly, she explains, her assignment is to gather signatures onto touchscreen forms and make sure people don't get too greedy about the water. She explains that the tent has been set up there by the Army to provide a Red Cross contact station and to support recruiting for CGC, the recently chartered FERAL Civil Guard Corps.

The Red Cross, Francine explains, just uses the recruiting post as a watering station and contact point for evacuees needing assistance. Many more like it are scattered along all major routes out of Gotham Seaboard. In conjunction with an overall FERAL organizing effort, Francine coordinates both volunteer and CGC recruitment. She signs up and transports to the local screening center anyone interested in "helping defend the nation against gathering perils". Transportation, meals and form processing are handled by GloBanc Services. GloBanc, she says in rote formula, also offers high-risk deferred payment credit cards to displaced persons in need of financial assistance. "Would you like to apply for a GloBanc card?" Francine smiles unconvincingly.

"Transport? To where?" Kristin asks the young woman; her hand impatiently declines the offered line of credit.

"Over to Watchung Reservation. To the FERAL staging area and Civil Guard processing center. And there's an Army field hospital there, too."

"Can we catch a ride there?" Kristin searches her memory trying to recall in what direction the Reservation lies, "I mean it's on the way to Harrisburg, right?"

"Well," says Francine, "it is in that direction, but we don't provide transport to Harrisburg, or even to Watchung. We just sign-up and haul prospective Civil Guard recruits. And maybe call FERAL if someone needs medical attention."

Francine starts and stares dubiously beyond Kristin where an improbably silver-haired, blue-eyed, and now luminously swathed apparition advances toward her, his tongue lascivious in serpentine writhing, impudent eyebrows wiggling quizzically in practiced Groucho Marx mannerisms.

"How 'bout you, me, Palm Springs?" Grasshopper urges in huskily sensuous implication. The iLink hums a sonorous sultan of smooth soul, Barry White, medley. Projected holofield evocations of golden sunsets suddenly flare among distant palmetto groves.

Taken aback, Francine stares blankly back at him, "Palm Springs?"

"Yeah. You know . . . your palms spring into my pants."

"Pete, you fool . . . stop it. Such a wise guy," Kristin snaps apologetically.

"Fool? Wise? Geez, make up your mind!"

"Hell, I want to join the Army," Kristin announces, turning abruptly to face Francine, "So does my dimwit buddy here. What do we have to do to get to Watchung?"

"Well," Francine counters Grasshopper's awakened erotic overtures by inflecting a serious mindfulness into her voice, "It's not really for the Army. It's the 'Civil Guard Corps'. Mostly they're looking for civilians with skills needed in this time of our nation's perils. You have to commit to participate in recruiting classes and go through initial screening. It's sort of like agreeing to attend a real estate sales pitch in exchange for a vacation or a gift bag or something. You have to sign this agreement to get over to the staging center. They won't let you just walk over there."

Grasshopper's disappointment at being rebuked builds into a full snit, muttering and lurching through surreally ominous griping, ". . . cast ratings perils be for swine . . . ," as the iLink distresses the air with imaginal vultures that swoop low to circle menacingly just beneath draped billows of tent fabric. On the ground between his feet, three little light-limned cute and cuddly cartoon piglets huddle, quaking in the shelter of his legs as they glance nervously up against voracious overhead threats. Over by the pallets of bottled water a rakish Big Bad Wolf, clad in ancient zoot suit, sporting a foppishly extravagant watch chain, is just barely visible, lurking as he watches Little Red Riding Hood, elegant in sleekly sheathing cocktail dress, join the water queue.

These apparitions quickly fade as Grasshopper's prurient scrutiny is diverted to engage an uncommonly attractive young blonde woman just now approaching the table next to Francine. Striding in feline grace and exuding an assurance that arouses Grasshopper's courtly display into that of a quixotic peacock, he blesses this newly arrived honey-golden temptress and seeks to indulge her fascination with plying veils of overtly and voluptuously sensual cinema. The young woman glances into his dazzling erotomania and laughs lightly, as she glances down at the limning iLink dangling from his belt like a flashing codpiece, her azure eyes equal to the indulgent radiance in his, "That's quite a light bulb you've got there, honey."

Francine, on the other hand, is decidedly taken aback. She starts to comment on this suddenly intimidating firestorm of sensually pictorial pique igniting into lascivious conflagration, but bewilderment overwhelms and she simply lets it go. She closes off the entire incident from her awareness. She has long since learned the wisdom of not letting happenstance dishevel neat priorities. Neither obnoxious overtures nor bizarre apparitions need affect her seeing to her duties. Taking a deeply sustaining breath she turns the touchscreen so Kristin can scan paragraphs of densely packed text above input boxes identified for *Signature, ID#* and *Date*.

Kristin takes the offered stylus, while keeping a wary eye on Pete who now has the blond woman huskily laughing, "Hey, I can handle anything that gets us closer to Harrisburg."

"Well, just sign right there. And put your natid there."

"Natid?"

"National ID. You have to have a valid natid. They won't process illegals."

"No problem," Kristin groans as she notices the fanfare he is directing toward the blond woman, "Here, Pete—Pete, stop fooling around and pay attention. Besides, you're wasting power. We're on the verge of no more powerpaks . . . here, sign this . . . and look into that port."

She shoves the stylus into Grasshopper's hand and watches him absently scribble a name, only vaguely legible, into the indicated space, then peer into the indicated scan port. Preoccupied with switching off his limner, Kristin does not notice that his signature might more easily be read "Grasshopper" than "Peter Dooley". She retrieves the stylus, signs her own name, glances into the green wash of retinal scan laser light and prints both their IDs . The touchscreen resolves her hastily drawn characters into neatly printed text numerals and blinks a lime green label next to each signature, *Natid Validated by Retinal Scan*.

She knows both Bill's and Pete's IDs better than ever they have. She negotiated their deal with the Amberson Agency. Since then she has taken over as manager, or, more accurately, as she often complains to the unruly duo, their would-be House Mother. Handling obligations, overseeing schedules, and ensuring compliance with Cornrath's strictures regarding the pair's public behavior has earned her their sarcastic sobriquet: 'Big Sis'.

Now she returns the touchscreen to Francine, who announces they are "good to go".

Francine looks around vacantly, as if she has misplaced something, but mostly she is still determinedly ignoring Grasshopper now sauntering back toward the entrance. He waves solicitously toward the young blond woman now departing, but still giggling back into his raucous banter. Francine reaches into a box and removes two strips of orange imprinted *CGC Prospect*. "These are valid for two days. By then you'll either have joined the Guard or decided to continue on. But while on the Reservation you'll have to wear these on your left wrist, fully visible. Otherwise they'll put you back onto an evacuation route."

"Evacuation route?" Kristin wonders aloud.

"Back on the Interstate highway. The official route out of Seaboard, I-78 takes you to the Relocation Center at Harrisburg."

"I was thinking maybe we'd get off the Interstate? It's so bleak and separate from everything. Who cares where we walk?"

"Listen," Francine's voice drops into an urgent tone, "It's really dangerous out in the back country. Law enforcement doesn't really exist out there anymore. There's too much going on. All kinds of . . . the residents, the ones who haven't left yet, are very suspicious of anyone they haven't known for a very long time. Especially roamers. If you get into trouble there won't be anyone to help you. And lowlife out there may kill you just for fun. Or make you wish you were dead."

Francine directs them behind the tent to a parked bus where several people have already boarded, "Don't wander away from official camps and routes," she reminds them, then marshals an offering of parting good cheer, "Stay safe. Good luck. And thanks for your interest. Have a nice day!"

Aboard the bus, Kristin and Dooley plop down at the front onto a side bench just behind and to the right of the driver seat. Kristin watches other passengers board. To her left is seated the same young blond woman she earlier saw talking with Dooley. About her slender neck is a pendant, ecSTACY, on a gold chain. The woman is accompanied by a somewhat older male companion who nods back graciously and acknowledges her flicker of a smile. Their manner, as well as clothing and gear, hint at prosperity long accustomed to leisure.

Kristin notes that quite a few prospective "recruits" on board are young parents accompanied by children flailing through various moods: fretful, happy, tired, noisy, drowsy. But all youthful attention quickly converges on the front aisle space when the iLink limner switches on. Animations play from programmed background routines working through teleview archives of *Aesop's Fables* and *Grimm's Fairy Tales*. Grasshopper fluidly leads virtual avatars through improvised variants of ancient roles, miming and interacting with them to mischievously usurp familiar plots, energized by young trills of delight. He feeds on their laughter. He is a child among children.

"I hope the CGC aspires to a younger force," she laughs aloud.

"Who is that, Jim?" the young blonde woman seated at Kristin's left asks her companion, "How does he do that?"

"I have no idea how . . . aha . . . that's Grasshopper!" the man suddenly recognizes a celebrity, "That's who . . . but I have no idea how. Especially on an old bus."

"You want the long spiel or the short?" Kristin volunteers, "It comes out of *BANG*. That stands for the converging technologies: bits, atoms, neurons and genes. Specifically, optronic media processors and nanobiotech . . . developed for military use in war-gaming, battle control and the like. But with him it takes media production to a whole new level. Peter and my brother Bill used it to generate the Ant and Grasshopper media spots you've seen."

"But how?" the two chorus.

"This particular implementation is called *ECHOES*. That stands for *Ethereal Computer-Human Onto-Experiential Synthesis*," Kristin laughs as a panic of tech-overload flits across their two faces, "Billions of nanobiobots—nbots. Little speck-like biomachines customized just for him. They're really, really tiny."

The couple stare back, uncertainly.

"Eighty-thousand nbots can fit in the width of a fine hair. Anyhow, a dose of nbots were cultured specifically for his tissue type and injected, just like a transfusion. Along with it another dose of so-called replicators and a third dose of custom referents interact like billions of little automated factories to multiply the nbots and accommodate them to their function in his nervous system. Then, when everything is set, the replicators are switched off and expelled from the body."

"Why?" asks the blonde woman.

"The nbots disperse throughout his body, homing in on neurons. Once in place each one just sits there and constantly signals by low-level radio whatever state its host neuron happens to go into. Other times they trigger a response state in the host neuron. There are billions and billions of them broadcasting and receiving at once."

"Broadcasting to whom? To what? Why?" the man asks as he watches Grasshopper's elaborately mimed antics within the surrounding theatrical pleroma of glowing imagery and sound.

She points toward a small fabric sheathed object, faintly reminiscent of codpieces sported in Renaissance costumes, worn on his belt, just below his navel. "His iLink—that little belly pack hanging from his belt is actually an optical computer—orchestrates all the signals from the nbots and renders them into a virtual entity—an nbTwin, they call it." Kristin elaborates, "The nbTwin is a mirror-image, a self-twin, an adaptive media avatar, of the person's nervous system activity."

"But how does he do it, make the little animals, and trees, and landscapes?" the woman is perplexed.

Kristin wrestles through her own shaky grasp of those more advanced technology layers gleaned from hearing Bill, Pete and technicians talk during early experimentation. "It takes a while to adjust to it. My brother Bill used to compare it to sharing space with a ghost.

"It was so disorienting at first that he had to re-learn everything 'with an extra person in the suit', as he said. Even when the iLink isn't turned on he says it's distracting. Walking. Thinking. Talking. Every kind of experience and sensation. And there's a little feedback echo that can be disconcerting.

"But eventually, after a lot of rehearsals and tuning and recalibration, the nbTwin becomes so familiar that the echo just dissipates into subtle tingling. They each learned to live in a sentient space shared with their nbTwin. It mirrors their neural processes and acts like an in-between for all sorts of digital media hookups. Kind of a 'front end' for their own self-presence. Once they got used to it, they feel out-of-sorts when it isn't switched on. Peter runs through power paks like you wouldn't believe!

"Production machines synch with the iLink across maxnet so there are incredible reserves of image and sound resources available, as well as computing routines. It all gets controlled through an associative spray of memory images, thinking and visualizing into the sentience shared with the nbTwin. The psychic spray of events is mirrored in the nbTwin, so there's no sensible distinction between his memories and thoughts and those synched into the iLink pipes, except . . . except the sheer enormity . . . and complexity . . . of what's available. It's a hyperkinetic stream of imagining . . . way beyond anything else. Active scenarios percolate up from the virtual generator, guided by his thoughts and reactions, and are coordinated through the iLink. New process streams are downloaded as bandwidth allows."

"What happens if he gets into something no one thought to program into it?" the man wonders aloud.

"Adaptive AI is built into the iLink and can handle just about anything. It tries to make sense of the senses but if a scenario is totally new it has to be worked into a story frame. It can get pretty bizarre, like dreaming on steroids. It constantly monitors home servers to download service updates or find additional resources. But mostly they use what's built-in or streamed from archives. I'm told there really are only about thirty narrative story lines, anyhow, so the built-in AI can handle most situations.

"The limner that projects images from the nbTwin's stream is for rehearsal. But things really get humming when the iLink synchs into a fully deployed rendering platform, like they use for media production. They don't use the limner for that. Along with screenwriters and special effects teams, musicians, it's 'plug and play' whatever their interacting imaginings concoct together. Bill and Pete could work together with all sorts of artists in different locations when they had sufficient bandwidth available."

"Amazing! I've never heard nor seen anything like it. Does it run on batteries?" the man asks.

"No, not really. The limner is the biggest power drain. If he doesn't switch that on, the iLink will sustain itself on just the energy of ambient light and acoustics. There's a photovoltaic piezoelectric sheath on the surface of its case that captures and converts the energy of both light and mechanical vibrations. It can sustain itself just bouncing around in the sidepack when he walks. You could recharge it just by putting it in a box and shaking it or setting it in bright sunlight. But of course, shaking it is probably not a great idea" she laughs. "Power gets stored in a gizmo kind of like a battery, an ultracapacitor. This model also has a sugar fuel cell to generate additional power used by the limner. Specially engineered bacteria feed on sugar syrup in the powerpak and convert it into electricity,"

"Where do you get this stuff? I've never seen nor heard of anything like it."

"From Eric Cornrath, the executive producer of the Ant and Grasshopper campaign. He's an advanced media systems expert. He also works with the Pentagon and probably the intelligence community; he's a Colonel in the Army Reserve, or something. Anyhow, he consults with them on media projects, for both publicity, war gaming and even combat applications," Kristin replies, "I guess that gets him access to use stuff for official purposes, such as making Ant and Grasshopper episodes. It gets really crazy. Once Bill and Pete agreed to the project it got so frantic and insane I didn't know if we would ever get through it all. So new. And strange."

"But why do all of that?" the woman, "They're just making teleview clips. Why go to all that expense and bother?"

Kristin pauses to search for a suitable simile. "It's like having an optronic studio scattered all over, creative people everywhere, anyplace on the globe, inserting their respective contributions at the proper times and places. This way of working is so much more powerful. You can do stuff that would be impossible any other way, and much faster. In some respects it all pays for itself in increased productivity. After staging media feeds and hookups from remote locations they rip through scripts with only a little post-production touchup occasionally needed. And that's usually because someone has changed their mind about an earlier choice.

"But it can get to be too much temptation to play, too. Bill and Peter used to ad lib through and beyond what was in scripts. Cornrath would get so pissed when he couldn't use a take. But they didn't care, they'd take off wherever their whim led them. A lot of stuff never gets shown. They could get really obscene and libelous—fast—unless somebody kept them on message." Kristin points to herself, "That became my job. But someday somebody will make a fortune by releasing those oprocks. Those bits were so damn funny. But they're too scandalous for right now!"

The woman frowns in consternation of too much to take in, "Will he always be like this? What if he wants to do something else?"

"No. Now that the project is finishing up they'll deactivate the nbTwin back in the lab by broadcasting a signal to switch off the nbots. Eventually the nbots will be flushed out by normal body processes. We were over at Carnegie-Mellon, working on the last re-cuts. Their nbTwins were already scheduled for decommissioning. Then Bill, my brother had a heart attack. He died . . . ."

"Oh, my dear," the woman exclaims, "I am so sorry."

"God, I miss him," Kristin sighs. "Pete and I came back into Manhattan. We had to look for some of Bill's papers and his security keys to start probate and deal with their partnership issues and all . . . then all hell cut loose and they said we had to get out.

"Cornrath had let us have a car and driver so we could dash in and out. But I don't know what happened to our driver. When we came back down to the street he wasn't there. The car either. And the maxnet was down, we couldn't get in touch with anyone to let them know where we were. So here we are." She smiles a tightly stoic bravado as the bus driver comes aboard, gestures snappishly for Grasshopper to be seated, nods curt greeting to his passengers, settles into his own seat, cranks the engine, and pulls out toward Watchung Reservation.

The man winces, as much to reflect sympathy for Kristin's loss as to rectify his social blunder. His voice rises above the roar of the engine, "I'm terribly sorry about your brother. And please forgive our not introducing ourselves sooner . . . I'm Jim Meerling. This is Stacy Madeleine. It's terrible, your having to go through all of this on top of your grief."

Kristin nods on past his discomfiture, smiling as she extends her hand to briefly clasp each of theirs, "Kristin. Kristin Banner . . . and of course, you already know him." Her head nods toward Grasshopper who has switched off the iLink and is gazing impishly at Stacy as the bus lumbers onto the roadway.

Passengers have debarked the bus at Watchung Reservation. Most mill about, idling until whatever comes next. Stacy lags behind, between the bus and main tent, waiting for Jim who has gone off in search of toilets. Grasshopper comes upon her and brazenly asks in sly innuendo,

"Woman . . . where is thy husband?"

"What? Thy? Why thy?"

"Why, thy thighs sigh to ply my prize size."

His finger tips nimbly gather hers and lift her palm to his lips as his free hand sweeps in extravagant musketeer flourish. His body lowers, leg extended athletically forward and slightly across, into a graceful bow. Lingering at her upturned palm his nostrils inspire her fragrance in rapt adoration.

"Sum ergo amor . . . I smell that you want me."

Typically there would be consequences. Now astonishment simply transfixes Stacy. Her perplexed trance dissolves slowly into a chuckle that quickly escalates into rich laughter. At a loss for how to answer his baroque overture she slides into effervescent giggling. Finally, lightly blotting tears with the edge of a blouse sleeve, she chuckles, "Peter. Which organ leads this dance? Your nose or your dick? This is so ridiculous."

"Peter? Nose?"

"Knows what?"

"To smell the now . . . the hour comes, now you want me," His nose quivers in little rabbit twitches.

"Don't do that . . . and stop saying such things," Stacy warns, still giggling, "Jim will break your neck. He's very jealous."

". . . you want."

"What I want or don't want is not the issue." Stacy sternly takes back her hand, "You can't just go up to a woman you hardly know and tell her you think she smells ready now!"

"Now is here."

"Good grief. At least bring flowers and suggest dinner first!" Stacy resumes cackling as she strolls away, hips swaying confidently.

The FERAL encampment is just off I-78 in the hills of the two thousand acre Watchung Reservation. A large circus-size arena tent, along with several smaller ones, has been drafted into Civil Guard service. Tents and several mobile homes for official offices occupy most of the picnic area along Sky Top Drive at the south edge of the Reservation. Only the predominance of olive drab color among vehicles and prevalence of camouflage combat fatigues counter an initial impression that either a fair or an evangelical revival might be in progress. Set up on the far Feltville side of Blue Brook—perhaps in deference to civilian Posse Comitatus sensitivities—the Army has established field operations headquarters in several impressively arrayed geodesic domes.

In a clearing nearby are several immense rigid-skinned "variable buoyancy" light-as-air VTOL skypods, each measuring over one hundred feet wide, two hundred feet long, and sixty feet high, each capable of lifting and hauling more than five hundred tons. They land at Watchung for inspection, maintenance and to be loaded with evacuation supply cargo for distant deliveries. Off to the side smaller gray surveillance skypods descend and arise for changing shifts of personnel.

Maintenance drones flit like dragonflies above and about the large skypods and their smaller sisters who solicit attentive care, perusing and inspecting upper and side surfaces of thin film solar panels that provide the bulk of their power for aerial maneuvering as well as recharging depleted ultracapacitors of deployed usher drone swarms.

Beyond the aerodrome field is a large plaque posted inside the *Trailside Museum* that recounts the history of the site:

A water-driven grist mill was built by original English settlers, Peter Willcocks and wife Phebe Badgley, who purchased land about Blue Brook in 1735. Local legend recalls the mill was converted to manufacture of gunpowder to support the American Revolution. The Willcocks and Badgley families and their descendants worked the land for more than a century.

In 1845 David Felt, a wealthy New York businessman and proprietor of thriving commercial stationery stores in New York and New Orleans, purchased the Willcocks holdings as part of a 760 acre tract. Drawn to the sparsely populated area by the ready energy source available in the mill pond of Blue Brook, Felt developed a residential water-powered mill factory complex. A company town housed managers and their families. Workers exercised a score or more skills as printers, sealing wax makers, bookbinders, paper rulers, stationers, paper colorists, and machinists. They lived simply in well-regulated, gender segregated dormitories constructed of wood in modified Greek Revival style. Company stores furnished ordinary needs and notions while a school and church catered to civil commonweal. In addition to housing were barns, a blacksmith shop, tailor shop and other storage and outbuildings. The census of 1850 counted 178 persons in residence, including twenty-seven children who attended a one-room school conducted by nineteen year-old Sarah Tobey.

Life in Feltville was orderly, as mandated by the owner who enjoyed his nickname, "King David", so dubbed by his employees. He was boss and therefore responsible for the safety and welfare of his community. Workers were expected to strictly observe all rules, both at work and in domestic bliss. Daily schedules commenced with a 7:00 a.m. bell to begin work; 12:00 and 1:00 p.m. bells bracketed the noon recess. A 9:00 p.m. bell demanded that each person must be within their respective domicile. Attendance was required at church each and every Sunday. Workers were paid in company script 'credits', redeemable only among various company stores for purchases.

King David's village prospered for fifteen years. In 1860 with Civil War looming to threaten his southern commerce David Felt sold his factory holdings, closed the mill and departed Feltville. During Reconstruction and on into the Gilded Age the abandoned community became known as "The Deserted Village".

In 1882 Scotch Plains investor, Warren Ackerman, bought the abandoned property with derelict buildings to renovate into a summer resort named Glenside Park. Train service from nearby Summit permitted men to commute into city occupations while their families remained at the resort. Facility comforts were upgraded by adding indoor plumbing and electricity. Guests were able to sit outside and enjoy fresh air on large, rustic log Adirondack-trimmed porches. A stable was adapted to house both carriages and horses. Tennis and golf courts were provided and the millpond opened to boating and bathing. Life at Glenside Park encouraged sense of community with carefully planned group activities and entertainments for guests.

Glenside Park flourished for thirty-four years; its quiet and sedate countryside recreations were observed within a rigidly embracing social structure. Times, however, had changed by its closing in 1916, on the eve of World War I: mass automobile transportation had fostered a taste for suburban living year-round, as well as whetting appetites for distractions less sedate than those of staid resorts.

The property was sold by the Ackerman family heirs in 1919 and eventually was acquired by Union County in 1927 to establish a public park, Watchung Reservation, where people could enjoy nature. Feltville-Glenside Park was preserved, although many buildings fell into disrepair over subsequent years. In 1980 The Deserted Village was listed as a historic district in the National Register and shortly thereafter efforts were launched to restore and maintain ten remaining buildings. On all sides, as viewed from the buildings, nature has returned the hills to a state much like that in which they first were seen by Peter Willcocks and Phebe Badgley in 1735.

After a bathroom break at ganged arrays of portable toilets the group of bus passengers is led into one of the smaller CGC reception tents for an initial welcome presentation.

"Hello. My name is Donald Ackerman. I'll be leading our little session today and hopefully answering any questions you may have about the 'Civil Guard Corps'. And some of the roles you might choose to play in these times of national peril."

Donald is fit, middle-aged, casually dressed, and speaks without notes in earnestly paced packets of words timed for easy assimilation. He stands beside, not behind, a portable wooden podium. His audience of approximately seventy adults is seated before him in rows of plastic folding chairs. To his left, over in the corner of the tent can be seen a large, muted flat screen teleview. It depicts a number of children in age-group clusters being led in a variety of activities by young adult counselors wearing CGC Pal T-shirts. The screen offers a silent window into daycare facilities provided on the grounds by FERAL so that parents can attend recruitment classes without distraction, but are able to check in a glance that their offspring are safe and happily engaged.

Kristin sits between Dooley on her right—already he fidgets restlessly—and the Meerlings on her left.

"Pete," she leans toward his ear to speak in a low voice; his head turns slightly as she urges, "Please. Don't turn on the limner. Just stay in the background. You know what I mean?"

"Okay. Background . . . got it."

She takes a deep breath, turns to smile at Stacy but pitches an eyebrow slightly upward to signal faint hope that he will behave.

Stacy chuckles and whispers back, "He's just a big kid."

"You know," Donald Ackerman launches themes like doves released well-groomed from fastidious cages, "back during the American Revolution, when colonists had been repeatedly whipped by the best army of that era, the British. Back at Valley Forge, shivering in the snow, bare feet freezing, poor folk all around dying of disease and starvation and begging their men to quit their foolishness and come home. Back then, in midst of a world falling apart, Thomas Paine, the author of *Common Sense*, wrote, "These are the times that try men's souls."

Donald pauses, leaning forward slightly as he takes the measure of this assembly. Heads nod back at him, returning and affirming implied sentiments.

"These, too, are such times," he repeats softly, "indeed, these are the times."

Straightening up, he smiles brightly, voice supple and pragmatic: "These are times for common sense, as well.

"That's where we have a lot to do. Applying common sense to problems that have lain uncorrected for a very long time. Working together to undo some of the snarls and pull apart tangles that now tie our hands and put America into a pretty bad bind.

"One of our former Presidents, Calvin Coolidge, a pretty common-sensical guy himself, famously said, 'The business of America is business.'

"Think about it. It's so simple. You can take all your fancy philosophers and put them in a room together and at the end of the day they'd have a pretty tough time coming up with a more succinct description of the American way. Calvin Coolidge saw that entrepreneurial spirit builds our success in the world. American spirit once kicked out British royal governors and set loose the little guy to build this great nation of innovative enterprise. They surmounted the challenges of their time. And we are more than ready for ours. We just need to get organized!"

Thoughtful nods endorse Donald's remedy for dangerous tribulations. He looks toward a large opalescent-surface white screen and then back to an optical projector situated in a rack behind the group. In his right hand is a remote control he thumb flicks to power up the projector. It flares large printed text, centered on the screen,

Federal Emergency Response Administration, Llc. (FERAL)
invites you into the
CGC: Civil Guard Corps

Kristin thinks she hears Dooley's iLink switch on. She looks to see if he is respecting her admonition.

"What?" he asks.

"Nothing." Kristin turns to glance back at the projection rack behind them. It is part of a Kurzweil media station, similar to those Cornrath had assembled in equipping their production studios. She supposes that the iLink has just automatically acknowledged a maxnet bus-polling signal. "Thank you for being good," she whispers encouragement in Dooley's ear.

"Good? I'm the best," Grasshopper brags, extravagantly mimicking her whisper.

She takes a deep breath and returns her attention to Mr. Ackerman. His left hand gestures back toward the screen.

"For a long time," Donald says, "Americans thought that government could solve their problems. They tried a New Deal and wound up building a big conglomerate mess that was supposed to solve the Great Depression a century ago.

"Only, what they learned, the hard way, is that government really doesn't know how. What really brought America out of the Great Depression was a world war economy. It was the spending that stimulated business, that charged up our entrepreneurial zeal and got us to focus efforts to figure out how to defeat terrors threatening across two great oceans. And in the process, how to generate prosperity for all.

"This time the response to crisis is different. And now we know better. Government is not the solution. Government is the problem. Government clogs entrepreneurial options and stifles opportunities. Government is a kind of paralysis imposed by self-important budget-squandering bureaucrats."

Heads nod approval.

"FERAL has been set up as a limited liability corporation. Why a corporation? Well, it's not a new idea, just a good idea," he grins his own personal endorsement, "A lot of people don't realize that for the past few decades even the Post Office has actually been a separate corporation, not an agency of government. And so has the Federal Reserve Bank, the institution that reliably governs your checking and banking accounts. In fact, the Fed has always been a government chartered private corporation. Ever since it was established back in 1913. Both the Post Office and the Fed are monitored by Congress and underwritten by the full faith and credit of the U. S. government.

"That is the true function of government: to guarantee full faith and credit. To keep the economy strong and to protect access to essential resources. All over the world. And that is the mission of our military. To keep us in the big global game that is the most cutthroat competition imaginable. Stay strong. Organize. Compete."

Donald waits as terse logic flows and settles like fresh water pooling and sinking into potting soil. Attentive expressions drink deeply.

"Do you know that the Pentagon has become the largest corporation in the world? The Secretary of Defense is its COO, that means Chief Operations Officer. And President Crown is the CEO. We call him 'Commander in Chief', but he's really CEO. Chief Executive Officer. And furthermore, Each and Everyone of You is a Shareholder. You are shareholders. You receive annual—no, actually, daily—dividends. Your wealth is built upon its service and protection of our economy."

Brows gather in dawning awareness, steeping and soaking upward into bemused tiltings of heads and quizzical expressions.

"Now, to help support the military and to deal with new challenges emerging in this time of national peril, FERAL has spun off a new corporate entity: The 'Civil Guard Corps', or, CGC. CGC solicits, underwrites and coordinates effective team-planning of civilian response to ensure achieving our national goals."

The screen blinks into a list of the Civil Guard Corps' most pressing concerns:

Initiatives:
national survival
threat evaluation
reorientation
technical education
& urban restoration

Unnoticed by Donald the word **restoration** suddenly splits in two, pulls in the ampersand and becomes **urban rest & oration**.

When the revised text is noticed someone giggles. Kristin has not yet seen the reason for laughter, but chuckles at the sound of suppressed giggling.

Suddenly, spellings of all the words explode and scatter. Letters scuttle about on each row as if high-octane ants are playing 'anagram tag'. Row by row the display settles into impromptu poetry:

it is ant I vie:
ill vain ant saviour
a travail unto thee
in a nite rooter
i launched one tactic
a bean & i turn rotors

"The world constantly changes," David continues to speak, unaware of the revised scramblings on the screen behind him. His eyes narrow slightly and scan to find several now impertinently giggling culprits, "New competitors arise. Old foes become friends. Faith falls into confusion. Lies seem true and former truths are unmasked as false posturing."

The level of mirth is rising quickly. Once the insouciant nonsense rendered onscreen penetrates her awareness, Kristin also laughs, then tries to restrain herself as she nudges Dooley with a chiding elbow.

Consternation knits Donald Ackerman's brow. He feels the meeting dissolve into a buzz of chatterers. He looks back at the screen. What are they laughing at? But, just before he completes the turn, all the letters scamper back into their original positions. Donald finds the display exactly as he expects:

Initiatives:
national survival
threat evaluation
reorientation
technical education
& urban restoration

Donald's voice presses onward through a well-practiced repertoire of verbiage. Meanwhile, other agents of his awareness search to explain the growing rambunctiousness. The little finger of his right hand discreetly reaches down and probes whether the zipper of his trousers is closed. (Check, all okay.) Casually his eyes wander down to see that his name tag is not doing something crazy. (Nope, it's fine.) Long familiar phrasings still work through his lips, but he is distracted in newly wary anxiety. His carefully honed professional manner is being seen as foolish. Why?

He stammers uncertainly, "This cutthroat world constantly challenges us to learn better ways to protect the full faith and credit of American prosperity. We're all going to have to pitch in and do our part in this new struggle. We're all going to need CGC. And CGC definitely needs us. Good solid American citizens.

His eyes burn with threatened tears as his voice weakens before the now avidly chatting, no longer listening, group, "These are the initiatives into which CGC currently is recruiting. Our purpose today is simply to introduce you to what we're about, why we need you, and how you can put your God-given talents and hard-won skills to best serve our national goals."

As he manages to veer toward a concluding thought the letters in reawakened mutiny cautiously tiptoe back into upstart verse:

it is ant I vie:
ill vain ant saviour
a travail unto thee
in a nite rooter
i launched one tactic
a bean & i turn rotors

The clamor gains fresh energy. Donald gives up. His voice falters and fails into tremulously mechanical rhythm as he bravely shouts above the din, "To that end, here's a brief spot that will give insight into what we're about. And give me a chance to take a sip of water."

Donald's pained smile does not subdue the now general levity. People prattle animatedly among themselves, debating the intentions behind what has been received as comedic warm-up. Donald retreats back to the edge of the tent and collapses onto a chair. He lifts a bottle to his lips and does not care that water is dribbling down his chin to darken splotches on his khaki shirt. His thumb flicks at the remote and the screen fades to black. An audio track opens into a spacious cavity of silence as the crowd suddenly hushes.

Scattered instrumental sounds of an awakening orchestra briefly stir. Symphonic hints work toward an opening theme, delivered in full orchestral surround sound. On the screen yellow text appears, centered against a black background:

WEBSTER DEFINES STRONG AS
HAVING GREAT PHYSICAL POWER,

Musical exploration continues searching:

AS HAVING MORAL OR
INTELLECTUAL POWER,

Music keeps searching:

AS STRIKING OR
SUPERIOR OF ITS KIND.

Musical hunt continues:

BUT WITH ALL DUE
RESPECT TO WEBSTER,

Musical pause anticipates:

THERE'S STRONG,

Horns announce imposing theme:

AND THEN THERE'S ARMY STRONG.

Long: A platoon of twenty or so soldiers in combat gear are gathered on a ridge, around a brigade flag, for a group photo. Only the flag moves slightly, stirred by a breeze.

Medium: Soldiers advance across a battle field, in the distance a helicopter lands to field more soldiers into fray.

Close: Face of fire control officer at control pad.

Close: Fire control screen shows computer map of terrain, deployed forces and target data.

Long: Cruise missile guidance video feed homes in on target.

Medium: Paratroops jump out rear bay of transport plane; ground far below is swathed in clouds, smoke.

Medium: Camera looks up toward a single parachutist descending through overcast sky.

Music sustains:

IT IS A STRENGTH LIKE NONE OTHER.

Close: Soldiers in battle gear slap 'high-fives', celebrate, and embrace manfully.

Music sustains:

Close: Profile of uniformed drill instructor yelling.

Medium: Seated novice recruits listen respectfully, wide-eyed and attentive.

Zoom close: face shot of strong pretty female face.

Music sustains:

IT IS A PHYSICAL STRENGTH.

Music swells:

Medium: Line of soldiers in combat fatigues work through obstacle course.

Zoom close: exultant expression on young male face during physical exertion.

Music swells:

Medium: Soldier rappels from hovering helicopter.

Medium: Soldier wearing cold weather exercise garb in marathon run up steep hill.

Long: Airborne camera looks down onto line of combat-ready soldiers trekking through forest clearing.

Long: Multiple amphibious assault craft streak across open water toward unseen objective.

Music sustains:

IT IS AN EMOTIONAL STRENGTH.

Music sustains:

Close: Soldier in combat fatigues embraces wife firmly and lovingly.

Music sustains:

Medium: Soldier in full combat gear with rifle walks beside three indigenous children whose hands are linked in solidarity of feeling secure.

IT IS A STRENGTH OF CHARACTER.

Music sustains:

Black &white: close: Helmeted soldier's face gazes into camera.

Medium: Officer and aide de camp pin medal on attractive young woman standing at attention, all are dressed in combat fatigues.

Black & white: medium: 'elderly veteran' in VFW cap, stands at attention, saluting, amid cheering and adoring civilians waving US flags toward off-screen soldiers.

AND STRENGTH OF PURPOSE.

Music sustains:

Night: medium across close: Profile of soldier looking down along ranks of assembled soldiers, distant light casts their long shadows toward him and camera.

Medium: Soldiers advance toward camera, emerging out of desert dust storm.

Music sustains:

THE STRENGTH TO DO GOOD TODAY.

Music sustains:

Close: Man talks into head-mounted mike.

Medium: Combat controller points to computer display.

Zoom close: Finger points at objective.

Close: Man talks into handheld mike.

Medium: Camera looks into controller vehicle; helicopters swoop low and are reflected in vehicle window where controller talks into handheld mike while looking up at them.

Medium: Helicopter, silhouetted against sky, flies over, so close as to fill most of frame.

Music sustains:

AND THE STRENGTH TO
DO WELL TOMORROW.

Music sustains:

Medium: Young civilian executive gazes from office window of an upper floor, out across cityscape of commercial skyscrapers.

THE STRENGTH TO OBEY,

Music sustains:

AND STRENGTH TO COMMAND,

Music sustains:

Close: Battle weary four-star general in sunglasses gazes resolutely into camera.

Extreme close in quick cut: T-shirt, "ARMY".

Close: Frames upper-half head shot of female next to lower-half of taller male, both wearing black woolen caps.

Long: Camera in grass looks up at distant squad trotting across training field, lead soldier carries flag.

Music sustains:

THE STRENGTH TO BUILD,

Black & white: long: Combat engineers deploy pontoon bridges, amid gigantic splashes where they drop into water.

Long: Soldier rush across completed bridge.

Music sustains:

AND STRENGTH TO TEAR DOWN

Medium: Tank lumbers and splashes through water toward camera.

Music sustains:

THE STRENGTH TO GET YOURSELF OVER,

Medium: Young female in combat fatigues climbs up rope ladder toward camera.

Music sustains:

AND THE STRENGTH TO GET OVER YOURSELF.

Medium: Camera views from below and behind her as hands reach down to help female soldier climb from rope ladder onto wooden scaffolding.

Music sustains:

THERE IS NOTHING ON THIS GREEN EARTH

Long: across silhouetted soldier standing looking out side door onto ground.

Medium to long: ramps deploy simultaneously, viewing down along line of landing helicopters, soldiers with weapons, rush into battle.

Music sustains:

THAT IS STRONGER THAN
THE US ARMY

Medium: Soldiers in battle wave helicopters forward, advancing across and above them toward camera.

Close: Helicopter pilot looks down.

Medium: Helicopter formation swoops and veers right, silhouetted against sky.

Music sustains:

BECAUSE THERE IS NOTHING ON
THIS GREEN EARTH

Medium: One soldier among several comrades waves at helicopter passing overhead.

Extreme close: Several stars and a few stripes are visible on billowing American flag.

Music sustains:

THAT IS STRONGER THAN A
US ARMY SOLDIER

(fade in)

SUPPORTED BY THE
CIVIL GUARD CORPS.

Music sustains:

Medium: Camera looks from behind a number of silhouetted figures, arms upraised to solemnly declare oath, background is gigantic wall-size American flag, a few stars and several stripes are visible in frame behind figures.

Close: Young male soldier's face, standing among infantry platoon, in body armor with weapons at ready carry.

Medium: Three soldiers advance toward camera, helicopters hover and advance just behind and above them.

Close: Soldier loads tank cannon.

Medium: Tank roars into full frame, advancing toward camera.

Music builds:

STRONG

(fade in)

ARMY STRONG

(fade in)

ARMY GUARD STRONG

(fade in)

CIVIL GUARD STRONG

(fade to black)

Music crescendo:

Fade in: close:

U.S. ARMY,

yellow/black/white star above

green **CIVIL GUARD** logo.

Fade to: green **CIVIL GUARD** logo.

Music silent while complete logo sustains briefly.

Fade to black:

Music rises:

Fade in: medium: Young civilian office workers, 2 male, 2 female, sit with notes, etc. in avid discussion around circular table.

Overlay, translucent red rectangle wipes right, superimposed by text wiping left:

CGC
Comfortable
Environment

Narrator (female voice): *Civil Guard staff praise their supervisors and find that they are often mentors.*

Fade out graphic as narration pauses:

Narrator (female voice): *They like the informal atmosphere in most offices and their positive working relationships.*

Medium head/shoulders shot of CGC Language Analyst, Loretta Brusco, speaking into camera:

Coming to work for the Civil Guard Corps as an entry level individual, you will have the opportunity to work in teams. Most of our work is done in teams. So you're not alone, there's a lot of support. The supervisors are very approachable and it's a great learning experience.

(fade to black)

(fade in)

Close: Yellow/black/white star above
green **CIVIL GUARD** logo.

Silence while logo sustains briefly, fade to black.

"That," announces Kristin, "just has to be an Eric Cornrath production."

"Who?" Stacy asks, "Why do you say that?"

"From the Amberson Agency, he's the producer behind Ant and Grasshopper."

"What makes you think of him?"

"Well, for one thing, his style in depicting women. Always dependent on men."

"That's every man's style, honey," Stacy looks puzzled.

"Every woman in that spot was shown among taller males." Kristin elaborates, "You'd think at least one female basketball player might wind up in the Army. And when that one poor little female soldier finally made it all the way up the rope ladder and was trying to climb onto the scaffold—she had to be helped up. Those were all men's hands lifting her up."

"I guess I didn't notice," Stacy muses.

"That's the point. They're so good at planting thoughts you never notice, like viruses. Exploiting patterns you already accept. In fact they even have a name for them. *Memes*. Eric Cornrath's one of the experts. People think they're in control, but media types like Eric want you to think that. It makes their work so much easier."

"Isn't that a little paranoid?"

"No," Kristin assures her, "It's very paranoid . . . and worse, it's true."

Left to her own amusements while Jim deals with some of his concerns, Stacy has found a nice private little space over at the far end of the encampment, away from the main processing area. She has spread a blanket she managed to 'borrow' from the daycare tent so she may relax alone, drowsing peacefully in the sun. Lying back, her eyes closed, she feels solar warmth penetrate her skin, suffusing its energies through to the very bones.

Her thoughts drift back to the previous summer; she persuaded Jim to take her on a tour of New Age sacred sites in the Americas, places renowned for properties of healing and energic inspiration.

They traveled first to Ringing Rocks Park, nearby, just over the Delaware River in Pennsylvania where they tapped out some tunes on those intriguingly musical boulders. Geologists, she learned, have no generally accepted explanation for why some of the rocks ring like a bell instead of answering a hammer blow with a typically stone-like thud. She smiles, thinking of the joke local to the site: the first 'rock concert' took place there in June, 1890, when Dr. J. J. Ott selected a number of 'ringing rocks' for their harmoniously appropriate pitches and then played a number of musical selections, accompanied by a brass band.

Next on their tour of hemispheric harmonic zones they traveled to Jim's favorite of the entire tour, Sedona, Arizona, because he could play golf. New Age mecca of the southwest, Sedona sits near an ancient settlement, occupied more than eight centuries earlier by the Sinagua Indians. Today Sedona is the reputed locale of mysterious happenings, spiritual possessions, and all sorts of astral mystifications: "harmonic convergences" and "spiritual vortices". A burgeoning tourist industry caters to wealthy New Age devotees enthralled by gateway glimpses into farther spiritual dimensions. Moreover, extravagantly appointed spas of the Verde Valley are lovely, as well: a great place to unwind and enjoy a day of pampering, meditation and soul-work while the significant other is out hustling natives on the links.

But Stacy's favorite of them all, and the most exotic, as well, was Machu Picchu, high in the Peruvian Andes. There she was privileged to work for several days with Oreone, Inca shaman and healer. Oh, his magnificent eyes! His transfixing gaze and electric touch.

Machu Picchu means 'manly peak' he had said; his magnificently muscled arms gestured out toward a suggestively rising shaft of rock. There he called upon profusely converging earthly energies to well up from deep beneath these ancient royal resorts, exclusive spa of Inca kings and their consorts. Ah, the spiritual journeys through which he led her!

She sighs now in longing, sun-blessed recollective transfixation, as from deep within her belly spreads a warmly tender suffusion of sweet ecstasy . . . is shattered by intruding footsteps swishing toward her through the grass. Opening her eyes Stacy cannot quite discern the identity of an impudently looming silhouette outlined in brilliant sun. Whomever, they appear to be thrusting a handful of hay out toward her.

Reflexively, she raises an arm to ward off any unwelcome overture. She shades her eyes with her extended hand, but still cannot see who has shattered her precious solitude.

"Jim?"

"No . . . it's me."

"Me?" Stacy repeats blankly as she pulls herself up to assume a seated position and change the viewing angle with respect to the sun, "Oh, Peter. I didn't recognize your voice."

"Grasshopper," he corrects her, "I brought these flowers I found over by the brook. And here's dinner." In one hand is a disheveled wad of bluets, bluebells, goldenrod, gentian and bladderwort. From his other hand dangle two plastic foil bags: military meals-ready-to-eat.

Stacy looks from the disheveled clutch of wildflowers to the rumpled bags and back again at the flowers. She rises slowly to her feet and stands for a moment to peer deeply into his fiercely ecstatic eyes. Their color is very different but the energies stirring therein remind her of the shaman Oreone. "Peter, you are just so sweet," she smiles appreciatively.

Taking the flowers, she lifts them to her nose, sniffs delicately at various petals of bright hue and then leans forward to kiss him affectionately, but lightly, on the lips. For a moment she ponders her situation, looks back toward the tents, and then arrives at a decision. A fey smile plays at her lips as she reaches out for the two meal bags, tosses them onto the blanket, stoops down and with her one free hand gathers it into a wadded mass, which she shoves under his arm. Winding her own arm through his other one, she takes his hand, tousles his hair with the fistful of wildflowers and says brightly, "Hi-yo, Silver. Let's go for a walk. Maybe we'll find a good picnic spot."

"There won't be any Ants, will there?"

Laughing, her hand reaches down to brush the variegated bouquet lightly across the manly peak bulging urgently beneath the iLink codpiece, "Nothing we can't handle, Machu Picchu Man."

Well into the forest, about half a mile east of the CGC encampment, birds, otters, deer, and perhaps even a few Ants are privy to a spiritual cleansing and twining of harmonic energies. Within a secluded and shadowy glade, the thrumping sounds of an iLink limner, doing duty as a surround sound karaoke, pound out musical rhythms and humping phrasings of the ancient disco classic *Macho Man*. There, before a reclining and beamingly buff and transfixed golden Goddess, a silver-tressed nude and ecstatic Puck dances in front of a spectral chorus line of luminously attired Inca dancers, arrayed in full-feathered celebration to rival Mardi Gras Indians as they hoof like confident veterans of a Las Vegas floor show, all singing in electric unison:

ecStacy...wanna make me bawdy?
ecStacy...such a thrill, my rowdy
ecStacy...wanna fetch my baddy?
ecStacy...it's too much, my waddy
Scope it out, my howdy doody.
Don't you doubt my ruddy booty.
Talkin' bout my gaudy proddy,
Check it out, my body.

ecStacy's man got to be a Machu Picchu man,
To live a life a'hoppin', Machu Picchus make a friend,

Find their own way and reveal,
Laugh with strong assurance, life's a reel,
You can always spot a Machu Picchu man
He's always good to go, always in demand.

Hey! Hey! Hey, hey, hey!
Machu Picchu man (Machu Picchu)
I've got to be a Machu Picchu man
Machu Picchu man
I've got to be a Machu Picchu! (dig the moment at its peak)

Machu Picchu man (see my 'Manly Peak')
I've got to be, a Machu Picchu man
Machu Picchu man
I've got to be a Machu Picchu! (Breath cosmic wonders)

Machu Picchu man (seethe vim and vigor!)
I've got to be a Machu Picchu man
Machu Picchu man
I've got to be a Machu Picchu!

Machu Picchu man
I've got to be a Machu Picchu man
Machu Picchu man
I've got to be a Machu Picchu! HEY!

Kristin has come to the mess tent, gotten a cup of coffee and sat down to look over assorted CGC brochures. Several tables away she notices Jim Meerling deep in animated conversation with a man who appears impatient but maintains a semblance of casual chat. Jim passes an envelope to the man who looks inside, then acts annoyed, as if something is missing. Jim almost leaps to his feet in exasperation, but catches himself and settles back onto the bench. A few more words back and forth, then Jim removes a plastic card from a booklet carried in his shirt pocket and hands it to the man. Following close inspection the card is consigned to the envelope. Just before he gets up the man points an index finger at Jim's forehead and mimics a firearm discharging. Then the stranger grins patronizingly and rises to quickly depart the tent. Jim remains sitting in a forlorn posture. He stares vacantly at the table top, lost in thought. Then he resolutely braces his shoulders, straightens his back, looks up and about.

Kristin has anticipated that Jim will notice her. Intuitively she prefers him to surmise that she has been unaware of his presence. She has turned sideways to sort through her brochures, her back angled cater-corner to where he has been sitting. Her cheek rests on her knuckles, hair draped as a copper red curtain obscuring that field of vision. Aware of his approach, she hears him speak, seeming delighted to encounter her.

"Ah. Kristin. Contemplating a career in the Guard?" Jim manages to seem jovial; he takes a seat across the table from her.

"Oh, hi, Jim," Kristin smiles brightly, looking up as if slightly startled to see him, "No, not really. I don't think the Civil Guard has much to offer this girl. I'll pass on this particular opportunity to get knocked up."

"Opportunity knocks . . . knocked up," he recites elements of the joke, prospecting for still more hidden gems, "That's very good." As if just reminded of an obligation, he asks, "Have you been interviewed and signed off on your screening commitment yet?"

"No," Kristin waves a hand to dismiss the notion, "Why sit through pointless mind games? Pretending to be interested? I'm blowing it off."

Jim nods, presuming shared intentions, "Just catching a ride farther along your journey, eh?"

"Exactly," a note of resentment strikes her voice, "You'd think FERAL or some other agency would devote more effort to moving people out. But no, these days everyone is on their own. So if I have to renege on a promise just to get some slight assistance, then screw them all! No wonder Crown has postponed elections! Now he's got a bunch of pissed-off taxpayers to deal with!"

Her head tilts in angry indictment but her voice is perplexed, "Why do you suppose they've set it up like this? I mean it seems so awkward and self-defeating. As if dealing with actual survival and well-being of millions of people wandering the countryside is beyond them . . . the last thing on their to-do list. It's insane . . . we have millions of wandering homeless and just as many foreclosed homes standing empty."

Jim lifts his palm to signal his own roster of complaints. "Don't get me started. You can be damn sure the bucks aren't stopping anywhere!" His teeth flash to appreciate his own jest, "Billions upon billions of them are changing hands as we speak. I know the GloBanc crowd. They're big backers of Ira Crown. Well, like I said, don't get me started."

"Well, it's a pretty abysmal performance so far," Kristin snorts contemptuously.

"Stacy predicted this. Sometimes it amazes me how some little offhand comment she tosses off at breakfast will ring true by dinner. But this was a real premonition. She started nagging for us to leave months before even the hurricane struck. I, of course, thought she was being silly. Then I came across an item in a financial blog, *TradingGoddess*, I believe it was. When I saw it, suddenly it all made sense . . . it's the same scams power brokers that run central banks have been pulling for centuries."

"What do you mean?"

"I'm quoting from memory, but it said, talking about how the really big financial players suck up wealth from everybody below: 'This is how it works folks, they herd the assets (and real estate is the most real of all assets) into pools and then they poison the pools, force everyone out, buy the assets up cheap and then miraculously (at great expense to the taxpayers) turn it around and end up holding the deed to most of the land in the country which they will then sell back to the people for 5 to 10 times more than they paid for it until the land is once again in the hands of the people, at which point they start the cycle all over again. This scam was run in the Great Depression and it took us 30 years to recover from that one!'

"When I showed it to Stacy she just laughed that that's the way it's always been, the rich get richer and the poor get children."

"How long have you and Stacy been married?" Kristin inquires, preferring to know more about them than about financial gamesmanship.

Jim considers the question for a moment, as if considering any of several possible answers, "Oh, we aren't married. Stacy is my, ah, assistant."

"Oh. I'm sorry, I didn't mean to be nosy."

"That's okay. It's no secret. Stacy is very resourceful and tenacious. My business involves lots of quirky details and scenarios with all sorts of people. Stacy manages to go with the flow and to keep it all running smoothly. Perhaps more importantly, she keeps me on task," his laughter seems rehearsed. "And you and Peter? Are you two involved?"

"Peter and I? Involved? Oh, dear God, no," Kristin gruffly dismisses any such insinuation. Her features express rueful recollection of first becoming acquainted with her brother's new partner. Bill Banner and Peter Dooley had discovered the comedic compatibility of their opposite styles, well-synched tastes and convivial timing during call tryouts for an improvisational comedy troupe. Initially there might have been some tenuous flickerings between Kristin and Dooley. At least she had thought so. But his coarsely obnoxious and persistently leering advances had quickly fouled any amorous prospects as far as she was concerned.

"It's strictly business," she assures Jim through a matter of fact smirk, "My brother, Bill, persuaded me to help them when they first got the Ant and Grasshopper gig. I had just abandoned any hopes for my own career in modeling and was looking for something more challenging. Their agent had just bailed so I agreed to handle contract negotiations with Eric Cornrath at the Amberson Agency."

"Did you say Eric Cornrath?" Jim interrupts.

"Yes."

"I think I've heard of him . . . he's a PR whiz, works with Amberson and all the other spooks, right?"

"The whizziest," Kristin nods, "but I don't know what you mean by spooks."

"Spooks, you know CIA, NSC . . . national intelligence guys."

Kristin frowns for a moment, "I don't know anything about that stuff . . . how well do you know Eric?"

"Oh, just through associates. More by reputation than direct contact. Anyhow, I'm sorry I interrupted. Please, continue."

"Then," Kristin picks up her narrative thread, "since neither Bill nor Peter seems to have any common sense genes I found myself acting as their manager, and banker, and housemother. Actually, more of a handler . . . you'd have a hard time finding a more irresponsible and self-indulgent pair. I didn't mind Bill calling me 'Big Sis', but I don't like hearing it from Peter. Now I think he calls me 'Sis' just to annoy me.

"You mentioned earlier that you'd lost your brother. What happened?"

"Bill had a drug problem. Crack. Meth. Majik. You name it. Anything to buzz him up. I begged him to get help, but he just laughed it off. Then when they started commissioning the nbTwin . . . ."

"Commissioning?" Jim is puzzled.

"That's what they call the process of creating and getting used to the echo feedback from the nbTwin interacting through the iLink."

"Oh."

"Anyhow," Kristin continues, "Bill couldn't give up the stuff. Or wouldn't try. He should have, of course. At least until decommissioned. I think it killed him. Between work sessions he seemed even more turned on by going into adrenalin overdrive. It escalated out of control. He had a massive heart attack as they were finishing up foley dubs. They said he died of *Bell's Mania*, 'excited delirium' I think they call it. It's a kind of extreme exhilaration brought on by the drugs. . . ."

"I'm very sorry," Jim says, reaching to lay his hand over hers.

Her own hand lingers only for a moment, then withdraws, finding something odious in his touch. Kristin feels curiously offended, but elides easily into a different topic, "What do you do, Jim? Banking?"

"We Meerlings have always been traders." Jim doesn't seem to care that she so quickly pulled her hand away. "There's a family joke that when Peter Stuyvestant bought Manhattan Island for thirty-two dollars worth of stuff, he used Meerling beads and trinkets."

"Trader? Like stock trading? Or trading cars, or what?"

"You name it, there's a Meerling trading it somewhere. For a while, after I realized college wasn't for me, I traded commodities. Then I tried my hand in currency markets. But when hurricane Moira wiped over Manhattan and Long Island I got involved in salvaged property transactions. I like dealing with actual stuff. And there's no place like salvage operations for opportunities to move real stuff. The US dollar has been sliding for a long time. It's getting so that working currency trades is like rolling dice. Do you double-down or just go bust? And it's really hard for an individual guy to survive swimming with institutional sharks . . . it's the most nerve-wracking tedious ordeal you can imagine. So I got out. I saw a massive debt bubble building and I decided to only trade tangibles. I just don't care for the whole speculative scene anymore, especially now that the dollar's fallen into the toilet. The powers that be, the banks, brokers, financial juggernauts have turned it all into the Red Queen's Race."

" That's *Alice in Wonderland*?" Kristin guesses.

"*Through the Looking-Glass*," Jim corrects, "the sequel to *Alice's Adventures in Wonderland.*

"But anyhow, they've built inflation into the system by being able to print money whenever the government wants it. What a scam! Money is devalued by arbitrarily printing more of it and so the prices of tangible stuff goes up, inevitably. It acts like a hidden tax. Your average joe is working on a treadmill and doesn't notice he's being cheated because he's scared of simple arithmetic. The government tells him everything is normal and okay, just work harder.

"Like the Red Queen says, in her race it takes all the running you can do just to stay in the same place and if you want to get somewhere else you must run at least twice as fast as that!"

"But why? Who benefits?"

"Politicians. Bankers. People at the top—Owners. It acts like a slow and steady wind blowing wealth from ordinary pockets below into the accounts of the wealthy. And they use growing riches to buy more power and influence to stack the odds even more in their favor. Hurrah for the Owners . . . a few families around the world who own and control just about everything that affects everyone else."

He glances toward the tent's open entrance where Stacy has appeared, bathed in an aureole of light, gazing beatifically toward them.

"So what are you two going on about?" Stacy joins them, smiling happily, her face lightly beaded in perspiration. She sits next to Kristin on the bench.

"I thought you were going to take a nap," Jim voice turns waspish, "What did you do? Run a marathon through it?"

She placates, "Oh, Jim, you are so clever. I decided to go for a walk. Up along the forest trail. There are such powerful harmonic convergences up there."

"Have you seen Peter?" Kristin inquires abruptly, suddenly concerned for what he might be up to.

"Last I saw, he's over at the daycare tent, playing games and drawing with some of the children."

"Oh, good grief," Kristin exclaims, jumping quickly to her feet, "I'd better go check on him . . . he's gotten so damn weird lately."

"Don't worry. Sit back down. He'll be okay," Stacy brightly reassures her, "A couple of counselors are there. I think they like having him around. And the kids just adore him."

"Hell, I guess so. He's a walking, dancing, circus," Jim snickers, "He brings out the kid in everybody."

"Oh, he's more than that," Stacy assures him.

Jim peers closely back at her pacific assurance, as if just seeing her, "Stacy, you are positively radiant. Where have you been?"

"Like I said, sweetie," Stacy smiles mysteriously, "In the forest. . . powerful energies are at work."

Jim frowns vaguely; his voice thins, "I can imagine . . . another Oreone?"

Kristin sits back down. To break suddenly stony silence she asks, "When did you decide to leave the city?"

Jim looks at her, back at Stacy, then answers, "Actually, a month or so ago . . . we started thinking it might be a good idea to go inland. Permanently. Back to Cincinnati where I went to school and then later worked for a while. I still have some contacts there."

"But it didn't go quite as planned," Stacy interrupts, "That was back before the general evacuation. We waited too long."

"What do you mean?" Kristin is puzzled by a disjoint she feels suddenly surfacing between them.

"We couldn't just leave everything," Jim protests.

"So Mr. Operator here comes up with a couple of humongous road vans and a crew of men to load and drive them," Stacy frets, "And somehow thinks it's all okay because he has spent a small fortune on the trucks and transport permits. Small thanks to some of his buddies. What's the phrase? Only cons get conned, right, honey?"

"What happened?" Kristin asks.

"At a checkpoint we were stopped just after crossing into New Jersey," Stacy continues, her voice rising in ire as she looks accusatively at Jim, "It was like they were waiting for us. Turns out the trucks had been stolen from the Army and repainted. The papers and permits were all forged. They arrested Jim and confiscated the trucks along with everything on them. And I spent three days bouncing from magistrate to jail and back again just to get Jim released on bail. I don't think we'll ever see that stuff again. Something's not right," her eyes accuse him of some secret mischief, "I should have left him there to rot." resentment washes across her mentor's faltering demeanor.

"But then the general evacuation started," Jim's voice is uncharacteristically meek, "Newark is inside the threat zone, so we left. Technically, I've jumped bail."

"Technically, my sweet ass," Stacy fumes.

"Actually," Jim admits, "that's the real reason we can't be screened. It would pull up the records. I don't want to have to deal with that crap here."

"But how did you get passes to come here on the bus?" Kristin motions toward the CGC imprinted orange strips they all wear on their left wrists.

Stacy's mood flashes into a light humor as she laughs, "You helped . . . I managed to filch them while Peter was ogling that woman—Francine, wasn't it?—you were talking to. She had laid two of them on the table, then got so distracted by Peter hassling her she didn't realize I'd palmed them."

As they queue for the evening dole of plastic-wrapped victuals all evacuees sporting orange wrist bands have been reminded that this evening is the last they will be allowed to remain at Watchung Reservation. Successfully screened CGC inductees have already been moved to another location for the next phase of training. Remaining orange-banded evacuees are to convene after breakfast tomorrow for a brief presentation to help prepare them for rigors of returning to the Interstate for the long journey toward Harrisburg.

Glum to return to the highway, yet anxious to get to the end of the ordeal, Kristin disinterestedly toys with her foil wrapped package of dehydrated food as Dooley sets two plastic bottles of water between them. At such moments he seems to be the familiar Peter Dooley. Although she cannot imagine a motive, it almost seems he has been working through improvisational scenes and now is taking a break.

"Pete." She gauges his expression for any response to his name, but he might as well be deaf; his attention totally is devoted to peeling away the foil seal of the food packet in a mini-charade of surgical precision.

"Pete," she repeats more firmly.

"Chicken", he grins avidly and as he lifts a wad of dried fibers for her inspection. "What'd you get?"

"Pete!" she repeats sternly, determined that he shall acknowledge his name. Grasshopper looks blankly back at her.

"Pete 'n' repeat, Pete 'n' repeat, 'Oh, pardon me," he whines, mocking her tone, "what did you say? Pete 'n' repeat."

Kristin flushes hot. "It's your name, damn you!"

Grasshopper pauses chewing momentarily, then resumes working his teeth against a coarsely fibrous mat of dried chicken. He watches her closely as if he needs some focus to hold his eyes during such jaw-grinding exertion.

"Why do you say that?" he eventually asks, only mildly interested in her answer.

Exasperated, Kristin angrily rips open the foil container constituting her food packet. Her fingers fumble and drop its contents on the grass. Annoyed, she picks it up and brushes off dried bits of leaf, "Because I know you," her tones are measured and specific, "you are Peter Dooley!"

"This is all Peter Dooley's dream . . . ," Grasshopper twists his lips into a smirk, ". . . of Grasshopper."

"Dream? What do you mean? You are Peter Dooley."

"Oh, no. No, no . . . nobody puts words into Grasshopper's mouth. No more lousy scripts . . . This one is on its own. This is for Grasshopper and Grasshopper is for this. 'Hic est iam . . . iam est hic . . . sum ergo amor."

"What?"

"Here is now. Now is here . . . now always, therefore, now love."

"Right. Got it! What the fuck does that mean?" Kristin fumes, "It's a role, a goddamned character you used to play. You remember, I know you do!"

"Peter Dooley is the role you want me to play," he snaps vexation, "but you can't trick Grasshopper. Grasshopper hoppens too fast! Hoppening too fast for you! Hoppening too fast for Peter Dooley!" He jumps up, bounds a few feet away, and pauses to watch her briefly. Then, suddenly cheery, teasing and playful, he darts in and lightly taps her shoulder.

"Tag," he cackles triumphantly, "you're it! You be Peter Dooley!"

Kristin watches him dance a victory step, enjoying his own joke. Then suddenly he is distracted by nearby people as they finish their meals and carelessly toss remnants and waste packets aside on the ground.

"Hey, hey, hey!" he caroms abruptly toward the nearest gaggle of litterers, "What do you think this is, you throwbacks? Your goddamn living room? We all live here. Don't clutter this place up with your crap! Now pick up that trash and put it where it belongs! It's people like you that give people like you a bad name!"

Grasshopper snatches up a large plastic bag that once held a gross of food packets. He shoves it into the hand of a man who at first stammers protest but then gives in and begins picking up trash lying nearby.

Darting to yet another cluster of people departing before their littered wake, Grasshopper soon has them enrolled in a cleanup detail. Kristin watches in utter amazement as this final proof of his wacky dementia aborts her hope to restore Peter Dooley's self. Only a lunatic would attempt to impart housekeeping skills to a throng of desperate and disgruntled refugees. And only a lunatic could succeed.

Within a few minutes all the trash has been gathered and placed into plastic bags neatly stacked at the side of the mess tent. Children have happily and enthusiastically joined Grasshopper's earnest project, helping cajole and wheedle parents and older siblings to join in policing the area.

Most remarkable of all, Kristin notes in stunned wonder, it makes a difference in attitudes she senses in the individuals about her, as if a little touch of normalcy has been returned to their control. People seem to relax a little, to smile and to sit about talking and laughing as if a demonstration of communal intent has relieved, at least a little, some deeply anxious chord.

As she watches Grasshopper, she finally admits to herself that he truly is a stranger. Alien to her. Perhaps alien even to himself. He seems exiled from all the things she knows and recalls. Even traits she once despised might now gladly be welcomed back.

And it seems she can no longer be certain that he has not changed physically as well. His small, lithe frame still wears Peter Dooley's denim jacket and trousers. The hair is the same startlingly natural silver dishevelment it has always been. A two day growth of beard is new, but is easily ignored in the community of similar growths on other masculine faces in the crowd.

His posture, his gait, the overall manner in which his body conducts its affairs of moving about, sitting and standing—all are vague reminiscences of Peter Dooley. There does seem to be an additional factor, however. She realizes she has struggled against it, as if her refusing to recognize it might hasten restoration of his departed self.

She finds it difficult to look into his eyes. When she does she must acknowledge that indeed there is some fey presence, as if there is an other inside him. She avoids looking into it. It threatens the sole remaining link binding her to a life from which they have been so unexpectedly evicted onto a wandering and aimlessly paved path of uncertainty.

Looking about, Kristin wonders if there can possibly be any haven for all these displaced people and the unimaginable hordes scattering even now outward from Gotham Seaboard. Where can they all go, except into death? Why has there been no plan revealed for dealing with rising chaos? Where is government aid? Or even encouragement? Can it only send soldiers to confine them to such narrow strands of helplessness, helicopters hovering above to tease them forward a little each day with plastic wrapped packets of tasteless food and fetid water?

Images shudder through her mind, fusing in a melange of memory: sounds of females screaming way off in the distance, corpses of old men and even some young, mauled and beaten, bodies slashed, pockets ripped open, tactile reminiscence of hands reaching down through tunnel dimness to paw, invade and abduct her. Why do people allow this to happen? Are they so immersed in such scattered insular concerns that no one cares about collective tragedy unfolding? Where is the we in U.S.?: a political campaign complaint uttered long ago renews in tragic irony.

But now Peter's transmogrification into Grasshopper seems the most insidious of all the blows dealt her. Simultaneously she envies and despises the ease in which dementia accommodates him to such public travesty, this senseless turmoil. He is part of it. He gathers from its useless and jarring rhythms an ease, a lightness of spirit that is at once callously inhuman and individually merciful in sparing him anguish she feels in losing all that has been precious and right. Especially her brother, Bill. They had lost their parents in a sailing accident just as they reached adulthood. He had consoled her; she supported him. Their bond displaced all other claims. Now he is gone.

And even the last faint link to him, through Peter, is betrayed in some raucous malady that styles itself Grasshopper.

Despite eyes burning in urgent emotion, it is impossible to cry. Impotence of ordinary release coalesces witnessed terrors and moments of grief into a relentlessly oppressive and brittle mirror. Whatever occurs now seems personally directed at her. The corpses seen on the highway are the possibility of her own violent death. They promise only agony ahead, relieved perhaps only by uncertain ending. The listless mob gives no succor. They populate an image of wanton uncertainty through which sinister helicopters escort uniformed shepherds who wield electric prods to herd them.

"Cattle," she murmurs to no one, "we're nothing but cattle."

Grasshopper has returned. He sidles back to stand close by, but unacknowledged by her, where now he chants an impromptu ditty about a cowboy herding bovine paramours:

There is a young cowpoke from Dallas
Who looks upon sheep with great malice
But never construes
Discouraging moos
While poking his cows with his phallus!

Gradually prospects of tomorrow, being forced to resume trek across unforgiving pavements, abrades recent memories and forestalls anticipation of anything. For a while she just stops thinking and is suspended in merciful trance among thumping helicopter noises rending distant skies.

~ 3 ~

In the CGC Daycare tent a young woman, just out of college and working as a volunteer, is sorting through various examples of children's art. Chloe Jernigan, herself a writer and illustrator of various child-oriented media vehicles, from books to animated entertainments, is showing a curious set of images to her new friend and companion, Jason Hood. Since their initial meeting upon her arrival at Watchung several weeks prior, Chloe has spent most of her free time with Jason, sharing experiences and brushing up on each other's biography.

Today Jason is distracted and gloomy. He has just been fired from his job as media technician with CGC. His supervisor, Donald Ackerman, has accused Jason, who was on duty at the time, of being either incompetent or mischievously complicit in sabotaging his welcome presentation on the previous day. From discussions with various members of the audience Ackerman has gleaned that the disruptive and, for him, humiliating laughter had been sparked when letters spelling the initiatives list had suddenly reshuffled behind his back into nonsense verse. He has fixed the blame on Jason and rejects all protests of his innocence, as well as supporting testimony by Jason's fellow workers, that none of them tampered with the media feed. The process log, they point out, shows everything as having been normal. Whatever happened during the presentation must have occurred on his side of the feed. But Ackerman is unmoved and remains adamant. Jason must go.

For the moment, Jason has lain aside his own tribulations. Chloe is mystified by a set of drawings and paintings produced by a new acquaintance, a young man with long locks of silver hair who has been showing up at the daycare tent to play among the children and join in various pastimes.

At first Chloe was apprehensive of the stranger, but gradually has accepted him as a benign and even helpful, if somewhat eccentric, presence. Indeed, he seems to elicit an unusually buoyant and receptive mood among her young charges.

Earlier today, her curiosity deepened as she watched planned art sessions unfold throughout morning and afternoon. She and her new 'helper' rotated among age-grouped platoons of children, leading them in pictorial renderings of day to day experiences, encouraging them to depict events and observations impacting recent family experiences.

"Do you see anything unusual in these pictures?" Chloe asks Jason. Before him she lays out a sequence of increasingly adept pictorial efforts.

"Just a bunch of drawings. Some look like they were done by really young kids, toddlers. Some older. Then teenage. And that one, by an adult." Then Jason points to a very accomplished portrait of Chloe, rendered in pencil on sturdy paper, "That one is by a trained artist."

"That's what I would think, too, if I hadn't seen it myself."

"Seen what?"

"These were all produced today, by the same person. That funny man I told you about."

"Why would he do those," Jason points to scribbles and clumsily articulated figures typical of very young children's pictorial imaginings, "when he obviously is capable of this?" He picks up the lovely portrait of Chloe rendered in a deft economy of line. "Was he just goofing around? Pretending, for some reason?"

"I don't think so," Chloe shakes her head slowly, "I watched him closely because there's something so earnest about the way he engages whatever he's doing. And he kept asking the children around him for advice about what he should do next, like he really didn't know and was learning from how they responded. And he kept asking their opinions of his progress. He is the most empathic person I have ever met."

"What do you mean?" Jason lays the portrait down and picks up one of the earlier, more naive depictions of a big bug surrounded by lots of children, some seem to be playing and laughing. Others seem serious, even crying.

"When I look over at him working with children, he just seems to become one of them. When they are working on a happy picture, he's full of energy, laughing and joking with the kids. But if they make a picture about something sad he asks them to talk about what the picture is about. Sometimes he's so affected that I see tears streaming down his cheeks. A couple of times I intervened to lighten the mood because I was afraid the youngest ones would think he had been hurt somehow and be upset by seeing an adult crying."

Chloe pauses for a moment, thinking over her recollections. Her voice seems distant, in some obscure stray place where curious thoughts might glide, "He seems to absorb everything they know . . . to hear about their worries, and fears, and joys and learn how they deal with them. But then it's as if he kept on developing, but much, much faster . . . until he was able to draw this portrait of me." Chloe's eyes are luminous as she recalls avid enthusiasm of all the children working throughout the day alongside the indefatigably expressive zeal of her new associate.

"So?"

"So. I think that on each piece of paper he drew a picture the way he did because at the time, that's all he knew how to do."

"But that's crazy!"

"I know," Chloe acknowledges, "It seems crazy, at first . . . but if you lay these out in the order that he created them, they are a progression through classic stages of development in childhood art, like a set of examples from courses I took in Art Education. Now aren't you the one who always says that any consistent pattern deserves a reason for being?"

"Yes, but this could be coincidence . . . maybe he's playing a joke . . . maybe . . . ," Jason shrugs his inability to penetrate the mystery she finds in the evidence spread out before them.

"To get from that," Chloe's voice is insistent as she points toward a haphazard raking of roughly concentric lines circling in wobbly orbit about the center of the paper, "which qualifies as a 'pre-schematic stage' typical of a four or five year-old, to that," she points to a scene of people at a picnic beset by gigantic ants, all rendered in shakily looping lines, "which has clearly entered the 'schematic" abilities of maybe an eight year-old, and on through the ever more adept pseudo-realistic and realistic stages, on up to a mature and nuanced rendering." She pauses and tilts her head in astonishment. "Typically, that's almost twenty years of assimilating experience, of learning to gather and express ever more complex ideas about space and time and appearances and how objects and people and events interact. He's accomplished it in one day!"

"Maybe he was . . . ."

"Maybe he was what?"

"I really don't know," Jason throws up his hands and chuckles, "It doesn't make any sense. Unless he's trying to show the kids something. Or show off."

"Oh, look, there he is, over there!" Chloe exclaims, excited to see the occasion of their mystification off in the distance. She points to where a silver-haired man hops about as he talks and jokes with a red haired woman who seems, unsuccessfully, to be trying to ignore him.

It is Jason's first glimpse of Chloe's new acquaintance, though she has talked about him earlier. Now, sauntering a zigzag tack before the red-haired woman, his arms wave abreast like wings or sails. He veers now with, now against, some imagined wind. Not too far away Jason sees that the pantomime echoes through a wake of dancing and delighted children emulating their hero. The red haired woman seems suspended between dismay and disconsolation.

Jason's impression of the distant scene is that of a wildly careening milieu of diverse youth, all sizes and demeanors. They bump and bounce among a crowd of only mildly interested adults still milling about after their evening meal. The swarming young chorus echoes barking incantations of nonsense chanted by a cavorting silver-haired Pied Piper.

Ant's Sister just won't wait,
This mister isn't late.
Work winners bind their fate,
Love losers grind and grate,
Sister's blisters blind her mate
Grasshopper fun is fate!

The silver tressed man seems very familiar. Then in a flash of recollection Jason recognizes the berserk air and jaunty, comically arrogant gestures that inflect his gyrations.

"Chloe, that's Grasshopper!"

"Grasshopper?" her head pulls back slightly. She looks puzzled.

"Yeah, Ant and Grasshopper," Jason reminds her, "you remember . . . those silly spots popping up everywhere for the past few months. They had all the neat animated effects . . . I wondered at the time how they were doing them?"

Chloe looks closely for a moment then frowns lightly. She does not watch nearly as much teleview as Jason. But enough for recognition to suddenly break across her face in dawning excitement, "Grasshopper? Oh, yes . . . is it really him? Grasshopper? Here?"

They watch for a moment, but eventually Grasshopper and the children prance off out of view. The red-haired woman lingers to watch their retreat, then gets up and strolls slowly off in the opposite direction.

"So, back to reality . . . what are we going to do?" Jason returns to the topic they have been avoiding. "We need to decide what's going down. I have to find a job somewhere. Nobody around here needs a hacker-programmer type."

"How long before we have to leave?"

"That's just it," Jason is bitter, "Ackerman won't give me a break. He and I have never gotten along. And now he's really pissed. He told me I had to be out of the quarters tomorrow. Otherwise he'd have me arrested."

"But can he do that? Just kick us out?"

"Jesus, Chloe," Jason explodes, "just look out there! The camp is crawling with people they're kicking out. They're all going to hit the pavement tomorrow. I guess I'll be among them. But I don't want to leave you here!"

"But can't we call someone. Maybe to come and get us?"

"Come and get us? Who? Most of our friends are in the same fix we are. And how? Nothing works anymore—except for the filthy rich who can get anything they wish. Do you know anybody like that? And since our bicycles were stolen we don't have any choice but to walk."

"Oh, mess."

Chloe and Jason have gathered their few possessions and piled them into backpacks in preparation for departing Watchung the next day. They return to the picnic grounds near the CGC mess tent. Chloe wants to introduce Jason to Grasshopper, especially now that a bracing dash of celebrity translates her earlier puzzlement into beguiling fascination. She, undaunted as always at the prospect of a new acquaintance, goes up to Grasshopper and asks brightly,

"May we join you?

Grasshopper glances first at her, then at Jason, narrowing his eyes suspiciously.

"Ants can't keep up. Is anyone here an Ant?"

"Heavens, I hope not," Chloe's laughter is musical and clear. Apparently she appreciates some implication that totally escapes Jason.

"What ants?" Jason wonders aloud.

"Ants are angry and confused but march around like little soldiers keeping everything and everyone in control. But they hate themselves for it."

"Why is that?" Chloe asks.

"Ants can't play because they think it's wasteful and evil," Grasshopper sighs as if there is little point in discussing the obvious, "so they feel bad and hurt one another trying to make their own evil feelings stop. But it doesn't do any good because they really are strangling themselves in their own unhappy imaginings."

"But that's so sad," Chloe protests, "And so unnecessary."

"Exactly," Grasshopper agrees, "Your children don't act like that yet. They're still hoppening."

"Her children?" Jason is puzzled.

"He means in the daycare. The kids we work with," Chloe assures him.

"They showed me a lot of interesting stuff," Grasshopper says, "And they aren't yet like their parents. But they'll get there, I fear. You can see seeds already sprouting toward greedy pretensions."

"Ants or parents?" Jason frowns.

"Same thing! Fathers of denial and mothers of empty wishes. Taunted imaginings they pass on to their children. Sucker moms . . . fatuous fathers . . . do unto themselves before they do unto others." Grasshopper's voice tinges in impatience with Jason's dense reluctance, "So they want to break me into little pieces. Take my parts back to their nests to keep them warm and pretend they're happy, so they can feel good through their winter."

Jason grimaces a fleeting warning to Chloe that perhaps it is not wise to promote familiarity with someone in such obvious distress.

She ignores him.

"That's just the way Ants are," Chloe agrees lightly, looking mischievously at Jason while she speaks, "Always got their nose to the grindstone. Never have any fun. Got to get this project done because they've already agreed to work on another one. The open-source committee's never satisfied . . . always one more bug to find. Hack. Hack. Hack"

"Exactly," Grasshopper seems gratified by her response, "Hack. Hack. Hack. Hack attack. Hack attack back . . . but I can't break. Or won't break out. It's all the same, can't, won't, canto, wanto . . . I am that I am . . . Aliswasmaybe." He nods, delighted to glimpse some depth revealed in Chloe's dawning enlightenment. "I can't break because breaking is Antward: Breakdown. Breakout. Breakthrough. Breakaway. Breakneck. Breakback. Breakbread . . . all Ant words."

"Antward and upward," Chloe exults in their rippling flow of nonsense. Jason falls farther into gloomy exasperation at being so puckishly left behind.

Grasshopper brightens. "Antward or hopward?" he laughs delightedly at an apparently profound pun, "Antward to hopward. But you can't go back and change Antthings. What is: once was: what shall be, so Antward to hopward goes onward to upward. And all the world swims below . . . ," he croons melodically.

"I like that," Chloe laughs, "Anything? or Antthing? 'Y', what a deliciously easy choice!" Her faery demeanor has relieved Grasshopper's sodden mood.

"I like you," he announces a newly gushing fount of admiration, as if she lifts some hateful and solitary burden, then Grasshopper directs a baleful gaze toward Jason,

"Are you Antward, hopward, onward or upward?" the question is nonchalant in evoking some fatuous compass of aimless direction but he might as well be asking Jason precisely where quantum entanglement begins.

"I don't have the slightest idea what you two are carrying on about," Jason stammers, disheartened at his to failure follow them through an unfathomable hinter world.

Grasshopper peers carefully at Jason, then expresses misgivings to Chloe,

"He seems an awful lot like an Ant."

"Jason? An ant?" Chloe chirrups in amusement, "Who knows? You may be right. But he's my dear sweet friend and I do love him so."

"Well," Grasshopper grudgingly accepts her endorsement, "You would know better than I."

"Oh, Chloe," Jason sighs, his perplexity strives to reconcile with her first affirmation of their mutual attraction. He realizes that he is apart from their renegade camaraderie. He is no more able to participate than he can fly, or hoppen. Whatever that means! Earthbound, he wanders alongside in gruff acceptance of abrupt divagations and dancing flights to subtle joys.

In that moment and often thereafter Chloe beams vibrantly, seeming to flit into and out of a dimension of light playfulness beyond Jason's grasp. At such moments he knows himself to be truly prosaic and pragmatic, absently regretting loss of whatever touches some mortals with childlike spontaneity. Perhaps once he had known such a realm, but it has long since been closed to him.

Maybe they are right, Jason muses. Antwardly he plods through dubious misgivings of their perpetual states of wonder. Then he smiles, realizing that he is musing over leavings of nonsensical play.

So he pushes along such thoughts until they are guided back into matters that must be dealt with. He tries to interest Chloe in going back to consult a map and plan their routes. They have talked about maybe finding their way back to southwest Virginia where Chloe grew up in the mountain hollows. It seems to offer sanctuary. Now might be a good time to go. There, persisting myths of self-reliance take on enhanced value nowhere available in these days of calamity in urban milieus. But now, unexpectedly their plans seem sidetracked in a new affiliation which Jason does not seem to share.

"Chloe," Jason interrupts the ongoing fusillade of banter, "we have to get ready."

"Jason, what's the hurry? We don't own enough to worry about."

"I just think we need to make some plans. Talk about where we're going and how best to get there."

"We will," she soothes, reaching out to lay a finger lovingly on his lips, "just be a little patient, please?"

"Let me know when you're ready," he stalks away to sulk, unable to pry her from this new and intrusive influence. He is angry that she resists. In Grasshopper and Chloe's easy celebration he foresees the unraveling of his own claims on her. Jason chews angry frustration of futures skewed by chance encounter with a madman.

Jason sits at one of the picnic tables and withdraws from his pocket a pair of vuspecs which he dons as he unrolls a touch key panel, spreading it before him on the table top. He taps a dot printed to the upper right of the efabric key panel's QWERTY array of printed squares labeled by alphanumeric characters. The vuspecs begin to glow and a virtual computer display painted by a faint laser directly onto his retinas hovers in space before him. Jason adjusts the display's apparent size and resolution, as well as nudging the vuspecs' polarized lenses to dim the background view so he can concentrate on the monitor's contents.

The computer is embedded in his iVest and automatically synchs into the camp's ambient maxnet signal. He checks his CGC account. Yes, as expected, he already has been locked out of the CGC and FERAL communications systems. He dexterously taps the keyboard and opens a link into one of his maxnet accounts, an open source project maintained on servers running in Houston, with backup mirrors in Dresden and Hong Kong, where he is team leader of the AiPAL Project.

His team comprises a number of fellow programmers and cognitive theorists who are attempting to craft a general purpose intelligent assistant and expert advisor they have dubbed an *Artificially intelligent Pro-Active Linker*, or *AiPAL*. The joke within the group is that the i is lowercase because they don't want the AiPAL to start getting big ideas about who's boss, you know, like 'I think, therefore I am'.

The linking apparatus of the iVest uses standard maxnet protocols. Jason notices that it has also synched with a signal that wavers in strength, but must originate somewhere nearby. Am I being cracked? he wonders to himself, as he looks about to see if he can determine the origin of the signal probe. Within normal signal ranges he sees only Grasshopper and Chloe pacing about, discussing joint misgivings about some dubious universe of Anthood.

Jason taps to awaken a system probe. He quickly isolates the interloper's beacon tag in a list of background system servlets running on his computer. Looking at its properties he sees it is polling the system to determine available resources, apparently looking for media applications. He is unconcerned since he has taken precautions of always running even known and trusted apps in auto-protected system sandboxes. This thing can probe all it wants, he smiles confidently, it will not get access unless he grants it.

Suddenly Jason's hubris shatters. He realizes that somehow the probe has already downloaded an application from some server out on the Internet. In the monitor's display appear full 3D renderings of an animated Grasshopper along with an immobile Ant. The latter stands stationary to one side of the display field, emblazoned with a large '?'. The Grasshopper figure paces to and fro across the display center as it carries on rapt and avid conversation with some unseen companion denoted by the question-marked static Ant.

Puzzled, and more curious than alarmed at this sudden co-option of his system, Jason lifts his head and opens the polarized lens to admit the scene before him. He sees Grasshopper strolling about with Chloe tagging along, both seem enraptured of some newly enlivened universe of playful prospect. The movements of the arms and legs attached to the Grasshopper with her are precisely those of the animated avatar seen on Jason's view screen; Chloe's Grasshopper friend is driving the motions of a Grasshopper media avatar! Further, Jason hears the Grasshopper avatar speaking in a fractured cartoonish voice relayed along vuspec aural eartips His rakish voice rambles on through a discourse of impromptu revelation of shortcomings of Ants and attendant misgivings of Grasshoppers. Following a sudden impulse, Jason leans closer to the key panel's built-in microphone dot and says in a low but urgent voice,

"Grasshopper, tell Chloe to wave at me."

He sees Grasshopper lean toward Chloe as if to share a confidence. To Jason's astonishment, she pivots gracefully and happily waves back at him.

And then, in a flash, he understands. He knows.

Gnosis transpires not in singular realization, but across varied prospects of comprehending sentience. It coalesces out of reorganizing eruptions where disparate moments of event-filled insight cascade to revitalize fundaments of being. Of having been. Of might have been or of going to be. Of making sense of the senses. Of universal intimations that underlie awareness in self being self. It brings new constellation, new paradigm to re-order experience borne in images, thoughts, smells, sounds, tactile factors.

Jason smiles, then chuckles in pleasure of insight so renewed. He bursts into laughter, marveling in the grace of newly glimpsed schema of how things fit and interact. Pointing, perhaps, even down into the innermost mysteries of the quantum universe and outward to the far side of the relativity continuum.

"The only physical realities, then, are the particles that result when observation collapses the wave function," Jason quotes aloud from ancient lectures, smiling as he adds Grasshopper's recent surmise, "It's all just hoppening." He shakes his head in wonder that such imaginal farragoes at the junction of self and world can seem so clearly pregnant with implication.

In the distance Grasshopper leaps in graceful pirouette and lands, feet astride, arms out-stretched. He seems to bow slightly back toward Jason. Chloe claps her hands and again waves to him.

Now, as Jason drinks from impetuous necessity animating emerging prospects, his newly seen template emerges to coalesce far-flung components into the 'self-sustaining and musing entity' his colleagues call an AiPAL. He and his teammates have heretofore been unable to provoke any cybernetic combination into responsive alacrity that might pass for 'awareness', much less, 'consciousness'. But, if he can intercept the intimate interface of soul and sensibility and impart into it enhanced inputs of information, an AiPAL may yet arise among humans. If insight indeed proves out, such an AiPAL may yet evolve into a self-organizing and auto-referential network of information traces that exhibits sentient self-awareness and coheres within a context of ever emerging new possibilities: consciousness.

As he later explains to Chloe, at the very moment she waved back at him, he suddenly grasped this new paradigm intuitively: the entire universe is alive because within it creatures live: extension expresses intention. In profound aesthetic our entire eco-system reaches out into and across galaxies of complex process interaction, even down to the quantum level and across relativistic event horizons. Spontaneously complexifying sets of feedback loops somehow are universally intrinsic. Through them, through all sentient beings, the highest-order universe that comprises all things, is itself comprehensively alive. And a profoundly wary process takes place across every moment even down into the lowest, most obdurate orders of all being, however complex their form.

Suddenly he sees, too, what she has been trying to show him in the progression of images drawn by Grasshopper among the welter of similarly profound renderings by his fellow playmates. What occupies years for a developing child apparently requires only a few hours in Grasshopper. Perhaps, Jason surmises, he is driven to a higher level of functioning by the iLink computer running adaptive inference algorithms. Somehow the iLink taps Peter Dooley's memories to massively re-construe them, along with whatever maxnet resources he is able to draw from, enabling a new, more complex way of interacting.

"Grasshopper is not simply weird," Jason tells her later, "He's like a maxnet marionette, plying archived media imagery into probing narrative acts, like some raw imagination of the probable. . . ."

"He doesn't act like a marionette," Chloe frowns, "what makes you say that?"

"He acts differently when the iLink is on, less scattered and aimless."

Chloe does not respond, but merely puckers her lips as if finding the whole idea distasteful.

That evening, Chloe brings Grasshopper's iLink to Jason for repair. It has simply stopped working. Grasshopper, she says, is very anxious without it. Can Jason get it working again?

He quickly determines that its power module, never intended for use outdoors, has been knocked ajar by bumps and blows encountered along the highway, breaking the circuit by which it drives the processing units. He repairs the circuit and then secures the module to better withstand rigors of refugee life.

In looking over the internals of the resuscitated iLink Jason finds a standard diagnostic port and attaches the unit to his work station. Its screen blossoms into code signatures, some unfamiliar, but many identical with open-source components making up elements of the AiPAL. A quick scan confirms that many iLink constituents, such as its app manager and remote communications linker, have been gathered from standard open-source projects, some of which he has even helped develop. In hope that he may find clues to help resolve issues they've been having with the AiPAL, Jason dumps the iLink process image and its memory cache to optical storage, a tinted transparent plastic cube called an 'oprock'.

Jason next uploads a copy of the AiPAL software kit he has helped build from a remote server via his iPak onto his euPhone, a device that may either be handheld as a traditional cell phone or clipped onto a shirt front where its projector beams a graphic interface onto any convenient flat surface, even if only the palm of a hand. He hopes to gain deeper insights into how Grasshopper navigates and construes oceans of information available to him via the iLink and maxnet. What software is running? How much of Grasshopper is legacy of Peter Dooley's sensorium? Just where in him is sense made of the senses? Especially when that sensorium interfaces with seas of heretofore unimaginable prospects of information and churning tides of data. It is an answer that thus far has eluded the AiPAL team.

Chloe is delighted that Jason has repaired the iLink, "You've made his day," she giggles.

"It may be mutual."

~ 4 ~

"Let me have your attention, please," Lieutenant Nathan Pomeroy calls to the seventy or so people gathered in the CGC Welcome Center tent. "People, come on, work with me. Quiet, please."

Gradually buzzing chatter hushes. Lt. Pomeroy's cordial smile briefly acknowledges their cooperative spirit.

"I know some of you are anxious about resuming your journey to Harrisburg. I want to do everything I can to reassure you that you are going to be okay. The US Army, working with FERAL authority, is available to help you every step of the way.

"I'm Lieutenant Pomeroy. I command one of several platoons of sentinels that monitor the route to Harrisburg. We operate out of a Command Center in over in Lebanon, about half way to the Harrisburg relocation center. I want to assure you that we will do everything in our power to see that you arrive safe and healthy in Harrisburg."

He pauses, watching for signs that his words are being absorbed, then continues, "The first question that always comes up is, 'Why do we have to walk?'"

Dispersed grumbling among the audience confirms that this is indeed uppermost in a number of minds.

"Well, long story short, what with all the efforts we're making throughout major urban centers on both coasts, and ongoing overseas commitments of troops and equipment, there just isn't enough FERAL or Army to go around. We don't have transport capacity nor adequate fuel to move everyone who seems to want to go somewhere else. The trains out of this sector can't haul everybody. There simply are not enough cars, trucks or fuel to do the job. So we have to reserve trains and other modes of transport for the physically challenged, the sick, infirm, aged . . . anybody who can't hike the distance."

"How about wealthy big shots? They don't seem to be on anybody's list of people not walking!" a disgruntled voice complains aloud.

"Well, actually, 'Wealthy big shots' are notorious for taking care of themselves," Lt. Pomeroy chuckles, "Frankly, they have their own resources . . . much more welcoming and convenient than anything the Army offers!"

A spray of caustic laughter repays the Lieutenant's affable humor.

"Actually, I'm here today to help provide you with the information and resources you need to arrive in Harrisburg fit, healthy and happy." A pause. "Yes, I said happy . . . even charged with energy and enthusiasm. How do we accomplish this?"

On the screen a map depicts Interstate 78 between Watchung Reservation, NJ and Harrisburg, PA. Along the line representing the route are house symbols labeled with 'Way Station #' and a mile designation.

Lieutenant Pomeroy continues, "The task of assisting and monitoring outward migrations on both east and west urbanized coasts is immense. No one officially ordered any evacuation. In fact, just the opposite, FERAL advisories have generally cautioned people to remain in their homes. But rumor, urban legend and media speculation have stimulated millions of people to spontaneously decide to depart major population centers. We cannot move millions of people at the drop of a hat. But we do have to deal with them."

Lieutenant Pomeroy sips from a bottle of water, then resumes, "As indicated on this map, FERAL has divided the Harrisburg evacuation route into thirty-two segments, five miles each, marked by way stations. Each way station is operated by a local business that has bid and won a FERAL evacuation sutler franchise. Certain basic staple items, food, water, first aid and toiletries, are rationed at no cost at all way stations. But each station vendor may also offer their own goods and services as well, which will vary from site to site. While FERAL emergency ration voucher cards cannot be used for anything other than designated necessities, you may use your own funds or credit accounts to purchase any of the goods and services available from participating vendors."

As a couple of privates distribute printed notes of the Lieutenant's presentation he clicks a remote. On the conference screen flashes the logo of the *Susquehanna Hiking Society*, beneath which is a list of precepts to guide happy hikers:

Why Walk or Hike?

To Stay Fit: Millions of Americans get their daily exercise by walking or hiking. It's a great way to stay in shape and reduce stress. Walking for 20 minutes a day at a moderate pace of three miles per hour (a leisurely 20-minute mile) can:

• burn an extra 120-150 calories a day
• promote fat loss and preserve lean body mass
• reduce the risk of heart disease
• increase mental alertness and memory
• increase energy levels throughout the day enhance one's motor skills

Hiking with a 10-15 pound pack provides all the benefits of walking, but also increases the calories burned by ten to fifteen percent.

"And there are further benefits," Pomeroy muses, clicking the remote:

To Take In the World Around You: It's only by getting out and experiencing the natural environment that one truly appreciates its wonders. Walking and hiking both have numerous built-in educational and entertainment possibilities:
• wildlife watching
• learning about and experiencing beautiful foliage and fauna
• historic, tourist walks
• themed day hikes with outdoor or singles groups

To Help Preserve the Environment: Walking and hiking on trails help millions appreciate the need to preserve our remaining open spaces and greenways. The less we drive, the better off we all are.

Pause. Click.

Just Your Speed
Whether you are walking to lose weight, walking around tourist attractions, hiking to enjoy a beautiful, scenic trail, or just to reach a FERAL Relocation Center the following guidelines can help you understand the appropriate pace or activity level for you. Walking is divided into three main speeds:
• strolling - 20 minutes per mile, low intensity, physical exercise

• brisk walking - 15 minutes per mile, the pace of most exercise walkers
• aerobic walking - 12 minutes per mile, generally for the advanced walker

"The trick will be to gradually build up your pace. Start by strolling. Then as you develop stamina, your pace will naturally pick up." Click.

Can You Go the Distance?
• Start out with short segments and build up to two-to three-mile walks.
• Take frequent breaks, don't overdo it.
• After building up your confidence and experience don't dwell on negative experiences; look around and experience the beauties and new experiences. Make this hike an adventure!

"There are some basic tricks to maximize your effort." Click.

Putting the Right Foot Forward
• Keep your breathing at a natural pace with your heart rate.
• Always maintain good posture, with your lower back flat and pelvis tucked directly under your spine.
• If you plan to walk or hike at a pace above strolling, it is important for your body temperature to rise gradually, So warm up for at least 5 to 10 minutes before increasing your speed.
• Stretch out after your walk, when your muscles are warm and flexible.
• Relax. Control, rather than tense your muscles.
• Take quick steps, not long strides, for the most natural stride.
• To determine your target heart rate: walk fast enough to notice your breathing, but not so fast you are out of breath or gasping; if conversing, you should have to pause regularly to breathe.
• Avoid blisters by choosing a properly sized and fitted shoe. Also try wearing synthetic fiber socks because they reduce friction and draw moisture away from the skin.

"And for you parents with children." Click

Walking and Hiking with Kids

• Take time to learn. Children have their own perspective, insights and ideas.

• Try incorporating games into the outing, like looking for loose change on sidewalks or different types of trees and leaves on trails.

• Take frequent breaks. Remember, kids do not always speak up. You need to keep asking them how the are doing.

• Caution children not to wander away. Have a plan in case you get separated.

• Involve the kids in planning the walk or hike. Go over checklists together and review trail guides and maps.

• Get them their own backpack and let them be in charge of certain items, like snacks or like the flashlight.

"I know it may be difficult, but we'll try to help you equip yourselves through the roadside sutler stations." Click. "Here are some guidelines."

What Gear Are You In?

Footwear – The proper shoe is vital to any walking or hiking activity. When walking remember to:

• find the proper fit – plenty of room for the toes with a snug, comfortable heel

• look for a solid support and good cushioning inside the shoe

• look for a firm, resistant heel counter outside the shoe When hiking remember to:

• choose boots about 1/2 size larger than your regular size; there should be plenty of toe room; the heels should fit snugly

• wear two pairs of socks for trying on boots and for hiking

• look for boots with a thin polypropylene liner to keep your feet dry and use thick wool or acrylic socks for warmth

Outerwear- Always try to dress for the weather and be prepared for the worst. Some tips to help guide you:

• In the winter, dress your upper body in layers to keep you warm and to prevent overheating; you should take layers off as you heat up (and store the extra clothing in your backpack.)
• Get yourself one reliable jacket that is appropriate for the climate, then wear different layers based on the day's weather.
• Always take along a waterproof jacket and hat in your backpack just in case it rains or snows.

"We'll just skim across the equipment list, it's all in the handouts you've received." Click.

Backpacks – The right waistpack, daypack or backpack is essential. Packs vary in size and fit; buy one that is suitable for the type of walk or hike you'll be doing.

Waistpacks, or fannypacks, are belted around the wearer's waist and come in varying sizes and shapes. Depending on what you are carrying, these versatile packs can be used for shopping, traveling/sightseeing, fitness activities or a day at an amusement park.
[Look for those that feature a separate pouch designed to carry a waterbottle – no matter how short the outing, you should always drink plenty of fluids.]

Daypacks, sometimes called knapsacks, are the most commonly seen and used backpack. Worn on one's back with two shoulder straps or simply carried by the top handle, these serve the widest range of activities, including as a book bag, for picnicking, day hikes, as an overnight bag, or as briefcase for work.

Click.

Important features to look for in a daypack include:
• a padded back and padded shoulder straps for added wearer comfort
• durable, weatherproof fabric
• thermoplastic buckles, zippers, etc. to avoid rust

• stormflaps over zippers to keep contents dry and the right size and number of pockets or compartments to suit your activity
• one-piece body construction so there are no major seams that could tear

Internal and External Frame Packs are much larger than daypacks and generally are used for more serious hiking and overnight camping, when you need to carry a large load including a sleeping bag, tent, cooking equipment, etc.
[Some internal frame packs are designed to be used as luggage that can be worn on the back or carried by a handle. These are very convenient for the traveler that is always on the go.]

Click.

How to Enjoy Your Stride!
The Essentials - FERAL Evacuation Packs will be provided for each evacuee: food, water, maps and guides, a compass. A small first aid kit, a pocketknife, matches, toilet paper, a flashlight, sunglasses, sunscreen and appropriate clothing for all weather possibilities.

Stay on the Route – Safety is vital. Realize that people in the back country are fearful of being overwhelmed and are extremely suspicious of all evacuees. Surrounding and outlying communities will NOT welcome you. They may harass, arrest or even harm strangers they perceive as threatening their well being. Continue on to the Relocation Center at Harrisburg. There you will find assistance in contacting family and friends, as well as assistance in regaining normalcy in your lives.

The Lieutenant earnestly stresses, “This situation is not easy for you, but neither is it welcomed by outlying communities. You will find no help, no comfort, no rescue in the areas surrounding the route. I repeat. Do not leave the official evacuation route. Those who do will be fortunate if their fates are only to be arrested and returned to the route. There are documented reports of everything from rape, plunder and murder to organized press gangs of human traffickers picking up stray wanderers and selling them into global slavery. Leave the highway and you could easily wind up in a worse kind of misery on the other side of the world."

He pauses for a moment, "Stay on the official route!" Click.

Don't Litter! – Take a zipper locking plastic bag along to carry any wrappers, empty bottles or miscellaneous garbage to a waste receptacle.

Snack Break – It's very important to keep your energy level up no matter how moderate the outing. Always take along water and food for recharging and drink plenty of fluids before and after. Ideas: apples, granola or trail mix.

Safety

• Help one another.

• Watch out for bicycles and occasional official vehicles. Step aside and let them pass.

• Obey official directives.

• Use only portable toilet facilities placed along the route for your convenience and public health. Do not relieve yourselves on the highway or its right of way, fifty feet either side. Usher drones will severely punish any immodest and unhealthy behaviors!

Kristin leans over and whispers sarcastically to Stacy, "Can you believe this spin? They're saying this isn't an evacuation, it's really a recreational marathon to lose weight. And we aren't evacuees, we're a niche market for tourism marketers! This is all a health spa and mall, from here all the way to Harrisburg."

"Could it be more 'thick and heavy' from your friend Cornrath?" Stacy asks.

"He's no friend . . . just an acquaintance."

"Well," Stacy whispers back, "as Jim likes to say, it's always the way money wants it."

"What kinds of goods and services?" someone asks.

"You name it," the Lieutenant replies, "Bicycles, room rentals, hot showers, clothing. You might even find a massage, who knows? Somewhere along the way someone probably offers it."

"But the FERAL cards can't be used for those?" a despairing sound tinges someone else's query.

"Only for food, water and other necessities," says the Lieutenant, "But you'll find a lot of goods and services available. And you probably can still get a GloBanc card if you hurry over there after we finish here."

"Where do we get the FERAL ration card?" Kristin asks.

"FERAL card? You got that when you completed your CGC screening."

"Oh, that's right," Kristin replies, feigning misunderstanding. Muttering under her breath she laments not having gone through the process, "great . . . after screening."

Stacy leans over to whisper, "Don't worry about it, honey. You can have mine."

"How did you get one? Besides, you'll need it!" Kristin protests in restrained surprise that Stacy has managed to get a card.

"Honey," Stacy laughs lightly, whispering, "I travel with Jim Meerling. He specializes in collecting all kinds of cards. You take mine, I'll get another." She reaches into a blouse pocket, pulls out a green rectangle of plastic marked FERAL Basic Voucher and hands it to Kristin.

"A lot?" Kristin asks as she looks over the card, on the back is a printed number and beneath it, Stacy's signature. "How does Jim get these?" Kristin marvels.

"By being Jim. Contacts, deals and friends . . . he knows everybody," Stacy assures her, "and everybody knows Jim when they want something."

"Contacts, deals and friends," Kristin echoes, as she recalls Jim's exasperated encounter with the menacing man in the mess tent. Then she adds, gratefully, "Thanks, Stacy, really. But are you sure it's okay? This has your name on it . . . oh, look, my middle name is Marie, too," Kristin smiles and taps the card lightly.

"So just pretend to be me," Stacy waves her fingers in dismissal of her concerns, "Nobody really cares."

"You're from Florida?"

"Yes. North of Tampa, from Lake Magdala. I met Jim working a pole down in Miami."

Looking around, it dawns on Kristin that Dooley is not in the audience. She is surprised to realize that she has not even thought of him. Somewhat guiltily, she marvels at how much she has enjoyed the respite from his unrelenting peculiarity. But out of dutiful habit, suddenly concerned that she has no idea what bedlam he might be provoking, she asks Stacy in a huskily hushed voice, "Have you seen Peter?"

"No, I haven't" Stacy shakes her head slowly, "But he's probably with that little camp counselor over at the daycare. He spends a lot of time there."

"God, I hope he doesn't get into trouble. That would be my worst nightmare . . . if he gets out of hand with children."

"He'll be okay, honey. With them he's like a child himself, growing up fast," Stacy flashes a wry smile, "But don't worry about it. He can take care of himself. You take care of you,"

Kristin nods, acknowledging to herself that Peter is not responsible to her. Nor she to him. She is free.

At daybreak all inhabitants of the camp not displaying a valid wrist pass are directed back along Sky Top Drive, past Seeleys Pond, onto New Providence Road and up onto the I-78 westbound access ramp. In a tinge of dismay Kristin remarks how little ceremony attends their departure. "Goodbye. Gone. History. Forgotten." She, Stacy and Jim mount the long ramp up toward the Interstate. In the distance usher drones hover and zip up and down to remind wayward souls that there is only one authorized route. Here and there are posted a scattering of soldier sentries, each equipped with the ubiquitous long shepherd staves for handling the unruly.

Kristin surveys surrounding groups gathering in reluctant journey toward Harrisburg, "How did we ever get into this predicament?"

"By not buying into their deal." Jim snaps.

"No. I mean big picture—the whole country . . . the whole world, for all I know," Kristin opens her arms in a wide arc.

"Too many people wanting others to take care of things. Then 'the others' get used to taking care of themselves first. Eventually it all falls apart," Jim twists his mouth ruefully.

"But when did it go wrong? This can't just happen overnight."

"Oh, that's easy," Jim grimaces a near grin, "Once there was an ant hive. For a long time they lived very well just by foraging and gathering. But then a few really bright guys figured how to buy and sell claims on future resources. And then some others created what they called options and contract insurance and still others invented even more clever deals that they bought and sold with other hives, some very, very far away. Then one day it all busted apart because it had gotten so complicated that nobody could remember what promises had been made to whom and all those hives that had invested heavily in things they didn't understand suddenly found themselves facing winter with only worthless receipts to depend on. Then all hell really broke loose as the newly minted have-nots tried to take what they needed from the haves."

"And everybody decided to go somewhere else, even if they had to walk to get there," Stacy truncates the parable she has heard countless times.

"Head 'em up! Move 'em out!" Jim drearily caricatures bright sarcasm.

"Rawhide!" Stacy yawns disinterestedly, then says to Kristin, "Jim is addicted to old westerns."

"Why? A bunch of stray exiles, outcasts and misfits steal land while they kill and abuse one another . . . I've never seen the appeal."

"Just like life," Jim testily defends his entertainment preferences, "Nasty, brutish and short. And so, here we go . . . again."

Kristin glances back to see that Peter is dawdling far behind the main cluster. He seems to be walking in company of the young woman counselor who holds hands with a young man Kristin has not seen before. Around the trio a number of children gambol, as if even ouster from Paradise would be just another fun outing. "Who's the guy holding hands with the camp counselor? I haven't seen him before."

"I think they're sweethearts. Jason is his name. She's Chloe."

"How do you know all this?" Jim asks testily.

"Grasshopper told me."

Kristin notices a defensiveness tinge Stacy's mood. Jim is a mite annoyed.

"Please don't encourage that nonsense by calling Peter 'Grasshopper'," Kristin tries to ignore their friction.

"Honey, who am I to say who he has to be? Grasshopper? He says it is. What's in a name, anyhow? It's still the same package!"

"Package?" Jim fumes.

"Just drop it, Jim," Stacy warns, then maneuvers in a few quick steps so that Kristin separates them.

"What is it with you two?"

"Stacy has suddenly become quite the naturalist. She's been going on a lot of nature walks with your whacked-out buddy. Why don't you ask Ms. Zoologist?"

Kristin looks at Stacy who rolls her eyes in disbelief, "Men! Now who do you suppose came up with the notion of an 'ownership society'?"

"What?" Kristin is taken aback by their sniping.

"It means some men want to own everything. Women, especially."

"That's not it at all," Jim snaps back.

"Go ahead, tell Kristin your favorite joke," Stacy challenges.

"I'm not in the mood for funny," Jim snarls as he takes a deep breath.

"What does a joke have to do with it?"

"Jim says his father used to travel on business out in rural areas. Way back when fuel cost next to nothing and people drove around in automobiles whenever they wanted to go somewhere."

"Okay."

"Well, one day . . . ."

"One day," Jim impatiently takes over the narrative, "Dad was back in the hill country of western Pennsylvania, traveling among the coal fields. He got turned around and couldn't seem to get back to Pittsburgh. He decides he needs to stop and ask for directions, but it's in the middle of nowhere and there's only woods and fields. Nobody around.

"He drives down a lonely road and eventually comes to a driveway that winds way back off the road. He sees a farmhouse back several hundred yards at the end of the path, so he turns his car into the driveway and bumps along until finally he pulls up to the ramshackle cottage. He stops and gets out of the car just as an old man dressed in dingy overalls shuffles out onto the front porch. Off to the side of the house he sees a three-legged pig rooting around in a weed-choked flower bed. Around the pig's right hip is a big bandage where its leg had once been attached.

"'Hello, there,' Dad says to the old man, 'I wonder if you could help me get back to Pittsburgh.' The old man nods, thinks for a minute and then gives him directions to the main road back into the city. Dad thanks him and then can't help commenting, 'Looks like misfortune has hit your pig pretty hard!' he says.

"'Yep,' the old man replies, 'It sure has. And that's a remarkable pig, too. He's really smart!'

"'What do you mean?' Dad asks.

"'Why a month or so ago the house caught fire back in the kitchen. That pig dashed in, woke up my wife and daughters and even dragged the baby's crib outside to safety. They got the fire put out pretty quick and everybody is glad to be safe.'

"'But what about the pig,' Dad asks, 'What happened to him?'

"'Why,' says the old man, 'a pig like that, you wouldn't want to eat him all at once!'" Jim delivers the punch line in howling, gleeful abandon as Kristin chuckles a little apprehensively.

"See?" Stacy confirms her earlier thesis, "A charter member of the ownership society. Men think that joke is so damn funny!"

"It is funny," Jim insists, wiping tears from his eyes.

"Not to the pig."

"God, you have gotten so weird and picky," Jim scolds irritably, "What's the problem all of a sudden? Everything was fine until . . . ," Jim pivots around to look back across the crowd where Grasshopper strolls along at a leisurely pace, holding hands, hopping occasionally, between two prancing children.

"Come on," Jim says, his tone suddenly becomes resolute as he faces forward again, "Let's pick up the pace. I want to get off this damn highway!"

~ 5 ~

Intermittent rains of several days pass, unveiling a sky of crisp azure, smudged only at the fringe of its eastern horizon by a haze of smoke. Summer shows slow retreat. Early autumn chill christens evenings. Kristin notes various tokens marking seasonal change in overheard chatter. One longs for a college football game. Another gauges the transient spectrum of color soon to be touching rolling hills farther along the way. Shopping might be entering a new season now, if only confusion and failure had not closed down so many malls and extinguished all but the most vital commercial pursuits.

Kristin smiles fondly, noting that nature maintains its rhythms despite human travails. At least, she ruefully muses, we will be spared this year's bout of drearily festive shopping where "Merry Christmas!" seems more like "Meretricious!"

In the space of a brief week she already has grown accustomed to rigors of the trek. She finds herself surprisingly competent at dealing with whatever arises. Her innovations make do with castoffs and litter scattered along the cheerless pavement. She and Stacy crafted abandoned dull green plastic sheeting, torn from pallets of rations and water delivered from skypods, into a makeshift tent during a prolonged downpour. Nearby Jim had sulked beneath his own fugitive scrap of plastic for the duration of the deluge, complaining of their ousting him from the warm comfort of their shelter. Even sleeping outdoors in open air has been made tolerable by simple, but surprisingly effective, inflatable pads with lightweight fiber-clad aerogel coverlets, gotten from an outfitter's impromptu store set up just outside Watchung and charged to their FERAL voucher cards.

A week earlier it had seemed to make little difference whether she reached Harrisburg or fell along the way. Now she itches to depart spare constraints of the freeway, to pick up such threads of her life as have not been prematurely pruned in the failure of Gotham Seaboard. Renewed vigor stimulates a keen discontent with their snail's pace. She is anxious to arrive in Harrisburg and resurrect her life.

Now that aching muscles have acclimated to daily exercise Kristin feels more fit, capable and resilient: able to deal with whatever comes her way. Walking and talking with Stacy, listening to Jim hold forth from his uniquely eidetic store of experiences, Kristin has gradually let go of anxieties that increasingly seem adolescent. And daily she grows more impatient with the pace of the larger tribe that has coalesced about Dooley's Grasshopper antics.

Peter, on the other hand, merely trails along, or worse, actually retards the tribe's progress. She cannot persuade him to pick up the pace. She notes that he has become some kind of informal focus for a growing number of clustering people with whom they travel. In a manner that she does not fully grasp, a gathering tribe of increasingly familiar trekkers seem drawn to Grasshopper's nonchalant incitements. A migratory clan of families has coalesced out of his persistent intrusions into affairs of those about him. He is like an irrepressibly curious child, popping up unpredictably to interact in any flare-up that threatens serenity of kith and kin.

Whether in some skirmish over rations, a family quarrel or sudden jealousy erupting among any of several tentative liaisons struck up during the journey, Grasshopper always appears to be drawn into unfolding melodrama as if he is crucial to its denouement. At such moments he at first seems overwhelmed in some immanent loss . . . of what? Kristin does not know. Perhaps abdication of his frittered Dooley persona. Then some opening into renewal seems to infect him with dread. Panic momentarily flits his eyes only to be released into Grasshopper's raucous enthusiasm. Whatever motives drive any antagonist into threat of violence, he channels his own rowdy energies to interact, distract and transmute their force into exorbitantly sarcastic characterizations that gather, redirect and defuse prospective brawling.

Inevitably in such crucial moments the iLink limner switches on and Dooley wields an antic profusion of luminous clichés swirling like comic whips of an archetypal drover to wrangle quarrelsome souls toward peaceful pastures. As Grasshopper he berates, inveigles, exhorts, or chastises by pointedly directing mounting laughter of the surrounding crowd toward whatever foibles have incited current dispute. Eventually would-be adversaries become distracted, comedically drained of animus and welcomed back into familiar folds of family and friends. Kristin has ceased to be surprised when two or more contending figures simply shrug sheepishly and walk away toward friendly faces in a community of good-natured laughter.

At first she thought Grasshopper merely to be Dooley's narcissistic performer-self, craving and garnering attention into his own self-wonderment. But now the opposite seems true. He appears reluctant to engage bickering clashes and holds back until the last possible moment, as if hoping trouble will dissipate of its own accord. His head tilts to one side, like a young pup suddenly intrigued by some playful overture. Then, as if drawn to mounting heat, he inserts himself into the fray. At such moments Kristin senses a powerful current gathering through his manner to suddenly and spontaneously project diversion into any altercation rapidly becoming public.

At first, just after they departed Watchung, among onlookers to these disputes would be an air of expectation that his upstart intrusion would be quickly dispatched by one or more combatants. But now, having seen Grasshopper referee previous bouts, the crowd relishes promise of entertaining diversion.

Fortunately, along the way, Kristin discovers that concentrated sugar syrups of discarded and stale soda pop can serve as fuel for the iLink bioconverter to recharge its ultracapacitor. Otherwise Dooley's prolific reliance on the limner would have quickly consumed the few powerpaks they brought with them.

The style of his intrusion always begins unpredictably. It may arise out of quiet chattering or explode as violent charade. But he seems always capable of catching foes off-guard, keeping them unbalanced as he assails and picks apart whatever issue is in dispute through some extravagant pantomime, one moment racing around in tight little circles like a fool chasing his shadow or in flight from luminous pursuers conjured out of the iLink limner, another hopping on one foot while reciting a madly invented rhyme in a surround-sound cinematic effusion of brightly jubilant invention. His pantomiming shrieks, chortles or exasperated outcries orchestrate familiar story setups and resolutions gathered from movie archives, teleview commercials or recollections of earlier incidents still fresh in the clan's memory.

In these antics he crafts his spell by which the focus of an attentive crowd is transmuted from anticipation of grief, misgiving, or regret into raucously jovial appreciation of slapstick narrative. Inevitably, mocked foes either relent and join in the fun or sullenly escape a jeering crowd.

In such moments of crisis he is exceedingly quick-witted. Thrusting himself between two large men arguing over one of the women, the largest of the two males grasped a fistful of Grasshopper's denim jacket and with one hand hoisted his fragile frame to dangle precariously in mid-air, the other fist drawn back to smash his face. Grasshopper instantly became a small playful mutt, tongue lapping effusive affection along the arm and hand holding him aloft. Puppyish innocence, gazing trustingly up into assailant eyes, froze the impending fist in ambivalence as his tongue, avid in noisy slurping, strove to reach out to the captor's face. A voice in the crowd, responding to the tableau of a bully about to abuse a defenseless innocent, sighed, "Awwh . . . no . . . please don't." The plea was picked up and amplified into a plaintive murmur of encircling sympathies. The protagonist blinked momentarily, set Grasshopper down, looked self-consciously at the entreating faces about him, shook his head in frustrated consternation and strode away.

Such incidents have transformed the group's initially wary disdain for Grasshopper's peculiarity into a now cheerful celebration of his presence as some fey token of singularity, a defiant mascot in this time of tribulation.

It is his signal notoriety that has become an increasing source of Kristin's dismay with Peter Dooley. As Grasshopper idles and dawdles along the freeway, so does the crowd. She watches new strangers overtake and pass them in patient dedication to forward movement. She envies their efficiently departing strides and wants to move faster herself. Often, however, she is aware that such arrivals are drawn into the Grasshopper clan, apparently basking in and welcoming its unexpected warmth of heightened humor and good will. The swarm about him thus slowly grows larger and more ponderous and seems unconcerned with ever getting anywhere.

Kristin resolves to set her own pace. She sees no reason for loitering and wants to reach Harrisburg as soon as possible. Turning to Stacy she suggests, "Why don't we walk faster. I'm tired of hanging with this group."

"Why, honey? What's the hurry?" Stacy laughs.

"Actually," Jim says, "I'm with you Kris. I want to get off this pavement, too."

"Jim, we're gonna be on this pavement for quite a while," Stacy observes, "and I kind of like being with this Grasshopper bunch."

Jim wheels about to walk backward a few steps, looking to where Grasshopper strolls as moving epicenter of a host of gamboling children shepherded by several parents chatting amiably with the young counselor Chloe. Jim almost stumbles, and awkwardly recovers his balance, "Amazing. Never underestimate the warm fuzziness of sheep."

"Or the nimble poise of traders, " Stacy cackles.

"Warm fussiness?" Kristin puzzles aloud.

"Fuzziness," Jim repeats, pointedly ignoring Stacy's slight, "Fuzzy logic. Fuzzy thinking. Fuzzy and wooly dreaming that everything is okay. You're right, Kris, this is ridiculous . . . let's pick up the pace."

"Okay?" Kristin looks questioningly toward Stacy who reluctantly shrugs her assent, "Okay by me."

Kristin is gratified by how quickly separation grows between them and the lagging crowd. Estimating her own stride in memory of gym treadmills she reckons that in an hour they will be at least mile or more ahead of Grasshopper's clan and not long after they should arrive the next station. A renewed exhilaration rewards her heightened effort.

"We're getting off here," Jim announces casually as they approach the next way-station, which comprises an old freight trailer parked behind a temporary pavilion decked out with beaten canvas covering and impromptu signs that offer a curious mix of goods and services from snack-bar, outfitter, and pawn shop under auspices of *Branson's Backdoor Trails Outfitters & Sutler.*

"What? Getting off—are you crazy?" Stacy exclaims.

"I cut us a deal back at Watchung," Jim replies in a tone of taking charge.

"Cut a deal? How? For what?" Stacy rolls her eyes for Kristin's benefit.

Jim ignores her probing to nod toward an old man seated in a strap and tubing lawn chair who watches their approach, "Hello."

"Well, I was beginning to wonder if I'd see anybody at all out here," the man replies, "not much point in sitting here if nobody comes by."

"There's a bunch of people just behind us," Kristin volunteers, "about a mile or so back."

"Good job . . . there hasn't been anyone by here all day. Last bunch to come through was yesterday afternoon. Nice group of maybe thirty or forty church families traveling together. 'The Host of Beulah', I think they call themselves. But nobody so far today 'til you came. Not even the army or those danged bouncy drones. I was beginning to think maybe they'd called off the evacuation," he grins befuddlement.

"Do you have a phone I can use?" Jim asks.

"No phones out here," the old man shakes his head. He reaches to pluck a small portable from his belt. "Just this radio. Who you need to call?"

"My attorney," Jim says, "or at least the law office my people said could help us."

"Mister, if you're out here the law ain't going to have much to offer you," the old man snorts.

"Well, just the same, is there some way you could help us contact them? I'll make it worth your trouble."

"Sure, no problem," the old man replies. Jim hands him a business card with an office contact number, "Tell them Jim Meerling is here at your station, waiting as arranged."

"All right, I'll see what we can do . . .," the man repeats absently, "Jim Meerling . . . ," and rises to go back into the tented pavilion to find a pair of spectacles.

"What's this all about?" Stacy insists in a low hiss of surprised suspicions.

"We can get the charges dropped," Jim explains quietly, " but we have to go before a judge willing to hear our side of the story. We need to clean up this mess and get back to business. We're missing out on everything!"

"I'm not leaving Kristin out here by herself. That other stuff can wait 'til we get to Harrisburg," Stacy's eyes search Jim's face for some hint of mischief, "Wait . . . are you putting me on?"

"No, this is for real," Jim declares, "Back at Watchung, while you were tramping about your 'nature studies', I got a message out to some people. I made a deal and bought papers that will get us where we need to go. We're both charged; we both have to go and plead the case. We can get it taken care of now. I called in some favors and we can't let this slip by. It won't happen again."

"Go where?"

"Over to Lebanon, to the regional evacuation command center. If we get the charges dropped I can get back everything they confiscated before it goes to auction. A chancery court there is set up to deal with civilian claims against the army."

"Why did you keep this to yourself?" Ignoring details of his scheme, Stacy is suddenly wary, "Just what kind of deal is this? You should have told me—especially if you expect me to make nice with somebody."

"It's nothing like that. But I know how you are. I can't tell you anything because you'll blab it to anyone who'll listen . . . this has to stay between us. It's really tricky . . . maybe . . . besides, everybody who knows will want to come with us."

"Can Kristin come, too?" Stacy demands.

"I'm sorry, Kristin," Jim glances sheepishly toward her, "I didn't get you papers. I didn't know," he shrugs.

"Well, I'm not leaving her out here alone," Stacy vows,

"Alone? For chrissakes, Stacy!" Jim explodes in exasperation, "In an hour or so there'll be a thousand or more people milling about here. Kristin's with Dooley and he's got a whole friggin' tribe!"

Stacy looks back toward Kristin, her eyes open slightly as if questioning what she should do."

"Stacy," Kristin urges, "you take care of what you need to do. Don't worry about me. I'll miss you. And maybe even be a little envious," Kristin laughs, "but don't make me the reason for your staying out here and miss a chance to clear up your problems."

Suddenly Stacy's eyes find inspiration among the farther contents of the pavilion's wares. She wheels about to face Jim, "I'll go with you on one condition."

"Oh, jeez . . . what now?" Jim snarls.

"A bicycle for Kristin. Get Kristin a bicycle so she can get to Harrisburg really fast. Then we can meet her there." She points to a couple of touring bikes chained to the trailer platform over at the side of the tented pavilion.

"Stacy, I don't know how long this is going to take us," Jim anguishes, "If we don't succeed with the judge we don't know what's going to happen. And if I do get my cargo back and charges dropped, I've got commitments I have to deal with. This is not a friggin' vacation, you know!"

"In the first place, you've already worked a deal with a judge . . . it's already done or you wouldn't take the chance. Second, I don't care what happens to your precious cargo. It's been nothing but trouble."

"All right, all right. Okay. Put a sock in it. I'll get her the damn bicycle. If it's anywhere within reasonable . . . but we're not waiting around if something else causes problems. You're coming with me, okay? Deal?"

"What do you think, honey?" Stacy asks Kristin, "Will you be okay?"

"Oh, gracious! A bicycle. That would be wonderful!" Kristin exults, "I'll pay you back, Jim. Really, it would be a wonderful help and I can't thank the both of you enough."

"And it's got to be a good bike, too. Fast and rugged," Stacy warns Jim, "No junky crap or throwaways."

"Is it a deal?" Jim insists, ignoring Kristin's warm delight as he glares back into Stacy's close scrutiny of his face.

"Okay," Stacy's voice loses its carping edge and becomes suddenly musical, "Deal. We just have to work out how to get in touch with you in Harrisburg."

"So how much for the bike there?" Jim points to a shiny bicycle poised with several others on a rack.

"This one?" the old man asks, adjusting his spectacles to see the price written on the tag. "Can't put it on that voucher card. Maybe GloBanc. What you paying in? Dollars? Euros?"

"Do you have an aftar clipper in here?" Jim looks about.

"Aftar . . . that's the gold receipt, right?"

"American Free Trade Asset Receipt, right . . . credits in gold transferred to your bank. That good enough for you?"

"Oh, yeah. We don't get too many people using them yet, but I've got one here somewhere. I'll just have to get it out and connect it to the radio," he rummages through a box of networking cables and equipment. "Let's see, the price today is $12,650 . . . that works out to about, oh, call it one aftar."

"The dollar's not doing too well, huh?" Jim shakes his head.

"Not much useful happening these days," the old man replies, "and it gets to be a pain having to change prices every day. Soon they say we'll be doing it by the hour!" His voice becomes strident in his disgust with errant vicissitudes of finance.

"Well," Jim shrugs, "I've heard that an ounce of gold will always buy a good suit; now we can add to that, a nice bicycle. I guess that's what it always comes down to." Turning back to Stacy Jim asks expectantly, "Is this going to do the deal for you?"

Stacy appraises the bike's features as Kristin removes it from the rack and throws a leg across to try the seat height.

"Seems about right," Kristin nods approvingly, flexing the brake levers and leaning back against the seat to test the braking action.

"Yes, Jim," Stacy answers, "I'll go with you. But we're going to get back in touch with Kristin once everything's taken care of."

"Okay, okay." Jim walks over to where the man has set up the aftar clipper, leans down for a retinal scan, keys in several codes and waits until the unit recognizes his login. "One aftar being clipped to Branson Outdoor Trails & Sutler."

The old man tentatively punches a couple of keys, reads the authentication message and announces happily, "Pleasure doing business with you, Mr. Meerling."

Kristin's voice sings girlish thrill as she reads aloud from a tag attached to the gleaming new bicycle, "Modeled after classic Dutch commuter bikes, the *Spectra Amsterdam* gives you everything you need for jetting around a city -- and strips away everything you don't. Its steel frame can handle the bumps and grinds of crowded bike racks and won't buck you off when you hit potholes," Kristin laughs, "Plenty of those out here on this ragged pavement," then continues,

"And its three gears are all you need: Even in hilly Santa Fe, we never wanted more. But it's the user-friendly niceties that make this such a luxe cruiser: full fenders to curb the spray, an enclosed chain guard so you won't grease up your slacks, and a burly rack for lashing your briefcase or groceries. And the generator powered headlight means you'll never be stuck in the dark with dead batteries."

"Oh, that's just the thing for your briefcase and groceries," Stacy cackles.

"Thank God I don't have to worry about grease on my slacks . . . but it will make hauling this pack easier." Kristin's giggly exhilaration brings tears to her eyes. " Really, though, Stacy, I can't thank you and Jim enough. With this I can be in Harrisburg in just a day or so!" She wheels the bike toward the front of the tented pavilion and out into the sunlight as Stacy places Kristin's back pack on the cargo carrier and secures it with two canvas straps.

"Honey, it's nothing, really. I feel guilty about leaving you out here. But now that I think more about it, I worry about you striking out on your own. Wouldn't it be better to hang out here and wait for the Grasshopper clans?"

"Oh, I'll be okay. They're monitoring the interstate all the way to Harrisburg. And it's only five miles between way-stations. That's nothing on a bike. Once in Harrisburg I should be able to make contact with Cornrath's office, find out where they've moved to, and wait for them to come pick us up."

"But where are all the drones and patrols?" Stacy worries, looking around.

"Patrolling nearby, probably," Kristin reassures her, "See? Up there's a skypod . . . looks like it's coming down to land farther ahead. I'll be okay. I'll just keep pedaling until I get to Harrisburg."

After an hour or so an official vehicle appears at the nearby access ramp and approaches the Branson's Sutler Station. Moving soundlessly, the electric coaster glides to a stop. A young man dressed in FERAL work slacks, shirt and cap gets out and approaches, "Mr. Jim Meerling?"

"Here," Jim answers and holds out an ID card. "I guess there's room for the two of us in there," he peers into the small interior.

"Should be room for two," the courier replies indifferently, glancing at Kristin and Stacy, "But that's all I'm supposed to transport. Two adults."

Kristin and Stacy embrace tearfully.

"God, I hope to see you again soon," Kristin frets aloud, "And I hope you both get everything worked out."

"Honey, it's you I'm concerned about. Trust me, Jim's already cut his deal for us. It's just a matter of greasing the wheels. Old story, well practiced. But when you get to Harrisburg, be sure and register with FERAL's contact center. We'll find you one way or another."

"Jim," Kristin walks to Jim and leans to kiss him on the cheek, "I really hope it all works out. And thank you so much for the companionship—and, of course, especially, the bicycle."

"Peddler to pedlar," Jim jokes weakly, "have good travels." To Stacy he signals his urgent wish to depart as he says to Kristin, "I'm sorry to leave you out here, but we do have to go." He glances back toward the official FERAL courier already impatiently sitting back inside the coaster.

A final sensation of touching fingertips with Stacy lingers as Kristin watches them settle into the vehicle and it accelerates smoothly and silently toward the exit ramp for the trip to Lebanon. Stacy and Jim wave back to her; Stacy holds a thumb up in benediction as Kristin waves goodbye.

Looking about as her two friends disappear back down the same access ramp from which the vehicle came, Kristin finds the old man is moving his lawn chair into shade of the tent.

"There's a big crowd headed this way?" he asks.

"Oh, they're a few hours back, but they're definitely coming this way," Kristin assures him, "They do take their time strolling along . . . that's why we three pushed on ahead."

"Well," I guess I've got nothing better to do than wait for them, "Can I get you anything? Something to go with that new bike?" he offers.

"No, thanks," I guess I'll just get on with it," Kristin sighs weary realization that now she truly is on her own, "The sooner I start pedaling, the sooner I finish."

"Well," the old man waves, "good luck to you."

"Good bye. Thanks for having the bike out here," she smiles back, waving as she straddles the bike and moves it forward.

She pushes off toward the western horizon. The sun promises a few more hours of full daylight. She expects to pass several way-stations before settling down for the night.

After a few leagues Kristin realizes she is pedaling much too rapidly. She cannot afford to let her anxiety or renewed discontents with Peter Dooley wear her out. It will be a long journey yet.

Taking a deep breath, she abates the rhythm pumping her legs and looks about. Gliding beyond the largely abandoned derelictions of the ghost-suburban outreaches of Bethlehem and then Allentown, in each direction as far as she can see are rolling hills, expansively undulating in late green light of an Indian summer afternoon. Peacefulness of bucolic grace blesses her hopes. Gradually a brightly invigorating calm displaces her annoyance with Peter. An abiding sense of estrangement gives way to her welcoming newfound liberty that infuse her spirits in avid lightness of being, whetting her awareness that she is truly alive and that it is so good to be alive.

Despite reluctance to dwell upon events of the past week or so, still she cannot prevent herself occasionally marveling that she has made it thus far with so little discomfort or real distress. The first few days were the worst. Now those tribulations seem increasingly remote as she zips along aging complaints of pavement, avoiding gapped segments and most of the worst potholes. Occasionally she dismounts and pushes the bike across bad stretches to avoid ruining a tire or risking a spill.

Looking ahead, squinting in the brilliant light of the sun dancing over shimmering waves of warmth arising from distant pavement, she realizes that while truly alive, she also is truly alone. The road ahead is totally devoid of any person; not even the occasional sentinels she has grown accustomed to seeing posted along the way. It confirms her conviction that they have been traveling far more slowly than necessary. Everyone ahead has outdistanced them and are that much closer to Harrisburg and relief from their straits. She recalls the migratory gathering of church families mentioned by the old man back at the way-station and wonders if she will catch up to them by nightfall.

~ 6 ~

The sun is poised a couple of hours above the farther high hills. Ahead, just beyond a way station pavilion that proclaims *Uncle Don's Tantalizing Treats*, Kristin sees a skypod has set down at one side of the highway. Between her and the distant airship confusion attends a scene of apparent tragedy: many people—men, women, children—lie fallen and motionless, sprawled in clusters across pavement and grassy shoulder. A few motionless teenage girls and one or two young boys, prone and lifeless in hospital gowns, on tubular medic tote frames, are being lifted by women wearing dark blue jumpsuits onto medical gurneys to be hauled over and into the skypod's cabin.

Drawing closer amid growing apprehension, Kristin slows to consider how she may help. Off to one side a crew of men wearing *Uncle Don's* t-shirts rearrange and off-load crates of snacks to make more room inside a delivery truckster. A couple other men assist the blue-suited women moving gurneys to and into the skypod cabin.

Between her and the pavilion a number of heavy electric motorbikes are parked. Two men with close-shaved bristle haircuts stand near the arrayed bikes, their backs to her as they survey mayhem and occasionally shout orders. Identically lurid ornament on the backs of their black leather jackets depict a crazed wolf. Blood-dripped lettering proclaims them to be BERSERKERS. Out among the fallen, other similarly leather-clad men prowl and poke among hapless casualties. In horror Kristin watches a man rip clothing from an unconscious woman's body and begin unbuckling his pants. Electric pangs of anxiety spur her to flee on past before she is noticed—these are not rescuers at work, but plunderers robbing and molesting hapless prey.

She darts toward the narrow aisle of pavement remaining unhindered by the skypod cabin. One of the men standing by the motorbikes sees her panicked dash and shouts, "Stop her!" just as three BERSERKER T-shirted men step from the ship's interior. One shoves an empty metal gurney violently against her bike and knocks her off-balance, spilling her awkwardly onto the grassy shoulder. In a writhing blur of fear Kristin is dragged protesting and struggling toward the cluster of bikes where the two men seem greatly amused.

"Put her with the other ponies," the taller man calls out, waving them on toward the tented pavilion where Uncle Don ostensibly hawks snacks and basic hiking supplies. There twenty or so young women are held captive, bound by dense polymer wrist cuffs attaching them like awkward pearls forlornly strung onto a heavy polyester rope tethered to the pavilion's center pole. A tightly binding plastic cuff clasps Kristin's left wrist and is locked with a keyed clip onto the rope alongside the other prisoners. The three men hurry back out, eager for plunder.

From Agnes and Justine, two young women closest to her on the rope string, Kristin hears what has befallen them. They are part of the *Newark Host of Beulah*, a church group trekking to Harrisburg. When first they arrived at Uncle Don's way station it had been attended by only a few uniformed middle-aged women, two with European accents and several Asians who rarely spoke. One of the blonde women welcomed them and chatted pleasantly as more pilgrims gathered around. The entire Host of Beulah was offered chances to win money, passage to Harrisburg and even luxury vacation spots if they would all agree to participate in a marketing study of new flavors of bottled water. After a couple of weeks wandering as a homeless tribe the offer seemed irresistible; everyone volunteered.

Divided into three groups they were each given a bottle of *Uncle Don's WonderWet* flavored energizer beverage. One group was made up of adults thirty or older along with children younger than ten. Another comprised pre-teens and teens up to age fifteen; the final category, sixteen through twenty-nine. They were instructed to keep their bottle caps handy because a number printed on the inner surface would be their entry into the lottery to be conducted after everyone had responded to the flavor quizzes.

But as they awaited payoffs from the lottery people began growing dizzy and needing to sit on the ground. Then they began passing out. As anxiety grew, the Host of Beulah groggily demanded to know what was happening. The blonde woman called on a portable radio to report outbreak of some unknown malady. She urged everyone to remain calm, that help was on the way.

After a few moments off in the distance an ominously droning noise had arisen. "It sounded like didgeridoos and bull-roarers," Justine volunteers, "you know, those weird instruments you hear in Australian movies. All the bikes were making that sound."

"Oh, it's insidious," Agnes shudders, recalling the droning, faint at first, then gradually growing louder and louder until finally a gang of bikers arrived on the scene, with anything but the medical assistance they had been promised. The bikers greeted the blonde woman, whom they called Bo Peep, and she obviously knew them, calling out as names of the lead bikers, Br'er Rabbit and his buddy, Georgy Porgie.

Bo Peep's manner switched immediately from one of sympathetic concern for her afflicted customers to one of ruthless efficiency as she directed these newly arrived henchmen to quickly overwhelm any opposition among the groggy Host of Beulah.

Wielding high voltage crowd control staves and using vaccine inhalers, groups of bikers forcibly administered spray into people's nostrils, starting with any males still standing—each immediately had passed out. Then they set about subduing and dosing the remaining women and children until all were comatose, except for the young women in the late teens and twenties group who had not been affected at all by their water.

These young unfortunates, now mostly sobbing hysterically, had been methodically rounded up and tethered to this rope inside the pavilion while the blue-coated Asian women were instructed by Bo Peep to begin inspecting all the unconscious girls and boys in the pre-teen and young teen group. These they undressed, closely examined, and culled to a final selection to be re-clothed in hospital gowns. Each one, still unconscious, was placed on a medic carry-frame, wrapped with restraining bands, and conveyed by biker squads hauling them on metal gurneys over to the skypod that had descended nearby. Once all the target victims were either unconscious or securely tethered the rest of the bikers began robbing motionless adults and molesting some of the still comatose women.

"But where is the army?" Kristin exclaims in rasping incredulity at their account of outrages befalling them, "Nobody seems to be in any hurry . . . it's as if they know they have nothing to worry about."

"I heard them say that the army was sent somewhere else," Agnes replies, "somehow Bo Peep's people got them reassigned farther back in Jersey."

Kristin fights off paralysis of horror. She must somehow organize her fellow victims if they are to save themselves. She reaches across to vigorously shake two young women clinging to each other as they sob entreaties to God and His Angels to rescue them from ravages of Satan. Kristin forcefully yanks their hair to shock them into attention and orders them all to get control of themselves. "These men are the worst kind of criminals," she urges them. "They will use your own fear to goad and bind you to do what they want. But if you give in you'll only fall more deeply under their control. We have to work together . . . otherwise we're all lost. Don't give in. Don't cooperate. Fight them.

"Remember who you are. Don't let them get inside your head. You want God's help? God helps those who help themselves . . . you must think constantly of those you love and who love you. Refuse to be afraid. Refuse!"

Glancing toward the two gang leaders now sauntering back toward them, she stresses even more urgently, "Do not cooperate, no matter how these goons threaten you. They will use you against one another . . . don't go along, no matter what they do or say. Resist every moment with all your strength. Vow to survive the worst. And if you ever want to have your own children to love, think of what you would do to protect them and do no less in taking care of yourselves."

Rabbit and Georgy's boots pound across the boards of the pavilion platform. Rabbit is tall, has an athletic build, and bears a horrifically violent scar across his left eye and cheek.

Following Georgy's initial gesture he glances briefly toward her, looks back at Georgy and says something. Other men come and go, receiving instructions or permissions or even judgments to settle disputes over conflicting claims of plunder. Kristin watches the duo closely as she racks her mind for some inspiration that may free herself and the other women. Gradually it dawns on her that the one called Rabbit seems very familiar but she cannot imagine why.

Rabbit strolls along the line of young women, prodding, probing and evaluating each through narrowed gaze of his one good eye but saying nothing. As he comes nearer Kristin steadies herself to maintain the degree of composure she has just urged upon her fellow hostages. Rabbit pauses to closely look at her. His left hand reaches out to cup her chin; his thumb lightly brushes her lip as she gazes steadily back, refusing to be intimidated.

"Look, Georgy . . . the smile that angels envy!"

Astounded, she recalls that very slogan from a long ago cosmetics campaign featuring her as its signature model, always brightly smiling among bevies of lithesome, scornfully pretending angels. Kristin suddenly remembers, from before the terribly scarring wound that has disfigured him and blurts his name in an incredulous gasp of recognition, "Reese? Reese Hatlo?" She is dumbfounded, so aghast in sudden revelation that she momentarily loses grasp of any sense of their horrific plight.

"So, KayBee, great to see you again . . . especially like this." He might have laughed had the impulse not simply hissed away. "Remember the last thing you said to me, bitch?"

"No, not really."

"You said . . . quote," in falsetto he caricatures scathing condescension, "For you, I don't exist." Georgy snickers.

"That's not what I said," Kristin protests, "I said 'I don't exist for you' . . . you acted like I was your property—and I hardly knew you!"

"That's what's so funny . . . now you are my property. One of my ponies. And you can look forward to being ridden hard and put away wet."

Georgy cackles, "Sounds like you may have to break this one, Rabbit. She keeps her own head."

"What? What's happening? How . . . Reese . . . how can you be doing these things?" she stammers.

KayBee was a working name she once had concocted from her initials while a still hopeful and young would-be model/actress. So naive, unaware and not yet disillusioned, she had envisioned a stellar career that might somehow become her own branded merchandising empire. Hearing "KayBee" spoken again by a long-forgotten nemesis she fails to locate any reliable path of logic and falls speechless. Haplessly she stares wide-eyed back into her abductor's triumphant umbrage.

Rabbit turns, motioning Georgy to follow him back outside where they find and confer with Bo Peep, who becomes very agitated by what she is hearing. Berating the duo with flailing gestures, she seems to be making demands of Rabbit, who merely stands patiently before her onslaught, nonchalantly resting his weight on his right leg.

"You know that monster?" Agnes asks in an incredulous whisper.

"No . . . not really," Kristin stammers, "He and some of his punk gang buddies were recruited off the street, supposedly to add menace and color in an edgy video by a really whacked-out director of an ad campaign I once worked, years ago. It was supposed to be some kind of eruption of hellish underworld, a bunch of 'bad boys' intruding into feminine heaven. During the shoot he came on to me . . . really strong, and really scary. He wouldn't take no for an answer. A really dangerous narcissist . . . you know the type? He was even kicked off the set. Somehow he found out where I lived and just kept after me. I had to leave town to get away from him." Kristin slowly shakes her head in distress of hateful recollections, "I quit modeling and moved to another city to live with my brother."

Outside, Bo Peep has gathered her staff and boarded the skypod, which lets go its earthly bonds as electric propellers begin nudging the light-as-air ship skyward. Cries that their sisters and brothers are imprisoned aboard the departing craft renew sobbing waves of prayers among the string of ponies.

The skypod rises gracefully beyond their view framed by the pavilion entrance as Georgy, accompanied by two other men, returns. Using a key to unclip Kristin's cuff from the rope, he says brusquely as he pulls her to her feet and shoves her toward the entrance, "Rabbit wants you outside."

With the other two bikers in close attendance Georgy ushers Kristin to where Rabbit waits, standing near the cluster of heavy electric bikes, each competing in garish display of ornament and souvenir. Scattered about, some standing, others already astride their mounts, are a number of the gang occupied in swapping and sorting spoils plundered from the Host of Beulah. They pause to expectantly watch Kristin's approach.

Rabbit turns to slowly appraise her. "Here's the deal, Kay-Bee baby," Reese flicks a scheming smile to predict she will not care for any of it, "Bo Peep doesn't like that you know me. She wants you dead. Now." He waits to gauge her response.

Kristin hears words but does not react. They flutter by like butterflies, sounds without implication.

"Now, just for old times sake, I might willing to buck Bo Peep if you decide it's in your best interest to play along with us."

"Play along?"

"Come with us . . . help with the ponies. It's easier breaking them if they think there's someone on their side, kind of like a big sister."

"Me, help break . . . ?" Her wits flounder for some reliable grounding of prospects. Words dart mystifyingly through her awareness like familiar but startled birds that find they have no place to perch. She strains to regain herself but wonders what does she know of ponies? Then she understands the pony reference. The other women.

"To do what?"

"You help get them ready for market . . . these are church girls, probably mostly virgins . . . they'll bring a good price when they finally catch on and give in. You can help speed that up with a minimum of training wear and tear."

He has the gall to suggest that she might somehow help him betray these other women? But then she gains an inkling that therein perhaps lies a chance to get them all out of this fix, a frail glimmer that gradually garners her emotions and intelligence into purposeful intent. She plays along with playing along. "I don't know what you mean," she stalls.

"It means you choose. Right now. Bo Peep's way? Or mine? You just play along as if your life depends on it . . . because it really does." He chuckles in cyclopean stare to let her know how much he enjoys the prospect of her subservience. She tries to ignore its implications, her mind racing for bargaining leverage.

"Bo Peep?"

"Yeah, Bo Peep. She can cause a lot of trouble. Xanadu doesn't like loose ends."

"Xanadu? Is that some other gang you're afraid of?" Kristin tries for a mocking tone.

"Xanadu," Reese replies, "is where the richest and most powerful go to play. The Owners. They do whatever they want to whoever they want for as long as they want. And they don't fool around pretending."

"Where?"

"Wherever. Maybe Rio . . . Macau . . . Dubai . . . Vienna. Don't worry, KayBee baby, you and the others Bo Peep left behind in there are already too old. You're just the poor man's product. Xanadu raises and trains their own ginch. Starts them young, impressionable, pampered and trained to love what they do . . . whatever you call them. The lucky cute ones that know how get to live the good life, taken care of by their Owners."

"Church girls?" she is baffled by the term.

"Prize, grade A, . . . practically raised in pews," he grins.

"But these girls have no choice . . . you're making them slaves. And what happens to the unlucky ones, as you put it?"

"The rest . . . who cares? Maybe organ transplants, maybe cheap labor. They'll be peddled off somewhere. Who cares about you? You're out here all alone, pedaling like crazy to get somewhere," he mutters, "By the way, why did you leave that big group back there?" He gestures eastward, "What's the hurry?"

Kristin blurts out, "I had an argument . . . I wanted to get away from him."

Reese cackles, "A quarrel?" He moves in close to gloat, "KayBee, baby . . . with lovely lips that angels envy . . . I think you just lost the argument!"

His fingers lightly trace across her lips, then down her chin and throat to find her left breast where they tease the nipple.

"Please don't," Kristin whispers plaintively.

"Do her, Rabbit . . . saddle up!" one of the bikers yells. Another flicks a switch on a bike, causing it to emit throbbing didgeridoo moans. Yet another rumbles menacing drones of whirling bullroarer, "Ride'er I deride rider I deride ride'er I deride rider. . . ." A rising chorus of anticipation summons fellow BERSERKERs to closely gather around.

Georgy grunts hoarsely from behind, "Let's all get a piece of this." His grip tightens in anticipation as he mutters, "This snotty bitch needs taking down."

"Boys, boys . . . what's the hurry?" Reese cautions, "This is just a little tease . . . why, a pig like that, you wouldn't want to eat all at once!" His audience suddenly explodes in lusty guffaws of appreciation.

Reese chuckles as his hands begin unbuttoning her blouse, slowly spreading it open. A finger plucks her sports bra forward as his other hand expertly flips open a large knife in a single flourish and slices it to fall useless, exposing her breasts. Kristin fights to keep her gaze in focus as a buzzing mutter of approval moves across the encircling tribe. Reese closes and pockets the knife as his other hand reaches for her belt buckle.

"All at once?" she echoes lamely, recalling Stacy's challenge of Jim over the same pig punch line that had at the time been mildly amusing, if a somewhat puzzling nexus of discord between her former companions. Stacy's accusations against male ownership now prods her into awareness. She begins to resist with all her strength. Georgy tightens his grip, threatening to dislocate her shoulders.

"It's purely political," Reese grins as his hands appreciatively massage her breasts, "If you're going to work the ponies with us, you have to learn your part. Set an example for the others. You know they're watching." He glances back toward the pavilion and then returns his gaze to contemplate her belly as his hands pull the blouse tail out of her jeans. "Besides, you've got to learn where your place is, one way or the other, KayBee baby," he laughs, "and now we're going to fill you up with learning."

Kristin tries to break Georgy's grasp, with little effect. The group of men tighten their cordon, gathering closer as their odors commingle with the sharp scent of her fear, their breathing electric in anticipation. Her eyes lose focus as she looks away from hungry eyes. Georgy consolidates her wrists into one hand to enable the other to paw at her exposed breasts. She is drawn backward against his knees and suddenly, lunging upward she begins fighting furiously. Her legs kicked out frantically and connect with Reese's groin, "Goddammit, Georgy, hold her!"

"Who made you such a monster?" she shrieks, loudly enough for the other women to hear, "A fucked up father? A beaten-crazy mother? You've rotted out your own soul! You piss-pathetic loser!"

Grimacing, Reese straightens to look viciously at Kristin for a moment, unwilling to let her know whether her kick truly connected, as if to stage a suitable response. Suddenly his hand grabs her crotch and pulls forcefully up, causing her to gasp in pain as he hisses into her face, "Nag and whine, bitch and moan . . . if it wasn't for this bloody gash," his hand clamps hard against her pubis, making her recoil again, "all you whores would have a bounty on your head. What do you know? You've been coddled your whole life . . . what now, bitch?"

". . . demon pig . . . there's nothing human left in you."

Reese backs off, tallying prospects. His voice lilts, "KayBee baby, remember? For me, you don't exist. Besides, if men are pigs, women are sausage grinders." His good eye narrows, boring steadily into Kristin, then on past her to Georgy.

"Cole," he raises his voice without taking his eye off Kristin, gauging her reaction, " . . . get your pan-cam ready. Go with Georgy. Spratt, you and Humpty, too. Go with Georgy."

"The rest of you load those ponies in the van. We're heading out soon as they're on board."

"Georgy, take this cocky little 'debu-tante' down there," Reese points to scrubby brush at the foot of the shoulder sloping down into a ravine, "I don't care what you do with her . . . just make sure she's dead when you leave. And let's hear lots of screaming on the way. It'll teach the other ponies' their first lesson. King Cole, make sure you get it all recorded, up close and in full sound. The ponies can see what comes of forgetting where they are."

Into Kristin's disbelieving eyes Reese grimaces the husk of a smile, "I said you were going to help break them. You will . . . and I'll watch your performance tonight while we all eat a pig in your memory."

Stunned, angry and terrified, Kristin disbelieves so pitiless a verdict. Georgy shifts, releases her left arm, but keeps the right one behind her as he grabs her belt to lift her weight so that she no longer has any traction. She is hoisted like a sack of flesh and bones toward the site Reese has designated for her demise.

Desperate, she jams the thumb and first finger knuckle of her free hand backward to where she feels Georgy's eye might be. Both thumb and dense plastic wrist cuff connect somewhere and transmit enough violence to make him scream in pain and throw her onto the ground. When he reaches down to grab her back her heel delivers a crunching blow to his mouth, splitting his upper lip and knocking out several front teeth. Scrambling backward away from his outrage and the now advancing trio of men, she suddenly hears a swoosh and feels a wash of wind. An usher drone interposes itself, bobbing buoyantly between her and the advancing bikers, who abruptly pause as an amplified female voice commands, "Stop. This woman is under protection of the U.S. Army."

Georgy probes tentatively at an ugly gash ripped by her thumbnail and plastic cuff underneath his left eye and across his bloody mouth even worse disfigured by her heel. He lashes out with the other hand as if trying to bat the drone to the ground. An electric flash knocks him onto his back where he lies momentarily poleaxed, grimacing and writhing.

Two more usher drones join the fray. They swoop and dash among the dispersing posse of bikers. Red capsicum clouds spray out across the backward stumbling crew to send them shrieking, coughing and clutching at their faces as they scatter, bumping against each other, blinded by the potent spray.

Georgy shouts, apparently answering some imagined accusation, "They're not supposed to be here!" Kristin watches in confused disbelief as he jumps to his feet and lurches toward where Reese already has mounted his bike to edge back toward the highway, pausing to shout back for everyone to mount up, scatter and rendezvous back at the farm. Then his lead bike spurts silently toward a far exit ramp. Georgy and other gang members scramble to follow in close pursuit. Their machines noiselessly disappear down the nearby access ramp.

"Ms. Banner?" the same disembodied female voice inquires from a bobbing spheroid drone, "Kristin Banner?"

"Yes," Kristin pulls her blouse back together as best she can, tossing the severed bra aside and feeling strangely forlorn amid a twinge of reclaimed modesty, "yes, I'm Kristin Banner."

"Help is on the way," the voice tells her, "Is there anything that you especially need?"

"I'm all right," Kristin chokes back tears as she waves toward still motionless bodies scattered about her, "but these poor people all need medical attention. A lot of medical attention."

"Medical personnel and army transports are being dispatched. The first squads of sentinels should be arriving shortly."

"And hurry," Kristin implores, "a skypod just left here with some kidnapped children. It was less than an hour ago."

"Can you describe the skypod?"

"It was . . . big . . . and silver . . . I didn't get a chance to look at it closely," her voice fails.

"We'll try to intercept it . . . Colonel Cornrath sends his compliments and apologizes for the delay in assistance."

"Cornrath? Eric Cornrath?"

"Yes, ma'm, he's been looking for you. When the sentinel squads arrive they'll convey you back to regional headquarters where he'll receive you."

The drone abruptly peels away to join its companions and scan the scattered field of unconscious victims. Watching them bob and weave, lifting, then zooming downward like buoyant starlings, Kristin relents of her self-imposed composure. Uncontrolled sobs attempt to flush away all residue of numbing terror. Stumbling back, away from its realization, gradually the release subsides and she turns to see what she can do to help the other victims.

~ 7 ~

From the command console of the FERAL skypod *Monitor* Stacy surveys a fully moonlit landscape passing beneath the airship as it glides silently through early evening. Down upon the freeway occasional campfires dot the route not far from where earlier today she bid goodbye to Kristin to go with Jim in attempting to resolve their legal limbo.

His legal limbo, she gloomily corrects herself. He got himself into that mess. I got snookered into that and now, this.

She glances back toward the bank of video screens where Eddie Minnow, her host and would-be suitor converses with a uniformed duty officer about some directive that has been mishandled earlier in the day. It's all so dreary and remorseless, she muses, and pointless. Here I am again, at sea without any place or purpose, having to play a horny male to make my way.

Jim had worked through the intricacies of cutting a deal with various authorities, which required pro forma testimony from Stacy before a disinterested magistrate. She supported his contention that the charges were an unfortunate byproduct of a set of misunderstandings and their own victimization by fraud among truck suppliers.

Outside, when all seemed to have turned a corner and Jim had gotten documented releases for his truckloads of confiscated goods he abruptly turned to her with an accusatory glint in his eye, "Well, Stacy. That's it . . . goodbye. I hope you and your next mark will be very fruitful."

"What are you talking about?"

"You and me. We're finished. I can't count on you any longer. You're a risk."

"What? What risk? Haven't I just helped you get through another scrape? What's going on?"

"You're going on. I'm calling it quits. You and Dooley deserve each other. You don't know when you've got it good and he doesn't know when he's got it bad," Jim grimaces to heighten the irony of the quip.

"You're going to just leave me out here?" She was incredulous.

"I already have." He simply walked away.

Later, as she wandered about the FERAL building complex pondering her options, an officiously eager, neatly dressed middle-aged man, Eddie Minnow, had struck up a hallway conversation. Their initial banter led to a lunch invitation, an extended tour of FERAL facilities culminating in a ride on the airship Monitor and now here she is waiting for his next move along seduction's pathway.

As she watches flickering points of light down in the darkness below, her peripheral awareness is awakened as though she is passing through some kind of high energy force field. A lightness, a luminous joy, an ethereal splendor beckons. The scene passing below now seems transparent, along with a failing tangibility of floor, walls and fixtures of the airship carapace.

Every nerve of her body tingles as she grows aware of some dawning presence, faint at first, but growing stronger . . . his presence . . . Dooley . . . no, not Dooley . . . pure Grasshopper-ness. It emerges into this buoying reverie of revelations, a flood of storied images arising from untold oceans, unfolding new prospects of understanding. Somewhere in some myths, she offers five golden rings or, just maybe, receives five golden rings from him . . . their exchange deeply fathoms five interlinked golden spans in which breathe opportunities of new life.

Gradually it is revealed that these golden rings are at the core of all that is, pouring forth narratives that make meaningful a worldly inventory masquerading as ordinary stuff. Of people. Of possessions. Of places. Of things.

Even simple tales may render ordinary into uniquely surprising extraordinary. Each story gives into a world of events ever-aborning in idiomatic artistry, each expanding into new prospects of relations among things, people, ideas, options . . . birthing, finding or losing, struggling, attaining, warring and peacemaking, becoming brings forth becoming . . . all that was or is or ever shall be.

Aliswasmaybe, whispers a Grasshopper avatar in her mystically pregnant pleroma of nbots plying dialogues to interact through her nervous system and inform a burgeoning universe of inter-braided possibilities and amplified resonances of replicating necessity and hope.

Grasshopper's presence and the luminous gratuity of his being grows more intense and insistent within. She shudders recollection of their momentary coupling where in exaltation he had cried out, punning her name into "O-o-o-h-h-h . . . ec-

Stacy! Now his voice reverberates within, calling after her, "ec-Stacy . . . Machu Picchu . . . peek-a-boo!"

She passes into bliss.

Chloe and Jason have been sitting on the slope of an embankment arising toward a bridge over the freeway. Fires are dotted about, each surrounded by people huddled and bundled against the late September chill. Way stations now provide disposable campfire canisters filled with some kind of congealed pinkish sludge that burns slowly with a sickeningly sweet smell. Chloe says the odor stifles her, that she prefers to shiver a little in fresh air.

They peruse bright clusters of stars, wondering what their official names might be. Chloe has discovered that dim stars at the periphery of her field of vision are visible, but vanish when she tries to look directly at them.

"Jason, why does it happen?" she asks, mystified at the inability to look directly at something she knows is there.

"That's the most light sensitive part of the retina, out on the periphery."

"Well, of course, silly!" she laughs at his non-explanation, "but why is that the most light sensitive? Why shouldn't the part where you focus be most sensitive? Couldn't you see better? How does your evolution explain that?"

"Maybe it's so your great- great- great- great- grandmother would notice some predator stalking her in the nighttime brush." Jason teases, startling her by suddenly grabbing her arms.

"Jason!" Chloe squeals. Her voice is plaintively, distantly, echoed by another wailing cry that lifts them both to their feet.

It is the voice of a woman in terrible distress. They rush toward the sound.

A group has gathered about a middle-aged woman sprawled on the pavement beneath the overpass. She seems to be trying to rouse a little boy lying outstretched beside her, his head passive in her lap. His eyes are closed in pacific contentment that ignores her pleas, "Wayne! Wayne! Wake up, honey! Please wake up!"

Another woman kneels beside her. "What's wrong?"

"Wayne . . . he won't wake up! I can't get him to wake up!" She looks up at the encircling group. "Where's a doctor? He needs a doctor!" she cries, her face is blotched and red from anguished tears. The other woman reaches over to feel the boy's pulse at his neck. Her hand is abruptly knocked aside by the distraught mother.

"He's alive . . . he's just asleep! He just won't wake up! Wayne! Wayne!" She shakes his shoulders violently. The boy's head lolls against her lap, his blissful expression serenely indifferent.

"Mrs. Jefferson," urges a man moving to help, "Please, let me . . . ."

"Get away!" Mrs. Jefferson's voice threatens, "He's just asleep . . . I know he is. why won't he wake up? Go get a doctor . . . now!"

A man's voice calls from the edge of the crowd, "I signaled for help. They say medics are all busy at a big fight back up the highway. They'll get here when they can."

"Jason," Chloe murmurs in anguish, "What can we do? Is he alive?"

"I don't know. She won't let anyone touch him."

The boy appears to be about ten years old. His mother gathers his arms across his chest and pushes them stiffly together in the force with which she shakes him. She grows even more frantic at news that no medics are available. Her eyes plead for help, but rebuffs every attempt to touch her or the child. She has already bitten one man who seems to know her and stands nearby holding his wounded hand, baffled as to what should be done.

Tense frustration mounts within the crowd. A mute debate seems to consider forcibly separating her from the unconscious child. Speculations gather to play off one another: is it contagious? Is the water contaminated? The evening meal tasted funny. People mutter in consternation and growing fear.

"He may be sick . . . don't get too close," someone advises, "We might catch it"

The crowd's frustration quickly translates into nervousness at being too close. A circle opens about the distraught woman and listless child. For a moment there is only silence, then a new urgency suggests, "See if you can wake him up . . . make her let him go . . . see if he's really alive!"

"These people are going to hurt someone," Jason mutters nervously to Chloe. People are beginning to shift and make gestures as their apprehension welds them slowly into a coalescing mob.

"Jason," Chloe whispers fearfully, "do something . . . stop them!"

"Do what?" he exclaims.

"Make them go over onto the shoulder!" someone yelled, "It might be contagious!"

"Yeah," someone else shouted, "we could all catch it. Get them the hell out of here."

"Wait a minute!" Jason yells, amazed even as he realizes that he is speaking, but no one listens as he continues, trying to move closer. They have become a faceless antagonist against which he shouts, "Don't panic! Don't lose your senses!"

The crowd's press, which initially had drawn back at the threat of contagion, now closes in on the woman and her child, resolute in the intent to ostracize them. She pulls the boy closer to her body, shielding him against their newly hostile mood.

Suddenly Grasshopper appears in the midst of the narrowing doughnut of coercion. He stomps slowly within the encircling group, the iLink limner festooning smiley faces and cartoon bluebirds of happiness. Gradually his pace accelerates until he is striding rapidly around and around the inner periphery of the startled mob. His mouth makes a sputtering noise, like a boy playing with a toy automobile. The limner populates the space with cartoon automobiles whirling about a miniature track encircling the woman and child. Startled, the phalanx begins pulling back as Grasshopper unexpectedly leaps into a mad gallop about the stymied circle.

Here and there protesting cries of pain announce that he does not mind stomping toes to drive people back. His sputtering noises intensify as the iLink limner unleashes surround sounds of a *NASCAR* pack racing about some speedway into his miming of a racing car dashing through ever widening circles, fiercely shifting and ricocheting off obstinate bystanders who are coerced back from his centrifugal onslaught.

Gradually the crowd backs away from Mrs. Jefferson as Grasshopper extends a slow, protective spiral outward. His eyes narrow in intensity of total commitment to the race, the limner thunders a cacophony of a stock car melee: tires screech precariously, engines whine at full tilt, even a public address system provide an account of the progress of some imaginary contest. When the centrifugal effect of his circular race has opened an adequate space, he slides to a stop in what seems to be a tremendous din of screeching tires and groaning metal. His hands wave away a mimicry of dust and smoke.

Grasshopper snaps his head down to peer closely at his right foot. Gazing intently for a moment, he slowly pulls it to one side by rotating it on the heel. Reaching quickly, he lifts up an imaginary coin for everyone to see, then tests it in his teeth.

For the first time he acknowledges the woman and child sitting on the pavement. He jerks back to glower briefly at her, looks at the coin and then strides to the man bitten while earlier offering to help.

Grasshopper seizes the man's wounded hand, holds it to his ear, and pushes the coin into his opened mouth. Before the surprised man can recover his dignity, Grasshopper taps his nose with abrupt flourishes of his index finger. There is a faint sound of a touch tones, followed by a distantly murmuring ring. Someone must have answered at the other end, for Grasshopper launches into a tirade, yelling into the man's forearm,

"Purnell! Are you trying to get me killed? Who let this damn' woman and her kid on the track?

"Well, never mind! Somebody better wake up around here or we're all going to the hospital . . . watch it, Purnell!"

Grasshopper slams the arm sharply against the man's chest where it holds as Grasshopper peers into the shirt pocket to see if he has gotten his coin back. Taking two steps backward, he spins a flawless pirouette and abruptly ends with a quick hop.

Both feet slap flat against the pavement as he brushes back his hair, continues the movement across his shoulders and down his torso to smooth his clothing. He examines his right foot, as if inspecting the polish of his shoes. The mother and child beyond his poised foot appeared to catch his interest. He speaks in a mock, vaguely Eastern European accent,

"Aach. Vhat haf we here-in?"

Someone in the crowd mutters impatiently. An answering hiss quiets the grumbler. All eyes are directed at the trio. The crowd tries to rouse itself from distraction, poised in an equilibrium of fear, confusion, deferred frustration and now relief that something seems to be happening.

A couple of hops bring him next to her shoulder. He seems to have brought a retinue of medical colleagues, resembling

Buster Keaton, Stan Laurel and Oliver Hardy, with whom he quietly discusses academic implications of this particular case. Even the mother has become distracted in his quickened kaleidoscope of gestures and characterizations as he and his phantom retinue elaborate the conditions of an invisible examining room.

"Oh, of course?! Grasshopper exclaims triumphantly, nodding in agreement with his pretended colleagues, "Der liddle child cannot breath so vell all ofer. Iss da shoose. Dey are binding hiss tootsies!" He quickly leans over and removes the boy's scuffed and scraped shoes. Pulling off his socks Grasshopper squints one eye as he glances back and forth from the small bare feet to the mother's anxious eyes.

"Dis liddle pigky vent to market . . . dis liddle pigky stayet home . . . dis liddle pigky hat vroast biff . . . and dis liddle pigky hass none."

Grasshopper pauses to hold the little toe between his thumb and forefinger as if it might be a convenient handle by which to hold the foot suspended in the air.

"And dis liddle pigky said . . .kitchy kitchy kitchy kitchy koooooooo"

His other hand begins tickling the sole of the small foot. Suddenly Wayne cackles with delight; his eyes pop open and he squeals, "Grasshopper! Don't . . . it tickles!"

Dropping the foot carelessly as he takes a step back, Grasshopper pretends exasperation,

"Young man, get your mother off the track before she gets run over!"

Suddenly Grasshopper is distracted by a thought from overhead where a monitor skypod glides noiselessly through the night, almost as if someone is calling to him out of its silent presence above dying images of a fading speedway. He tilts his head slightly, listening expectantly.

"ecStacy!" Grasshopper suddenly bellows. His head rolls back and arms open wide in exuberant welcome toward the surveillance skypod overhead, its electric vanes rustling lightly like a wind in distant trees, "ecStacy . . . Machu Picchu . . . peek-a-boo!"

The skypod continues drifting northeastward and Grasshopper dashes down the embankment in pursuit of the departing airship.

Gradually the crowd regains its composure. People blink, look at each other and make conversation, "The kid was just asleep . . . happens sometimes to children . . . I've never seen anything like it . . . he looked dead to me . . . I don't know . . . that guy's really bewhackers . . . the boy seems all right now . . . ecstasy? Machu Picchu? Peek-a-boo?" scattering commentaries buzz across the dispersing crowd.

"Jason," Chloe gasps in wonder of the moment, He healed him! He really healed him! You saw it."

Jason sees her eyes glisten with sudden tears of joy and inspiration. A hollow suddenly opens in the pit of his stomach as he says, "Chloe . . . that's crap! The kid only just asleep. Maybe a fever, or something."

A brief gulf has opened suddenly between himself and a true believer. She smiles wanly and slowly shakes her head, "Poor Jason can't let go. No, no, we saw it . . . it happened . . . come on, we need to stay with him."

Without knowing why, he fails to protest. Jason grabs both their backpacks and jogs after as Chloe scrambles down the embankment in pursuit of her newly given mentor.

~ 8 ~

Until Kristin saw Eric Cornrath she was consumed in aftershock of mindless havoc and savage depredation, on a level of viciousness beyond anything she has ever imagined. Shattered, she gratefully had stood aside as squads of uniformed army sentinels and medical support personnel triaged and debriefed victims. They were efficiently directed by the same young Lieutenant Pomeroy she recalled meeting briefly at Watchung Reservation.

Eventually she had been escorted away by a pair of female MP's. A brief trip in a coaster brought her to the front door of a lavishly styled Greek Revival country house located near the *Harrisburg Corridor Regional Command Center* outside Lebanon, Pennsylvania. The dominant impression she gained along the way was of numerous signs boasting convenient access to the *Pine Meadows Golf Complex*.

"Kristin," Eric Cornrath rushes toward her reaching out his arms as she steps through the doorway, "thank Heaven you're here!"

As they embrace Kristin realizes that he is trembling like a startled puppy. She steps back to break a hug that lingers too long. His manner has visibly failed since he let Dooley and her take a chauffeured coaster into Manhattan. Usually poised, suave and erudite, his persona now seems to have imploded, fractured and scattered: vagrant now among rambling queries and disjointed comments issuing from beneath dreary eyes.

"I've been at my wits end!" Cornrath exclaims, leading her into a tastefully appointed sitting room. He is in pajamas and dressing robe. "Can I get you something? Would you care for a drink?" His hand waves indifferently toward a table bearing a bottle of brandy and several snifters, one already pressed into service.

"No, no . . . thank you," Kristin says, "just some water, please."

"Oh, certainly, my dear," Cornrath disappears to quickly return with a glass pitcher of ice water, which he dispenses into a clean goblet and hands to her.

"Whatever happened out there? Are you all right? Where is Dooley?" Questions tumble forth eagerly, as if they have been pacing about, muttering to themselves and now are breaking free to declare themselves.

She recounts her ordeals, starting with the most recent and working back, as if leafing in reverse through sheets of disconcerting scenes penned by a spiteful playwright.

"It was surreal . . . like it wasn't really happening . . . then those drones appeared and knocked him down and they all just scrambled to get away. When the voice said 'Colonel Cornrath sends his compliments' it all seemed so insane that I really wondered if anything could ever be real again," Kristin falls silent.

"But where's Dooley now?" Cornrath frets impatiently, "I sent a squad to find him among that large cluster farther back on the highway. He's not there. He's disappeared. And the evacuees are all worked up, claiming that Grasshopper works miracles, that he's a healer. Apparently did something to heal some little boy. Then he ran off into the night yelling 'Ecstasy! Ecstasy!' They seem to find some kind of divine revelation in that." Cornrath's voice is gruffly bemused as he adds, "It's crazy! We've got to find him!"

"He IS crazy! He's stark, raving mad! That's what I've been trying to tell you," Kristin strives for a remnant of her former self-assurance. "It started even before we went into Manhattan. I told you how he was acting so weird, hissing and spitting, carrying on about ants. And ever since the Holland Tunnel it has gotten worse. Now he insists that he is not Peter Dooley. That he is Grasshopper. He . . . it . . . whatever . . . spins through manic gibberish with no pretense of ordinary common sense. And it's like he's addicted to that damned limner!"

"Can it be?" Cornrath ponders a thought, then mutters to himself, "Oh, those fucking idiots!" He tosses down a shot of brandy.

"What? What is it?" Kristin asks.

"The nbTwin. Could it have overwhelmed Dooley's persona? Actually become Grasshopper?"

"What? What do you mean?"

"Back when we first commissioned them. You remember how carefully we stepped through ramping up the nbots. Proto-replicators were deployed throughout the nervous system, to build biotic interface?"

"Yes. And when the nbots saturated the neuron sites the replicators were switched off, to be flushed out."

"Well, that's just it. They neither switched off nor flushed out. They just went dormant for a while. I don't know if lab techs misinterpreted test results or just goofed off. Maybe there's a glitch in code somewhere. Maybe, who knows?

"Somehow the replicators activated again. When Bill overdosed his replicators magnified and enhanced the opiate hit. The amplified exuberance fried his nervous system . . . killed him. Now the replicators must be working up Dooley's system into God only knows what!

"It sounds like his nbTwin has displaced his sense of self, which I always found to be pretty shaky to begin with. Now, it may be controlled by neural net auto-adaptive routines in the iLink—not the limner, that would be a side-effect. The iLink gathers information into the nbTwin. Its neural nets self-organize while exploiting resources from maxnet. Who knows to what end? We have to find him and turn off those replicators. Before they kill him—or make him really run amok!"

Kristin glares back into Cornrath's speculative dismay. She is stunned by grim implications of his rambling premise. Unable to think of any practical reply she asks evenly, on a totally different topic, "How did you find me? Why did those drones suddenly appear out of nowhere? And where was the army all along? How did that gang get control of the highway? They acted like they knew they had nothing to worry about!"

All the suppressed energies aroused and lingering in vestiges of terror drive her questions percolating toward a shrill zenith. Her eyes have rekindled and she flashes outraged disdain across a simple observation, "And if you're really a Colonel in this damnably incompetent outfit don't you have to wear a uniform, for Christ's sake?"

"But that's just it," Cornrath pleads defensively, "it's all a terrible mistake—happening at the worst possible time!"

"A mistake? Ha! Let's see you sell that load of horse shit, Mister Madison Avenue!"

"But it's true," Cornrath insists, "It really is . . . you know, they assigned me a temporary rank of Colonel. But that was just to give me enough clout to expedite ECHOES nanotech applied to the Ant and Grasshopper campaign. I wasn't really supposed to be in the actual military."

"So, what's the connection to all this, then?" Kristin waves to include not only his very comfortable present accommodations, but also to implicate the entire evacuation support effort. "How did you wind up here, responsible for the Harrisburg Evacuation Corridor?"

"When President Crown declared a state of emergency, the Army was given the mission to assist FERAL operations. But they're already stretched thin by two wars. They don't have the manpower. So Crown ordered activation of all reserve forces, along with whatever National Guard units are available to be assigned to the effort for the duration. Somehow someone somewhere saw that a reserve Colonel was administering a publicity project—that was us—and decided, without checking, to shut it down and reassign all assets—me—to supervise the Harrisburg Corridor. I've been trying to get someone to listen, ever since. It's been horrible! I don't know anything about how the Army works or even what we're supposed to do out here."

"You didn't realize that it's important to protect people out on the highway?" Kristin explodes, "I see that you did find a way to be near a golf course!" Her scorn scrapes incredulity that this man, formerly a beacon of efficient accomplishment in complex PR media projects now confesses that he has no clue what to do next.

"Golf? Oh, that's Eddie Minnow. He's FERAL liaison officer. And a golf nut. He got us these houses for administrative quarters. His place is just over there. It puts this place to shame!"

"Shame?" she repeats for effect, "Shame?—how would you know?" Kristin sneers.

"Besides, I turned all the day to day stuff over to FERAL. They take care of operations. I just sign papers when they say something is needed. Meanwhile I've been trying to get someone to pay attention to the mistake. But no one wants to hear it. They've got enough to worry about just covering their own backsides! Mostly I've been looking for you and Dooley," he offers hopefully.

"And how did you find me?" She reminds him of the yet unanswered question.

"I've been frantic! The driver that took you into Manhattan was beaten nearly to death by a bunch of thugs that took over the coaster. I never should have let you go back into that madness. I had no idea how bad it had gotten!"

Kristin's head twists slowly side to side to deflect his excuses.

"Then, I thought the two of you probably would head out along the northern corridor," he probes for conciliation, "that's mostly where we've been looking. But just a few hours ago I got a phone call. Some woman identified herself as Stacy Madeleine. She got through because she told the operator that she had information regarding Peter Dooley and Kristin Banner. If there's anything they know at the operations center, it's those two names." Cornrath smiles thin claim to having accomplished at least something. "There are alerts plastered all over every surface and database!"

"Stacy? Stacy called you? Where is she? How is she?"

"I don't really know anything about her. It was a very hurried conversation. She said she was in an office somewhere taking care of legal matters. She had seen my name written down with a phone number. She didn't elaborate. Anyhow, she told me that you were bicycling toward Harrisburg and that Dooley was in a large evacuee group back near the Branson Sutler station. I pulled sentinel teams from where they'd been assigned to look for you and Dooley. Drones were dispatched from the monitor ship, way back up the line. It took them a while to find you. They were almost depleted when they did arrive on the scene. There was just enough energy reserves left to scatter that gang of hoodlums and hold the ground until sentinels arrived."

"Did you catch them? The gang? Reese Hatlo?"

"No, the sheriff and federal marshals are out still looking for them." His brows knit in puzzlement, "What's a Reese Hatlo?"

"It's his name. The gang leader . . . it's so bizarre . . . I actually knew him once, years ago. What about the skypod? Did you find it . . . with the children they kidnapped?"

"No . . . at least not anyone on it. The skypod was found abandoned, floating in Delaware Bay. FERAL called in the FBI, according to whom it may have set down to transfer the abductees to a ship. Or they could have been taken to some unknown location and the airship left in the bay just to confuse the search. They're still trying to track them down, but it doesn't look promising, I'm afraid."

"They all seem to be working for some underworld organization . . . Xanadu? Have you heard of it?"

"Xanadu? Well, yes, I've heard of it," Cornrath curtly dismisses its import, "it's just an urban legend . . . PR for the underworld. People want to believe that kind of crap. It makes the world seem more understandable, a familiar and reliable good-against-evil struggle. But that's just more superstitious conspiracy nonsense than anything else."

"Reese talks like it really exists. He makes it sound like a reawakened global franchise of Ottoman harems, resurrected for the superrich."

"He's just fantasizing," Cornrath sighs, shaking his head despondently, and pours another drink, "I think that particular myth is more about Hollywood's vision of the world than it is about reality. It's a meme that persists in modern societies."

"Eric, please don't try to solve all of this with alcohol," Kristin pleads, "you'll only make things worse."

"They're already worse," he mutters and defiantly tosses down the drink. She watches him start to reach again for the bottle, but then he takes a deep breath, changes his mind and pours the snifter full of water, instead. "Or at least I am. But you're right, of course. And the very fact that you're here does bring a scintilla of hope. I could always depend on you, Kristin. I didn't realize it until this insanity began. Right now, you seem like a beacon of grace and promise . . . and I'm sorry. Truly sorry for what has happened to you. I'm just thankful that you're alive. I had just about given in."

"Given in? To what?"

"To defeat. The entire nation is falling apart. Washington is crippled in corruption, lobbied by criminal money bartering for pieces of whatever is left. It's almost impossible to find a line anywhere separating organized government from interlocking schemes of corporate crime.

"Xanadu would be such simple explanation," he chuckles ruefully, tossing his hands in the air, "That everything is falling into Xanadu where the powerful regard the weak as chattel playthings, disposable trifles. But our cities really have become hellholes. New York isn't the only one in trouble. And two wars are going badly. The country is bankrupt after years of borrowing to pay ongoing expenses. The dollar is useless.

"And Ira Crown, that valiant little southwestern District Attorney who was elected to clean up Washington is finding that corporate interests, especially those with roots in the underworld, have far more clout in Congress and among the two major parties than either he or God. He's resorted to chasing about the countryside, leading cohorts of marshals and local sheriffs to conduct raids and sudden crackdowns as full media events with teleview cameras in attendance. He thinks he can rally a turnaround with such antics. It's laughable, really. Or laughably tragic."

He motions for her to take a seat on an oxblood leather Queen Anne sofa. He plops down nearby in a plush lounge chair.

"Even the Army is on the verge of mutiny. A shadow group calling itself *Morning Star Militia* is said to be operating secretly within the ranks. Nobody really knows what they're up to. Maybe mutiny? Who can blame them? Who can blame anyone for feeling pissed on by a wealthy few families who own and control everything. Remember the 'ownership society' . . . well, here it is, just not as originally envisioned.

"Even FERAL. You can't trust anything or anyone. Anything may happen, at any time, it seems. When I looked to see what had happened to the sentinels assigned the near Branson station I found that they'd been reassigned and sent to deal with crowd control during one of Ira Crown's expeditions out among the evacuees. And he's just spinning his wheels, Crown is, running around out there like he's still a DA—trying to show that he's on top of things. Someone somewhere took advantage of the opportunity. That's how a skypod was able to land, bring in that gang of hoodlums to rob, molest and kidnap those poor people. Somehow it was arranged within FERAL and I was the poor sap who signed the redeployment orders."

"Yeah, sure, Eric. You're the victim here."

Cornrath shrugs feebly, "I'm just waiting for General Michaels to show up and have me taken out . . . shot for dereliction."

"Surely not!"

"Why not? That's how it looks. Nobody wants to hear the whole story. It's too confusing, or too revealing. They'd rather find a scapegoat. And I'm the most convenient. Meanwhile

Dooley roams the back country, an upstart nano-mutant maniac vectoring technology of unknown potential."

~ 9 ~

After a restless night trying to wallow a comfortable cleft for his back into clods of dirt, matted leaves and sticks, Jason awakens to sunlight dazzling filigreed clouds stroked along the eastern horizon. He feels chilled; rheumatic pains twinge stiffened joints and muscles ache from the previous night's aimless dash through the woods: he, chasing Chloe; she, trailing her guru; Grasshopper, pursuing fugitive ecstasy. In the dark Jason hurt his ankle.

While mentally cataloging all the ills of his frame, he sees that Grasshopper also is awake and is noisily poking about the roots of a large elm tree with a broken tree limb. Chloe sits up, glances briefly toward the rattling stick and then peers over toward Jason.

"Good morning, Jason," she says sweetly. He gathers from her airy mood that she has slept peacefully and in relative comfort.

His grunt is non-committal, muttering, "Morning." Blinking several times, he adds, "I suppose breakfast is going to be a problem."

She giggles, "A big problem, Jason, if you expect eggs and ham with hotcakes and a side order of . . . ."

"Stop it!" he snaps. Already Jason regrets leaving the freeway. There, at least they ate regularly and sometimes found tossed off prizes along the way: snacks, toys, useful clothing, even a basketball he'd dribbled for several miles until Chloe complained about its constant pinging against pavement. Sitting cold and hungry in alien woods, the evacuation route seems lavish in his recollection.

"Why the hell are we out here?" he asks, more as commentary than question. But Chloe answers, anyhow,

"Grasshopper followed it out here."

"Followed what?"

"Ecstasy."

"Ecstasy?"

"Something about Machu Picchu."

"Jeez! . . . does this look like Peruvian ecstasy to you? Chloe, this is ridiculous . . . come on, you realize that, don't you? Jason's eyes implore her to reconsider their options.

Jason's entire life has been in cities. While he has seen teleview and read about people living in woods or even camping on vacations, no sense of outdoor life has ever penetrated his urban expectations. Back on the freeway, despite rigors and hassles with others, perhaps even because of such bickering and complaints among evacuees, he still felt at home: the way is paved and rules are enforced by people with guns. Hordes of people moving along the freeway are, like himself, urban spawned. They seem like a city on the move. When they complain, he intuitively grasps the motives and habits from which their grievances arise.

Here, however, in this unkempt wooded thicket crazy with growing things, he is alien. He has no conception of what opportunities to seek nor threats to avoid. He suspects there will be snakes but really doesn't know even what to dread in that regard. Give him a predatory band of surly punks any day! At least they are somewhat predictable—cough up some pocket change and usually it's okay.

In the distance he hears occasional staccato thumping of helicopters back toward the freeway. He listens and even after Chloe's breakfast fantasy finds himself in a reverie over reminiscences of stale biscuit, compacted egg loaf and enriched fiber caplets that, taken with water, expand in your stomach to make you feel comfortably full.

"Anyone ready to go back?" Jason adds listlessly. Chloe merely throws him one of her don't-be-that-way looks. "I didn't think so," he sighs.

Grasshopper still pokes at the tree roots with a stick, totally absorbed.

"What is he doing?" Jason asks Chloe.

"I don't know," she shrugs, then rises and goes over to ask, "What's so interesting?"

Grasshopper does not answer immediately. During a long silence his stick pokes, drags and occasionally taps against elm roots. Finally he observes, "Ants can't be alone. Drag one off to the side and it scrambles back to rejoin the others. They like to march in little lines like soldiers . . hup, two, three, four . . . and they gang up on bugs much bigger and tear them to pieces." He shudders apprehensively. "They seem very resentful of other creatures."

Chloe looks back at Jason, smiling as if to champion keen insight. He fights an urge to derisively roll his eyes. It will only annoy her and at the moment he does not want another huffy exchange. "Gee, that's . . . really . . . weird," he vents his annoyance into fading sarcasm.

"Look," Grasshopper's stick guides Chloe's gaze, "there's a beetle they're breaking up into little pieces to haul back to their little underground bungalows." His dismay seems steadied by profound realization. "It could be these," his fingers flutter lightly above the swarm, as if tempting the tiny creatures with a prospective cadaver.

"Social insects of the family *Formicidae*, along with wasps and bees, they belong to the order *Hymenoptera*. Ants evolved from wasp-like ancestors in the mid-Cretaceous period between 110 and 130 million years ago and diversified after the rise of flowering plants. Like wasps, they've a reputation for an ornery temperament," Grasshopper declaims haughtily. "Today, more than 12,000 species are classified with upper estimates of about 14,000 species. They're easily identified by their elbowed antennae and a distinct node-like structure that forms a slender waist."

"Wasps?" Chloe jokes, "Do you mean all those White Anglo-Saxon Protestants back on the highway?"

"Hold on," Jason interrupts suddenly, peering at Grasshopper in disbelief, "how do you know all of that?"

"From Wikipedia," Grasshopper replies, "it's embedded in my cache."

"You can't access maxnet out here."

"That information came up when Ant & Grasshopper were still in production, while building up character traits. Want to hear about Grasshoppers? It's in the pool, too."

"Please, spare me. But . . . ," Jason challenges, "if you know so much, why don't you tell us where we can get some breakfast. I'm hungry."

"Well, ah, let's see. Oh, yes. There seems to be a store over that way," Grasshopper points toward the northwest.

"How do you know?"

"Annotated maps . . . satellite view."

"Maybe so," Jason looks quizzically at his silver-haired rival for any excuse to discount his suggestion. "Come on, Chloe. Let's go see what they have to eat." Jason shoulders his backpack and moves off in the indicated direction. Chloe lingers as Grasshopper further taunts the busy ants by wiggling fingers of both hands above them. Then the two follow Jason.

After walking for about an hour, mostly in silence punctuated by Chloe's exclamations over incipient autumn leaf color and luxurious cloud formations, they cross several small hills and a meandering stream, eventually to emerge at the edge of a recently burned-off field. A faintly lingering odor of stale, scorched earth and charred leaves hovers to accentuate Jason's appetite.

Grasshopper detours about the perimeter of the field, muttering gibberish about a curse of fire being upon the fiery or fairy or fury or ferry—or all of them. Jason has given up trying to reckon any sense in such helter-skelter squibs gleaned from maxnet archives. His impromptu and disjointed soliloquies foster such tangles of irate illogic that finally Jason just tries to ignore whatever Grasshopper says.

Eventually they arrive at a path winding through woods that border the burnt field. Along the trail, trees gradually thin and open into small groves between acres of field peas that now are tinged with autumnal hues. Beyond these ripened fields a few small wooden buildings are just visible through a variegated screen of younger trees.

One of the buildings soon is revealed to once have been a modest service station and general store. Rusted signs and torn posters proclaim bygone eras of bountiful stocks of gasoline, groceries, fertilizer and hunting equipment. A general dilapidation of building and grounds testify to diminished prospects of commerce. Only a roof recently replaced with solar shingles counters its impression of dereliction.

"So much for those satellite views," Jason mutters acidly.

Rounding the corner reveals ancient gasoline pumps, broken and battered, their hoses askew across paved fuel islands like ramshackle sentries guarding a padlocked entrance. A screen door juts awkwardly from broken hinges; its wire mesh has long ago rusted out in onslaughts of weather and flies. Hanging from the sagging shed that extends from the store out over derelict gasoline pumps is a ravaged plastic sign barely able to identify *HERSHEY'S STATION*. The plastic sign has been shattered by rounds of bullets so that only careful attention to lettered remnants makes it legible. Below, a metal sign similarly punctured by bullet holes proclaims through ballistic filigree *QUAKER STATE MOTOR OIL*.

Warily looking about the premises they see that a number of signs, old cans and even an abandoned dull green fender of a pickup truck have similarly suffered long barrages of what seem to have been random shooting sprees.

"Gee," Jason mutter, "somebody needs a maxnet game to occupy their trigger finger."

"They're all red and green!" Chloe suddenly exclaims, "Why would anyone shoot only red and green things?"

"Who knows? Maybe they're color-crazy . . . I just want something to eat," Jason complains, peering into a grimy window. The shelves inside display haphazard groupings of disheveled articles, but nothing edible. "Looks like a flea market or pawn shop or something. But I don't see any food in there."

"Maybe somebody doesn't like Christmas," Chloe continues her detective speculations as she fingers jagged extrusions about holes in the red pumps. Rust has long ago eaten away sharp edges, but Jason still cautions her,

"Be careful, Chloe. You could get tetanus from jagged metal like that."

She jerks her finger abruptly back as if microbes might jump out and attack it. A *MERITA BREAD* sign draws her attention, "Me, too, Jason. I'm hungry also," she agrees.

"Food is on its way," Grasshopper announces as he peeks and pushes at the perimeters of the windows.

"How do you know?" Jason overrides his vow to ignore Grasshopper.

"It always is."

"Not always . . . people starve, you know."

"Ants maybe. Not Grasshoppers."

"You live in a weirdly complacent little world, don't you?" Jason snaps bitterly, "Why don't you let us in, then we can all be happy and carefree . . . ?"

"Okay," Grasshopper smiles exuberantly, looking directly at Jason for the first time all day, "You're in. All around us. Now. It's just now. Now . . . isn't that neat?"

"Neat as never—it looks the same to me," Jason grouses stubbornly.

Grasshopper peers through the windows at clusters of furniture, clothing, household appliances, books, and toys, "Ant tracks. Their stuff always looks the same . . . sad old turds of wasted ways. But now is always new but still is as old as the universe."

Suddenly Chloe gasps. An older middle-aged man in faded denim overalls and a flannel shirt has silently come upon them and now stands watching them. A dirt-gray pit bull glowers from just behind the old man's left leg; a slow growl rumbles ominously. Cradled in the man's right arm is a worn and beaten double-barreled shotgun. Its muzzle points toward the ground but is ready to come up level if provoked.

"Ah," says Grasshopper, as if expecting the new arrival, "there you are."

The man eyes him warily then asks suspiciously, "There who is?" Both eyes squint, the right one so much that it nearly closes. He is small in frame, but sturdily built, and wears a dirty plaid shirt under faded overalls pulled over scuffed high-top boots with leather laces. From his left shoulder hangs a dilapidated canvas bag; small bulges hint that it might be filled with apples. His face is grizzled with whiskers; his eyebrows and thickly grown hair also are grey with clumps of white against deeply tanned wrinkles of his face.

"Lloyd Hershey," Grasshopper grins affably. Jason has grown accustomed to Grasshopper suddenly "switching on", as if donning a persona appropriate to the moment. Now it seems as if he and this elderly newcomer are old buddies. Or at least are destined to be old buddies.

But the man is not buying into the act, "How'd you come to know my name?" he demands, the shotgun twitches slightly as he steps closer. "Do I know you?"

"No, I don't think so," Grasshopper chuckles and points to a small sign over the door that reads Lloyd Hershey, Proprietor beneath a bullet-hole riddled Dr. Pepper.

"So . . . how'd you know it was me?" the man persists testily.

"You walk like you own the place."

"Ah, well, probably so . . . so what d'you want? Are you some of them damned refugees?"

"We're just three hungry people," Jason interrupts. He is unnerved by the man's tone, the fact that he is armed, the dog's menacing rumble and the way man has assumed that the trio are refugees. "Can you sell us some food?"

"Store don't open 'til noon," Lloyd Hershey growls, "that ain't for a while yet." He eyes Grasshopper's long hair, then looks at Chloe, apparently amused that her hair is cut shorter.

Chloe asks, "Do you really have food?"

Lloyd Hershey smiles enigmatically, " Don't do much food business these days . . . folks hereabout generally grow their own."

"From the looks of the place you don't do much business," Jason observes.

"That just goes to show you that peekin' in other folks' windows won't tell you everything about them." Lloyd Hershey snaps, "I've been opening up for longer'n any of you can remember and I'll be opening up long after you're gone." He returns his attention to Grasshopper,

"How come you let your hair grow that long? It's longer'n hers."

"Grasshopper is a mercurial child . . . mythically lunar in nature . . . silver locks testify to a wealth of untapped implications . . . ," Grasshopper's voice is musical, as if captioning a teleview presentation of cultural icons.

"Ah-h-h-h, what do I care?" The man quickly loses interest in the soliloquy rippling from iLink cache of script notations. He turns back to peer accusingly at Jason, as if the trio has transgressed some universal code. The smudge-gray dog steps ominously forward; his master's shifting stance promises violence. A thrill somersaults up Jason's back and he reflexively retreats backward as the dog presses forward to whiff stranger-scented air.

Lloyd Hershey watches the three interlopers for a moment, considering options. Finally he says, "Well, you can't stay here —you'd best come with me. If the girl can cook and you two can chop wood, you can eat up at the house. If not . . . well," he grins wryly, his upper front teeth are missing and the others don't seem too secure, "you can watch me eat." He chuckles at the prospect and clucks to the dog, "C'mon, Peery."

"Ah ha," Grasshopper waves to his companions, "brunch has arrived." He smirks triumphantly at Jason, "Now— neat . . . you see?"

Wending a slow arc through the trees, they are led to an old wood frame house sitting in what once might have been an open clearing but now is overgrown with underbrush wildly rambling among thickly clustered trees. The two story house is fronted by a balustraded porch extending its entire width. Like the store, it once was painted white, but now meandering patches expose weather-textured wood, making the exterior seem like sunlight has cast sporadic gray and splotchy shadows over its surfaces.

As they mount steps onto the porch, Grasshopper veers off, unnoticed, to disappear around to back of the house. Jason and Chloe follow their host, their feet making scraping wooden sounds as they mount the steps. Lloyd Hershey opens a tattered screen door, the center of which is adorned with a dirty puff of cotton batting.

"What's that for?" Chloe asks, pointing to the cotton.

"Keeps out the flies, girl," Lloyd Hershey answers as if the question is silly.

"Keep out flies? But how?" Jason asks.

"Hell, I don't know. Ask the damn flies."

"It couldn't," Jason argues to no one in particular as the man turns the knob to open the glassed-paneled entrance door.

"You may as well come in." Looking around he frowns and asks, "Where's the long-haired guy?"

Jason shrugs dismissively, "He just wanders off like that."

"No work, no food," the man snorts. "Whatever. Good riddance, anyhow. I don't like him. Too smart-assed weird!"

The door swings open into a dimly lit but neatly kept hallway. To the far right is a staircase running to a floor above.

"What a nice old house," Chloe remarks as she examines very old but carefully maintained furniture. The old man watches her for a moment, then observes,

"Everybody that ever sat in these chairs is dead now—except for me. And I don't live in this part of the house anymore."

Chloe's hand draws back quickly, uncertain how to respond. Lloyd stands the shotgun in a corner by the door, then walks over to rub the back of an elaborately carved rocker as if it were his prize for surviving all other contenders.

"I never sit in it now," His eyes imply that none of them should presume to occupy it either. Returning to the screen door, he snaps its hook into place, removes the bag from his shoulder and says gruffly, "C'mon to the kitchen."

The duo follow him through a wide hallway to the rear of the house, onto a disastrously untidy porch and back into a large kitchen off to the right that is in even worse shambles.

"This is where I live, now," Lloyd announces, "Never seems to be any reason to clean it, though. It just gets dirty again." He beckons to Chloe and shows her a pantry stocked with food as well as brooms and rags. "You can start cooking just as soon as you get the place cleaned up . . . there's water over there. He points to a dingy chrome swivel faucet rising above a large double sink overwhelmed by dirty dishes, piles of dirty overalls and shirts and shriveled remnants of vegetables and stale bread.

Chloe gasps at the overwhelming assignment. Then she looks at Jason to say, "Okay. But I'm going to need some help."

"Not this boy, Missy," Lloyd interrupts, "He's going to split wood. Ain't long 'til winter and it promises to be a cold one."

"Don't you ever throw anything away?" Chloe asks in dismay at random scatter of filth and junk.

"Cleaning-up ain't my kind of work," Lloyd says spitefully. "Had a girl clean the place for a while, but she run off. Don't matter."

He unsnaps the cover of the canvas bag and empties it onto the already cluttered table. Five freshly killed squirrels tumble onto the table, their gray fur matted in specks of darkened blood drawn by shotgun pellets. They have already been gutted, the cavities packed with dry straw.

"God, that's gross," Chloe protests, turning away.

Lloyd laughs derisively, "Hell, girl. There ain't nothing worse than a squirrel dyin' of old age—it's a waste. You ain't one of those vegetarians, are you?" he accuses.

"No," Chloe says, "I'm not vegetarian." She still is unable to look at the motionless creatures without wincing.

"Then hush your fuss—you can't very well eat'em alive, now can you?"

"I guess not," Chloe murmurs. She takes a deep breath to gather her energies, looks at Jason, raises her eyebrows and smiles bravely. She begins to assault the heaped debris strewn along the counters and cluttering the sink.

Lloyd Hershey opens the firebox door of an ancient cooking stove, tosses in several lengths of wood and stirs at glowing coals with a short poker to coax them into flame. He hands a kettle to Chloe, saying, "Get the water really hot, it'll make the dishes clean easier." Then he retrieves a knife from the clutter and tosses it onto the table by the squirrels, "I'll clean these critters. I don't want nobody screwing up the pelts." To Jason he says, "You go out back and you'll find an axe by the woodpile. There ain't no use in your wasting your time in here. You'll want to bring in some wood for the stove so she can get to cooking. The pieces need to be no thicker than your lower arm. And if you see your buddy, tell him to get to work or he can find his own food!"

Jason looks inquisitively at Chloe, but she smiles brightly as if to reassure him she will be all right.

Lloyd has just begun to remove the skin of the first squirrel carcass. He looks up at Jason, squints his right eye in renewed appraisal, and then says with a contemptuous snort, "Hell, boy, don't just stand there. If you're so hungry . . . go split some wood to feed that stove so it can feed us." He chuckles as if he has just told a joke, then proceeds to dismember the rodent carcass.

Jason sighs in resignation, "Okay. I'll split some wood for breakfast." He turns, walks out the back door and down several wooden steps. Presently the sounds of an axe can be heard splitting lengths of log into smaller sections.

Lloyd remarks, "You folks are from the city, ain't you?"

"Not any more," Chloe replies as she vigorously scrapes dishes crusted in dried food.

"Well, I guess that's true," Lloyd Hershey says casually as he peels a furry pelt from another carcass. "From what I hear there ain't much left of New York City anymore . . . lots of people wandering about in these parts. A man could make a pretty good living off bounty hunting."

Chloe drops a dish. It clatters in the sink as she looks warily at their host. He returns her gaze with a wry chuckle.

"Hell, Missy, I won't turn you in," Lloyd Hershey assures her. "I don't care what the law says—people got a right to go where they need to. Me, I just hunt squirrels and rabbits and leave the people-chasing to others."

"Is that all you do? You're able to live by hunting?" Chloe glances toward Lloyd's hands deftly transmuting slaughtered wildlife into the makings of a stew.

"No, I buy and sell stuff. Or trade, is more like it since the dollar ain't worth crap any more. But kind of like a pawn shop. People need something, they come by my store and if I don't have it, like as not I'll know where to get it. I've got the only maxnet feed around here. I do a pretty good business across the web when shipping can be worked out. If I can't get it, I tell them, then probably they don't really need it." He grins slyly.

"Do you have any family?"

Chloe immediately regrets the question, recalling his pronouncement that everyone else who ever sat in the front room furniture is now dead. "Oh, I'm sorry," she quickly adds, "I shouldn't pry."

"No," his voice falls, leaden, "just me. And Peery."

Conversation whimpers into blankness. Chloe returns to cleaning with renewed vigor, noisily punctuating a now leaden silence. Lloyd pulls the squirrel torsos and limbs apart, skillfully severing joints with the knife. Occasionally its point flicks a shotgun pellet out of the meat, clicking against the china platter. Presently he sets the pot aside, gathers the blood-specked pelts into a bundle for further working and tosses visceral remnants onto the floor where Peery snuffles and crunches through their final oblivion.

"Cover that about an inch with water and put it on the backside of the stove to simmer. Toss in some onions, carrots and potatoes from the pantry, along with a little salt and pepper."

"I'm sorry, Mr. Hershey," I shouldn't have . . . ."

"Never mind," Lloyd snaps, "it ain't your concern." He opens a drawer and pulls out all the various butcher and carving knives, gathering them, along with the one just used to render the squirrels, into a bundle which he carries across the room and places into a bedraggled wicker basket. "Time for these all to be sharpened. They're so dull you couldn't stick'em in the ocean." All but his eyes smile weakly.

After a brief pause he asks, "Where are you folks headed?"

"Jason and I are going back to where I grew up," Chloe says brightly, looking for some cheerful note. "Southwest Virginia. Out in the country, between two little towns, Hilton and Maces Spring."

"Never heard of them," Lloyd shakes his head disinterestedly.

"It's not far from Bristol, near the Tennessee border, but back toward the Clinch Mountains. "Have you ever heard of the Carter Family? Old time music? They came from there."

"That's a long walk," Lloyd muses, "and it promises to be a rough one, too."

"Well, we don't seem to have much choice."

"You could settle down around here for a while . . . wait till things calm down some. Maybe I could use some help around here. You and your boyfriend could work and save up for your trip."

"Oh, thank you, Mr. Hershey, but I guess we should keep going," Chloe thinks for a moment then laughs, "besides, I don't think Grasshopper would be much help."

"Grasshopper?"

"He's the one with the long silver hair . . . you know. Grasshopper. Peter Dooley is his real name. He was in the teleview spots Ant and Grasshopper."

"I don't watch that kind of stuff. Too much crap for too little payoff. Besides, I don't want him around here, anyhow. I don't like him."

Jason enters with an armload of firewood which he drops noisily into a wood crate next to the stove. Lloyd watches him cross the room, then says, "Take a break and help me bring up a table from the cellar. I sold it to a man yesterday and I need to get it ready to pick up."

"All right," Jason agrees, glad to take a break from swinging the axe, appraising the progress Chloe is making toward breakfast and approvingly eyeing the promise of squirrel stew for later. He follows Lloyd Hershey downstairs.

Looking out the window, Chloe sees no sign of Grasshopper. Hoping that he has not taken off without them she redoubles her efforts to get through the cleaning chores so that they can get something to eat and be on their way. In the distance it almost seems she hears muffled sounds of furniture being moved, then a distant door slamming. After a while Lloyd Hershey reappears and comes to the pantry, saying, "We need a light down there." He walks to open a cabinet between the sink and stove. Then before Chloe realizes what has happened, Lloyd Hershey quickly seizes her right leg, pulls out a chain lying hidden beneath and behind the stove and slaps a nickel-steel ratchet manacle onto her ankle.

"What are you doing?" She is so startled that she merely watches him check the chain.

"Just making sure you stay to do your work," he replies in a dead flat voice.

"Where's Jason?" Chloe suddenly is horrified.

"He's okay. He's down in the cellar. You get some breakfast made and then he'll get his meal. And as long as you behave yourself and do as I tell you, he'll stay healthy." Lloyd Hershey is breathing deeply and standing much too close. His index finger brushes across her lips. His foul scent is overpowering. Chloe moves away from him.

"Mr. Hershey, let us go. We don't really need any food. Please, just let us go and we'll leave you alone. We don't want any trouble."

Lloyd turns and works his hands along the chain yet again to ensure that it is securely fastened to the heavy cast iron stove.

"Please. This is not right," Chloe pleads, tears welling, "You can't treat people like this. I thought you were a good man."

"A good man may be hard to find," Lloyd Hershey rasps, "but nothing like looking for a good woman in hard times . . . looks like I just got me a bride." He grimaces a pretend smile. "That'll hold you 'til we get used to each other."

From his overalls pocket he takes out a black revolver and checks that it is loaded. His left hand opens a drawer to scoop a handful of cartridges from a box marked *Winchester .357 JHP*, which he shoves into his overalls pocket. "Now we'll see what that silver-haired devil is up to." Striding to the door he pauses to instruct Chloe, "Get hold of yourself, Missy. And get to work. Remember—the only thing keeping your boyfriend healthy is you doing as I tell you. Get some food ready. Likely there'll only be the three of us for breakfast." Again he mocks a smile.

"Come on, Peery." They disappear out onto the screened side porch.

About fifty yards farther to the rear of Lloyd Hershey's house Grasshopper sits in the upper branches of an aged walnut tree. On the way to it he picked several apples to assuage his hunger from a heavily burdened tree and then quickly ate them, tossing the cores back under the tree.

He has gone off by himself to find silent space and private time in which to resolve a new flood of information borne via the revived maxnet signal. The iLink latched the signal as they first approached the house. Last night it had latched a signal from a skypod floating slowly over the highway, a signal that had spun his nbTwin into a giddy debut-encounter with a new nbTwin: ecStacy! It had been her. He was riven, startled, in the initial encounter within and then quickly welded whole again in a sudden rush of blissful exhilaration as their nbTwins blended and harmonized in a playful rush of intimation.

Now, aching to recapture that moment, Grasshopper plies the maxnet via the iLink to extend searching tendrils of network process, probe global servers for a trace route to reestablish contact with ecStacy. Ping! Ping! Ping! Not since Ant departed has Grasshopper known the uniquely melding familiarity of nbTwins entwining prospects and possibilities to mutually enthrall and direct ongoing events. He shudders in heady anticipation of again plumbing depths of nbTwin-sharing that ecStacy opens between them. With Ant there had been self-indulgently stimulating banter and counterpoint, along with much laughter, but never the profound bliss of being-within-all that she arouses. Never has he known such gently profound necessity of promise.

Probing maxnet tendrils glean disturbing images from link-spiders crawling archives of nearby servers. A new resource has opened . . . the iLink spans an image-map, a collage of images and video across his ken, dynamically shifting layers of overlapping intimation and visualization. Young women captured, abducted, bound, abused, tortured. He recoils against the intrusion and tries to ignore it. Yet it persists.

He shakes his head, trying to force past such detritus, longing for some sense that ecStacy may respond to his probes. Now, he feels an actual intruder. Not an nbTwin, but another presence. Ah, it seems to be on the director channel, like the one named Jason . . . it is Jason . . . saying over and over, "Help Chloe." But that makes no sense . . . Chloe is in the kitchen and Grasshoppers don't belong in kitchens.

Now images map into a dynamic cartoon frieze of Snidely Whiplash binding poor Nell on a railroad track while Dudley Do-Right of the Mounties has been waylaid and is unable to rescue her. Nell's visage dissolves into an overlay of Chloe's smiling face from an old snapshot, which slowly grows contorted in an exaggerated semblance of panic. Grasshopper pushes past these annoyances and tries to more fully open into the global maxnetwork. But the access point suddenly collapses into local area network service only. Now there is only a relentless barrage of silent movie clips of helpless damsels in distress, of leering grasping villains, of heroes rushing to rescue.

Through the tree's dense foliage Grasshopper sees Lloyd Hershey open the back door and descend the steps from the house, scanning the yard and fields beyond. In his right hand is a revolver. He walks to a shed and searches all around it, the gun held ready. Then he heads toward a separate, sturdier concrete block building farther away from the house. It has a heavy wooden door made of heavy timbers, bolted together and padlocked. Its only two windows are fortified by wrought-iron bars.

Suddenly, two electric motorbikes appear back near the house, gliding silently up the drive from the distant road. Hershey puts the revolver into his pocket and stands waiting as the two cyclists pull alongside, turn off their bikes, and push down kick stands to dismount. One man pulls a pair of saddlebags from the back of his machine, waves his right hand expansively as he explains something, then hands the saddlebags to Hershey. They talk briefly, then walk toward the concrete building where they remove the padlock from the massive sliding bolt securing the heavy wooden door, open it and go inside. On the back of each newcomer is a picture of a snarling wolf beneath the blood-dripped word BERSERKERS.

"Help Chloe . . . help Chloe . . . help Chloe," Jason's voice persists through the iLink, "Lloyd Hershey will hurt her. Stop him. Help Chloe."

Grasshopper scrambles down the tree, trots over to the concrete building and on impulse shoves the heavy door closed. Inside, Peery barks savagely.

Grasshopper rams the bolt into place and latches it with the padlock. Inside three men yell outrage, punctuated by the barking dog. Several gunshots rip through the heavy oak door. But Grasshopper already has trotted back to the house, bounded up the steps and gone inside, where in the dimness he walks surrounded by a luminously swarming host of would-be rescuers: Buster Keaton, Charlie Chaplin, Mighty Mouse, Superman, the Green Lantern, Batman and Robin.

Chloe is banging a skillet against the cast-iron stove, attempting to straighten out a small sliver of silvery metal. She sees Grasshopper and shouts, "Thank God! I thought he had shot you!"

"Here I come to save the day!" Grasshopper sings, in exuberant Mighty Mouse voice.

"I've got to get out of this cuff," Chloe anguishes, "I hope this really works. We always played like it did in Dungeons and Dragons, but that's just pretending. I never actually tried it for real."

"Tried what?" Grasshopper seems baffled and a little miffed that she is not more desperate.

"Opening a handcuff—er, ankle cuff—with a shim. I pulled this metal clip off an old ballpoint pen that was in a drawer." She is bright with enthusiasm of her project as she sets the skillet back on the stove, leans against the sink and begins pushing at the manacle. "Do you know anything about how to do this?" she asks, without looking up from her task.

"Grasshoppers don't do locks."

"Well, today they do," Chloe glares at him deliberately, "go help Jason. He's down in the cellar. Maybe he knows how to get this damn' thing open. Go! Now!"

Startled by her uncharacteristically brusque urgency, Grasshopper backs out the door, his virtual retinue of derring-do rescuers fading forlornly as he mocks derisively, "Help Chloe . . . help Jason . . . help yourself, is more like it."

Chloe continues poking at the mechanism, trying to think it through: the ratchet is held by some kind of spring thingy. A pawl I think they call it. And if the ratchet teeth angle that way then the pawl has to pivot this way. So push at it from the other side, Chloe, . . . plickety-plickety-slip-p-p—it magically opens!

"That is so easy," Chloe giggles in astonishment as the manacle bracelet falls from her ankle. All the energy of her fright, gathered just moments ago, now dissipates in surging spasms of laughter at how piqued Lloyd Hershey will be. Mirthful tears stream down her face as Jason dashes in. Grasshopper nonchalantly follows along behind.

"Are you all right?" Jason pleads against his worst imaginings.

Still giggling Chloe nods as she proudly holds up the opened manacle with the little piece of metal she has made straight. "See?"

"Great . . . we've got to get out of here," Jason yells impatiently. "You wouldn't believe the hellhole that is downstairs. This guy is a real pervert. A sadist. I think he even traffics it. Grab some food and get your pack. Let's go. I sent a text message and an email, too, to the sheriff's office when I heard shots. But we don't want to be here when they show up. It's our word against his."

"How did you send a message to the sheriff?"

"Maxnet . . . with my euPhone," Jason pulls a small device from his jacket pocket, "thank God I had it with me. When I saw all the porno crap in that room down there . . . you wouldn't believe it, racks, chains, leather masks, bondage pictures like naked women tied up and strapped to motorcycles, men wearing BERSERKER insignia doing unspeakable things . . . I tried getting through to Grasshopper, like that day back at Watchung when he told you to wave at me. I kept saying, 'help Chloe,' over and over."

"I think he locked them in a building back there," Chloe points out the window above the sink.

"Chloe didn't need help," Grasshopper observes snidely, "you did."

"Them?" Jason ignores Grasshopper's barb.

"Two men on motorbikes, wearing BERSERKER jackets came into the yard just as Lloyd went out back looking for Grasshopper," Chloe says, "He spoke to them and they gave him a big bag, which they all took out to that block building. See? That's where Grasshopper trapped them."

Jason rips off a piece of brown paper bag and scrawls in giant block letters—the ball point pen works furiously back and forth to make each character large and bold:

LOOK IN CELLAR
&
IN BUILDING
OUT BACK

~ 10 ~

Soon after welcoming her on the previous evening, Cornrath showed Kristin to a bedroom with a closet of clothes that he suggested might "suit her needs." She luxuriated in a bath for the first time in what? Weeks? Then, from a night of anxious dreams—biker predators hunting Agnes, Justine and herself—Kristin awakens suddenly into a refuge of alien luxury.

Several moments pass as she regains her bearings and recollects how she came to this place. Tossing on a dressing gown she heads downstairs where she finds a breakfast of toasted bagel, cream cheese, jam, tuna, orange juice and coffee waiting.

"Good morning, Kristin," Cornrath's voice is steady and even ebullient, "I trust you slept well."

"Well enough." After a moment's reflection she observes, "Your mood seems much improved."

"Oh, yes, well, let's say matters have taken a more promising turn."

"They found the kidnapped children?"

"Well, no, not yet. We're still looking."

"What's happened to so brighten your day, then?"

"Well, first, I've been assured that General Michaels has full confidence and support for my efforts here."

"And second?"

Cornrath pauses, smiles knowingly and then says, "Let's just say that events seem to be taking unexpected turns . . . I think the heat is off, or soon will be."

Kristin looks at him closely. Last evening he had seemed about to self-destruct. Now, cocksure, even arrogant, he has retrieved the Eric Cornrath she has long known.

"You're up to no good," Kristin says simply.

Cornrath laughs pleasantly.

"I want to go on to Harrisburg?" Kristin announces.

"I'm afraid that's impossible."

"Impossible . . . why?"

Cornrath stares into flaring green eyes for a moment, then says, "Because your country needs you."

"Ha! Sell that load somewhere else." Kristin's voice firms irony into its challenge, "Eric, you've set yourself up here as some kind of warlord. You live the good life while millions depending on you barely stay alive." She pauses for a moment; her lip wants to curl into a sneer as she watches his self-contentment steep in its own reflective esteem, "This country needs for you to get on the stick, to pay more attention to helpless and hungry refugees than to yourself."

Cornrath glides past her accusations, "Just think for a moment. If Peter is as deranged as you say then you are the best person to help restore his sanity. You know how antagonistic he always has been with me. Only you and Bill could ever talk sense into him. When we find him, he'll need you as a link to his former self."

Kristin dismisses the prospects, "That's assuming you find him—and that his sanity can be restored."

"We'll find him," Cornrath assures her, "eventually he'll stop hopping around. Every time he switches on his iLink we're listening. Eventually we'll bring him in."

"But why are you so determined to find him? There's more to it than you're saying. You don't seem to care much about the fates of all the people out on the highways. So why make a big deal about Peter? You barely tolerated him before . . . and then only because he could deliver what you needed for Ant and Grasshopper."

"Right now Peter is many things at once." Cornrath leans forward, studious in laying out his argument, "He's a menace because we don't know what that technology may make him capable of. But, and this is an important 'but', he's also a breakthrough . . . he's survived reactivated nano-replicators. We need to know how that happened in him . . . and what kinds of new innovations it makes possible."

"So you can use it on other people?" Kristin's voice is faint in both disbelief and discovery. "Peter was right . . . you are Joseph Goebbels on steroids!"

"Hardly," Cornrath dismisses the slur, "but think of what this could mean for public health, for the common good. We've already seen nanotech facilitate interaction between the human psyche and communications media. What if those activated replicators could actively engage, accommodate and rectify disease vectors without harming the hosts."

"Think of the tons of money it would bring in," Kristin counters acidly, "and who knows what other applications you'll find for it. What? Super soldiers? Super workers? Happy and willing servitude, anyone?"

"Whatever," Cornrath signals impatience with her stubbornness, "we need you here to help with his debriefing and reconciliation. So here you'll be staying."

"You can't keep me here against my will. That's kidnapping."

"Actually I can. And I will. Only it isn't kidnapping. You and Peter both have defrauded the U. S. government and can be held until you make restitution."

"Defrauded? How?"

"You both signed up to participate in the Watchung Reservation programs, then failed to show up. You owe the government community service to reimburse for your food, lodging, transportation and other conveniences provided."

"How dare you!" Kristin seethes at the triviality of the charges.

Cornrath seems to be enjoying himself. "In any event, you'll remain here under house arrest. There is an electronic perimeter beyond which monitor drones will not allow you to venture, so please stay in the yard unless you want to be gassed unconscious or given a nasty shock."

"I won't help you. You can't make me."

"Is that a bet?"

Later, in the afternoon, Kristin watches from the window above the kitchen sink as outside Eric Cornrath converses and exchanges animated gestures with a companion. The two are slowly edging toward the house. Whatever the other man is attempting to persuade Cornrath, neither seems to want anyone to overhear. Repeatedly they glance about, as if concerned that someone may have wandered undetected within earshot. Cornrath wears a field uniform with tacked-on doodads and adornments; Kristin has no idea what the stripes or badges ornamenting his camouflage fabric mean. She wonders if maybe he just makes them up. The other man is dressed casually in slacks and sports jacket.

The two men mount the back steps and enter the kitchen. Cornrath seems surprised to see her. "Eddie," says Cornrath, "this is Kristin Banner. Kris, meet Eddie Minnow, Regional Director for FERAL."

Kristin barely nods acknowledgment of the Regional Director.

"She looks nice, but she doesn't say much, Eric," Eddie banters, gauging her reaction.

"Leave her alone," Cornrath says curtly, "Kris is an old friend. You remember me speaking of Bill Banner, don't you? Kris is his sister."

"Oh, I'm pleased to meet you," Eddie replies smiling back at her in his most charming manner.

Kristin turns and strides quickly from the kitchen into the living room, falls onto the sofa and assumes what feels like a guise of calm poise. While the rigors of the evacuation route seem very far away, she still has no idea what to expect. A part of her even wishes to be back on the freeway with Peter and the faces and primitive routines that had eventually become familiar. Another part of her tries to settle back into billowed comfort of the overstuffed sofa, pretending to peruse an old fashion magazine.

She is astonished to see her own image on the cover—a visage that now looks so implausibly varnished and fake. Kristin cannot, at the moment, recall what it felt like, but she does remember that, just as now, time never seemed to finish its confusing cascade of details, that she had rushed and rushed, bouncing from one assignment to another, never paying much attention to anything that did not lead into another batch of equally dispossessing concerns.

She hears the kitchen door open and close and then a solitary set of footsteps approach down the hardwood floor of the central hall. Kristin is irked that Cornrath goes directly to the bar and makes a great show of pouring himself a glass of Perrier. In cool detachment she has made an equally great show of not being the first to break the silence.

Moving to a large and vaguely modern chrome and black leather chair opposite her, he takes a leisurely sip of bubbling water, sets the glass on a cast resin block coffee table, massages his face with his hands, as if to emphasize the brutality of the morning's regime, and leans back to gaze contemplatively at her. Kristin meets his eyes, thinking how silly such contests have always seemed, but she refuses to look away.

"That wasn't very friendly," Cornrath observes disinterestedly, "What must Eddie think?" Kristin declines to respond. Silence opens and abounds until he adds, as an afterthought, "You look well-restored."

For a moment they both wait to hear what she will say. "I want to leave," Kristin finally reiterates in a voice held intentionally flat.

Cornrath winces slightly, sips his water, then eases himself back against the chair cushion. "That's impossible—at least for the moment."

"Why? Why is it impossible?"

"Kris . . . you're here because I brought you here—I'll tell you when you may leave. In the interim, I am in charge of everything that . . . shall we say, keeps you comfortable. When you realize and accept that, the more enjoyable your visit will be."

Kristin's sudden peal of laughter startles and interrupts his lordly poise. He waits for her to explain such sudden mirth. Her voice again purposefully deflates, "Meanwhile, we'll sit and stare at each other, I suppose."

Cornrath taps the rim of his glass with his forefinger. "You may suppose what you will. However, this is a base of military operations in a region under martial law. You should be grateful to whatever destiny brought you in from that chaos."

"Your chaos," Kristin's voice is quiet but firm above rekindled disgust mounting in her eyes, derived from memories of corpses, abandoned children, wasted and dreary hopes dragged along by tens of thousands.

Cornrath watches her for a moment, seems to toy with several possible responses, and then finally chooses. "I don't own the world, Kristin. I didn't create its problems. Like everyone else, I started out on my belly, then I learned to crawl and . . . ."

"You didn't progress much farther, so let's just cut to the chase . . . suppose I just leave," Kristin challenges him.

"Suppose you just leave," Cornrath speculates. "Suppose you manage to get by the drones and off the base without being knocked down or perhaps even shot—sentries do get edgy with rumors of rogue militias and even terrorists about. Suppose you make it through the woods without being recaptured by a patrol. Suppose you avoid being taken by one of the renegade bands or vigilante committees roaming these back counties. Or by the bunch of outlaws I rescued you from. Suppose you don't freeze to death. Or starve. Suppose whatever you will. There is no other place to go, Kris. There is simply no other place to go. Hopes, dreams, ambitions . . . what are they now? What do you think you would do even if you did leave and manage to survive some desperate journey alone? Your best future is here, now."

She glares at him. "With you?"

"If you'd just accept that I'm the host with the most . . . ," he attempts humor, but somehow chokes. Several droplets splash from the glass onto his uniform. He tries to blot them with a used paper napkin taken up from the plastic coffee table.

"You want to be 'straight?'" Kristin arches an eyebrow. "Get this 'straight'. I don't like you. I loathe you. I despise everything you love and love all that you fear. I didn't like you three years ago and that was before I fully grasped what scum-soaked evil you concoct . . . I want you to leave me alone. I want you to let me go to the relocation center at Harrisburg. There, at least, maybe I can help someone who really needs it."

Cornrath watches her briefly. Kristin expects for a moment that he is about to remark how beautiful she is when angry. Why, she wonders, do tired old movie lines come to mind in this man's presence? "Too bad about Bill," Cornrath blithely observes, ignoring her tirade against him."

"Leave my brother out of this. You'll never be even half of what he was."

"Get real, Kris," Cornrath chuckles snidely, "by your own account . . . Bill was greedy. He lusted after celebrity—you know that. He was weak and his weakness killed him." He smiles self-satisfaction that he has managed to penetrate her facade. She flushes and blurts out,

"Bill was a human being."

"Oh, come on," Cornrath savors the role of conversational critic, "surely you are more clever than such limp conceits. Where's the wit? The fire? The discerning Athena, rising above all challenge . . . I think that's what Dooley used to say about you."

"Peter? Said that about me?" Kristin feels a strange exclusion from an aspect of her world she has somehow ignored. She recalibrates the trajectory of her own animus toward Peter Dooley. It had blossomed even as she tried to ignore Bill's faults. Had she gotten it backwards? Has she been angry at the wrong person?

"Does that surprise you?" Cornrath evidently is pleased that he has managed to put her off-balance.

Kristin is suspicious of any so personal bridge into continuing dialogue that has unexpectedly opened. Her voice regains its icy edge, "No. It doesn't surprise me that he said it. It is, however, totally out of character that you noticed or even seemed to understand what he meant."

"Kris, Kris, Kris," Cornrath holds up a plaintive hand, "what have I done to you to elicit such hostility? I have always considered you a friend. I always thought you were among the most promising talents ever to work with the Amberson Agency. Give me a break, will you? I know you've had a rough time, but surely there's a chance we can continue to appreciate each other's friendship."

Kristin smiles brightly. A certainty has welled up within that now breaks in new conviction across her lips, even as her eyes brim with tears for all the turmoil and depredation she has witnessed in the past days and weeks. "No, Eric. There is absolutely no chance whatsoever."

~ 11 ~

Walking along broken pavement of a perilously deteriorated road that winds generally northwestward, conforming to the contours of rolling terrain, Grasshopper is in an exuberant mood; he chatters relentlessly, in agonizing detail about whatever rises to his lips. Adding to Jason's annoyance at her persisting volubility is the insistently chanting rhythm in which he speaks. He seems devoted to a career of nonsensical cheer leading.

Chloe, on the contrary, is wholeheartedly engaged. She skips along at Grasshopper's side, joyfully, joining the carefree and vaguely prodding rhythm, "R--it's rudimentary. O--it's opportunity. A--it's accidental. D--it's a direction! Put them all together and what do they spell?"

Grasshopper whirls about, leaping to land with both feet spread far apart, his arms extended in waving expectation.

"Road?" Chloe offers.

"No!" Grasshopper's shout is raucous, "Hoppening! Rudimentary - opportunity - accidental - direction!"

Turning abruptly to resume his sauntering gait, he intones the words in a piercing wail,

"Rudimentary - opportunity - accidental - direction! Rudimentary - opportunity - accidental - direction -hey! Rudimentary - opportunity - accidental - direction - hey! Rudimentary - opportunity - accidental - direction - hey!"

"Shut up, for Christ's sake!," Jason complains futilely.

"But that doesn't make any sense," Chloe ignores Jason's discontent as she explores implications of his effusive chant, "What is 'hoppening'?"

"You aren't listening," Grasshopper sniffs haughtily.

"But where are we going?" Chloe persists, suspecting that Grasshopper is merely teasing.

"I told you," he giggles, "It's along our rudimentary-opportunity-accidental-direction."

"Hey!" Jason adds, relishing Chloe's sudden perplexity. Then he shrugs back into her baffled gaze to let her know he does not intend to be drawn into inane exchanges.

"What is?" Chloe inquires haplessly.

"Where we're going."

"Then why don't you just say we're going nowhere?" Chloe persists.

"Because we're not going nowhere," Grasshopper chuckles at a ridiculous notion.

"Then we're going somewhere," Chloe insists, tentatively.

"Why, yes. Of course!"

"Do you know where?"

"Now that is a road of another horse-sense," Grasshopper muses gravely. He taps his lower lip with his right forefinger as he considers the options.

"Don't, please don't," Jason pleads with Chloe, "you two are driving me batty."

"I just want to know if he knows where we're going," Chloe answers huffily.

"It's a process of elimination," Grasshopper beams suddenly, inspired by a wayward thought, "Just like Sherlock Holmes says, 'Eliminate the impossible and however improbable, what remains must be the truth.' Where are we now?"

Jason frowns to himself, curious to know how their companion might delve into Arthur Conan Doyle when he does not even accept his own true identity.

"I don't know," Chloe confesses while she scans the surrounding countryside, "just here, I guess."

"Okay," encourages Grasshopper, leading her through some skewed Socratic tutorial, "so we're not going here." Grasshopper nods his head to prod the next step.

"If we're not going here, then we must be going somewhere . . . ," Chloe presses on into perplexity.

"No," Grasshopper sighs, realizing that she still does not perceive some fundamentally lucid tenet, "If we're going some where, it means some place in particular that isn't here yet. If we are going any where, it means that any old place will do. Neither of those will do. We are going awhere . . . not just anyplace, not someplace, but aplace."

"Awhere?" Chloe repeats, intrigued by the newly coined word.

"Yes," Grasshopper continues professorially, "'A' stands for 'Ant'. Antwhere is not anywhere, but it is somewhere."

"Antwhere," Chloe frowns in deep contemplation.

"Yes," Grasshopper intones seriously, "Ant goes antwhere."

"But we're not ants," Chloe objects, straining to grasp the thread he seems to be spinning.

"Oh, but you are!" Grasshopper shrieks gleefully, chortling that she has fallen into his trap, "Otherwise you would know what I'm talking about!"

"But you're not an ant," Chloe challenges the implication, "you said you weren't."

"That's very true, yes, indeedy-do!" Grasshopper winks in amusement at such an outrageous notion, "I'm just tagging along with you two Ants!"

"Why me?" Chloe asks in bewilderment, "And Jason?" She looks to Jason for assistance but he shrugs it off as if to say she is on her own.

"So you'll take me Antwhere," Grasshopper crows triumphantly, skipping in a circular celebration at the progress of their continuing education.

"But I don't want to be an ant," Chloe whimpers, sounding like a ten-year old unhappy with some role assigned on the playground.

"Oh, I don't blame you one bit," Grasshopper commiserates, "When you stop going Antwhere, then you'll stop being an Ant. You'll find it's all hoppening differently, even if it often seems similar." He smiles bravely to encourage her pluck.

"I can't believe you're taking this seriously," Jason's scorn will no longer be still, "Can't you see he's just free-associating anything you say; you're arguing with your own thoughts turned into nonsense."

"But Jason," Chloe protests, "when you listen carefully, it makes sense . . . until you try to pin it down. Oh, I don't know."

"It's what you know when you say you don't know and what you don't know when you say you do. It's like art . . . you know, that stuff that people tend not to throw away, even after its time has passed." Grasshopper tries to help sort out her confusion. He sticks his tongue out at Jason, squinting his eyes in a mawkish disdain of any contribution he might make.

"Even if it did make any sense," Jason attempts to ignore Grasshopper's fingers mocking his lip movements, "what's the point out here?"

Chloe persists, trying to dispel dismay invading her hero's opinion of the duo, "If we learn to stop being ants, will we be like you?"

Jason snarls in exasperation and vows not to intervene again.

"Do you mean 'be Grasshopper'?" Grasshopper asks for clarification.

"Yes, I suppose so," she answers.

"Oh, goodness me, no, dear," Grasshopper chortles in condescending amusement.

"Why not," Chloe begs to be included among the elect.

"Because there can be only one Grasshopper, dear, that's why." His tone suggests an elementary fact of the universe.

"But what would we be, if not ants?"

"It would depend on you," Grasshopper asserts gravely.

"Then what would I be?" Chloe presses on, sounding as if she were planning for a costume party.

"I'm not really certain," Grasshopper ponders possibilities for a moment, "I think probably something very light and graceful and musical."

"How very nice," Chloe trills as her spirit suddenly brightens. Jason glares briefly at Grasshopper—why hadn't he thought to tell her that? It pleases her so!

"It's very hard to say, of course," Grasshopper continues in the tone of a seasoned professional, "so much depends upon circumstance."

"How did you become Grasshopper?" Chloe breaks in, hoping for a clue to a more certain route.

"I don't remember, really," Grasshopper replies, gathering his brows into extra effort, "First they commissioned the nbTwin into Dooley, who was with Ant's Sister—she wanted him to stay in charge. But he left back at the tunnel. She became quite upset that I wouldn't be him. I didn't believe her then, but now I think maybe she was right, somehow . . . I tell you, it's not easy being Grasshopper." He invites their commiseration, sighing at a confused welter of opinions invading his composure, "But it's not important now, anyhow."

"Why not?" Chloe wonders aloud; it seems crucial to her.

"Because I *am* Grasshopper. I like being Grasshopper. If I weren't Grasshopper, I might be a lunatic!" He shoots Jason a dark glance as if fending off some slur. Then, looking back to Chloe, he smiles in a triumph of reasoning, "I would be going Antwhere and not knowing it."

"But where is Antwhere?" Chloe lapses into another round as Jason throws up his hands and surges ahead of the two quibbling dawdlers.

Grasshopper sighs, profoundly overwhelmed by her bewilderment, "At the end of a rudimentary-opportune-accidental-direction."

"Which is?"

"Ant . . . or Ant tracks. However you choose to think of it."

"Be quiet!" Jason suddenly turns. Not only is he exasperated at the prospect of another bout of inane discourse, but he thinks he has heard a loud motor approaching. Could it be a helicopter or patrol vehicle?

"I will not!" Grasshopper sulks, "I'll just get worse and worse and here comes a hearse to take us through another verse," he sneers.

"Be quiet!" Jason hisses at him again, "Someone's coming!"

"Certainly," Grasshopper shrieks in bated excitement, "it's a hearse!"

Jason stares blankly at him for a moment as the distant ruckus of an engine racing at full throttle gradually grows louder. Wheeling around, he is astonished to see a ramshackle and ancient Lincoln Town Car hearse lurching spasmodically along the fractured slabs of pavement, its modest progress strangely out of synch with the breakneck screech of its engine, wafting plumes of french fry and peanut flavored smoke in its wake. The black hearse bounces angrily toward the trio, making no effort to avoid mowing them under.

"It is a hearse!" Jason marvels at the implacably approaching specter. Then it occurs to him that maybe these are vigilantes or bounty hunters! Maybe they are going to run them down to collect bounties! "Look out!" he yells as he grabs Chloe and drags her into the scrub brush at the edge of the broken highway.

Despite Jason's animated display Grasshopper merely stands in the road, nonchalantly supervising the hearse as it slams from hole to hole, his head nodding to mark each lurch. He looks back at where his companions cringe in the brush, then stares back at the raucously smoke puffing apparition struggling to clear another jutting mound of pavement.

"What's the matter?" he yells at Jason, "It's not your funeral!" He shrieks happily at the joke. Turning back to the battered vehicle that now lurches to a stop beside him, Grasshopper leans down to listen as someone inside yells above the engine's din. Then the engine scream drops in pitch, falling into a precarious clatter that the driver intermittently keeps alive with flourishes of the accelerator.

A middle-aged man is at the steering wheel. He wears a tan windbreaker jacket and a plastic white rain hat. Obviously agitated, he leans out of the window to yell at Grasshopper,

"Hey, buddy! This is an emergency! Can you give me a hand?"

From our sanctuary of brush both Chloe and Jason gasp in unison as Grasshopper grins, looks at his two hands, and then holds them forward as he says something that sounds like, "Which one?"

The driver is at first confused, then becomes angry. "Look, Mac!" he yells in a tirade of exasperated anguish, "My brother's in the back. He's been shot! He's hit bad and he's started bleeding again. I can't take care of him and drive this thing, too. Can you help me out?"

"Sure!" Grasshopper yelps, reaching for the door handle to help the man out.

"Oh, no," Jason groans as he jumps from the brush. "Come on, this guy's in no mood for that kind of nonsense." The driver has just pulled free of Grasshopper's 'helping hands'. Confused, but now thinking that Grasshopper has offered to drive, the man curbs his anger and yells, "Thanks, fella!" He rushes around behind the hearse, opens the large rear door and jumps in, pulling it shut after him. "Well, let's go! What're you waiting for?" the man shouts hoarsely.

"No, wait," Jason calls out. Running to the rear he sees lying on a pallet inside the vehicle, his chest swathed in bloodstained bandages and lying amidst smeared splotches of crusting red, a young man his own age moaning, eyes closed against pain and shock, his head lolling slowly from side to side. Jason's stomach heaves at the sight of so much blood. The elder brother attempts to staunch the flow from his sibling's ugly crimson chest wound. He has ripped off blood soaked pads and now swathes the wound with a new bandage unrolling from a large plastic box on the floor.

"You know anything about first aid?" he shouts to Jason, attempting to shift his wounded brother to a different position to put pressure against the gauze pads.

"Not really," Jason apologizes, "no more than what you're doing." He berates himself for having had an entire life to learn about such things, and somehow failing an unknown challenge.

"Move, Jason," Chloe pushes him aside as she leaps into the rear. "You drive—you know Grasshopper can't." She quickly estimates the severity of the victim's condition, then takes over as the elder brother whirls about to yell at Grasshopper who now kneels laconically on the front seat, watching the drama in idle curiosity.

"Come on, Jack," the man orders belligerently, "get this heap going!"

"I'll drive!" Jason shouts as he dashes around to the front where Grasshopper already has plopped himself on the passenger seat. Jason clambers beneath the steering wheel and checks the instrument panel and gear shift.

Grasshopper grins sarcastically, "Goody! Now you finally have someplace to go . . ."

Jason ignores ghastly humor as he fights with the gearshift. The man yells something from the rear.

"What?" Jason yells back.

"I said the damn' thing won't shift out of low. It may throw a bearing, but gun it all the way . . . Grady won't make it unless you do!"

"Okay!" Jason assents and shoves the lever into drive. The old vehicle jumps, clanking hideously forward and aft, and then noisily resumes its plodding, screeching negotiation of battered roadway. Everything shakes ominously. Jason doesn't bother to ask where they are going because he doesn't think the heap will make it across the next hill. Grasshopper is miming uproariously, his hands clapped over his ears to muffle the din. A sudden rage erupts; Jason is fed up with his idiocy!

"Shut up," Jason screams at him, "just shut up!" His fury is no match for the wailing engine, however, and his sudden exasperation gradually melts into the chaotic miasma of horrific noise and smoke emanating beneath the black hood in front of him. Grasshopper just keeps shrieking and laughing and stamping his feet against the floor.

Then Chloe leans over the seat back to pull Grasshopper's left hand from his ear. She yells something to him that Jason is unable to hear. Grasshopper springs about onto his knees, perches up on the seat as he looks back past Chloe. His head turns slowly from side to side, as if he were considering flavors of ice cream in a parlor case.

"What is it?" Jason yells, his foot aching from the pressure of holding the accelerator against the floorboard. The engine noise is deafening. Even at their snail's pace he has to fight the steering wheel to maintain control.

"That man is dying!" Chloe shouts above the racket, "I asked Grasshopper if he could help him!"

"He can't do horseshit, Chloe!" Jason yells back in anguished dismay, "You can't let him back there. That guy will kill us all!"

"He can! He can!" Chloe shrills defiantly in Jason's ear and shifts to one side to make room for Grasshopper to climb over the seat back.

Grasshopper looks back at Jason indignantly and taps lightly on his shoulder as if to say, "So there!"

In the mirror Jason watches in anguish as Grasshopper scrambles nimbly over the seat back to quickly crawl where the stricken man lies. The elder brother squats beside, anxiously watching for signs of hope. Jason grits his teeth in dread of what is coming. He has glimpsed two rifles lying alongside the wounded man. He dreads that if Grasshopper launches one of his burlesques, egged on by Chloe, grief-stricken outrage could explode into murder. Theirs.

Jason shoves his painfully sprained foot even harder against the accelerator as if to wring yet another increment in the pitch of the engine's scream. The speedometer does not move much, however. It refuses to exceed thirty-two miles per hour, however hard he grits his teeth and presses against the pedal. It is just as well, he reflects darkly. He can hardly keep it on the road as it is. The wheels, all four of them, seem driven in different directions and the shattered pavement grows only worse. A faded road sign from a long bygone age advertises air-conditioned night's rest with color StimuVision in every room at *Brougham's Motor Lodge* in Lebanon, only thirteen miles away. He recalls staying in such a place, years ago, with his parents; it had been so cool and new and exciting. Only they wouldn't let a teenager turn on the StimuVision, of course.

Shaking his head to fend off plaintive reveries Jason tries to see what transpires in the rear of the hearse. Chloe is blocking most of his view in the mirror and he keeps having to look at the dismantled roadway to avoid crashing into ditch-like gullies. From what he is able to see in the mirror, Grasshopper appears to be sitting cross-legged by the wounded man while the elder brother kneels nearby, distraught in his indecision at not knowing whether to intervene or to allow this bizarre ministration to continue. Jason can hear him asking repeatedly, to no avail, "Are you a doctor? Are you a doctor? Is he a doctor? Is this man a doctor?"

He winces as he hears Chloe's asserting reassurance, "Kind of . . . he's a healer!"

The man's posture suddenly tenses, then sags as he loses all hope. With both hands he starts to push Grasshopper aside, then relents as he apparently decides it makes no difference now. Slumping disconsolately against the right rear wheel well, his head inclines at an angle as his eyes glaze into a remote and haggard stare. Jason cannot see Grasshopper clearly, but he presents a macabre blood-smeared spectacle. It is impossible to say whether he is kissing, or licking, or spitting repeatedly on the dying man's wound as his hands push blood across his chest, trying to coax and herd it back into its human vessel.

"Oh, no," Jason croaks in advent of certain doom. Transfixed by the bizarre scene before him, the elder brother's eyes already are murdering Grasshopper. They move slowly over his features and gauge the violence sufficient to snuff out life here and there until . . . the hearse slams down suddenly and then a crash and grind announce disaster through the sudden spasm of the steering wheel that violently slings Jason's hands away. The engine leaps terrifically and seems about to jump from its mounts. Blue black smoke explodes from beneath the hood and sweeps by both side windows as Jason agonizes in the shearing sounds of metal pummeling metal and bitter odor of scorching french fries.

Then it stops. It just dies. Where there had been the painful shriek of an engine at full throttle, there are now only sizzling, popping whispers that bubble and gurgle. Jason coughs and coughs in the acrid blue-black miasma pouring from beneath the hood while in the background Grasshopper sings happily,

Go tell Aunt Rhody,
Go tell Aunt Rhody,
Go tell Aunt Rhody,
The old gray goose is dead.

Yanking open the door, Jason dashes around and pulls open the hood. Black, pummeling smoke assaults his nose and lungs and blinds his eyes with tears. Choking and coughing in paroxysms of aching breath, he backs up to wait for the steam-mixed smoke to clear. For a moment he stands just looking at the sizzling, hot and useless clump of metal. Then he gives it up as hopeless and goes around to the still closed rear door where he simply stands, listening to Grasshopper sing the same mad song over and over. He stands waiting for a gunshot to announce his own demise. It is all over. He no longer even cares what happens.

He is weary of struggling against the times. Obviously the times want him and practically everyone else dead. There can be no question of justice or anything remotely like reason. It, whatever it is, is just something that happens and we go with it wherever it takes us: somewhere, anywhere, Antwhere, awhere.

He lies back against the ground and closes his eyes, allowing the sun to anesthetize him before the onset of violently delivered pain. He is certain that the man will kill them all. Just now he wants to kill Grasshopper for leading Chloe, and him, into this obscenely ludicrous impasse.

Behind bright blankness of his eyelids screening sunlight, he pictures the rear door of the hearse bursting open and the elder brother cutting them down in murderous bursts of gunfire. It is sad, he acknowledges, but only because he once had speculated upon so much to do, so much to accomplish, so much to taste and to try. A sad weariness grows from disappointment, from realization that there has never really been anything to do but constantly revise one's estimates of people, of the world, of self, of things to pursue, to accomplish . . . to taste. It's all a perennial train of revelations: new guises for old problems, a lifetime of disposable hypotheses.

He had warned Chloe about falling under the spell of this madman. He had chastised her childlike credulity, her unflinching willingness to accept the implausible . . . nay, impossible! . . . notion of this lunatic being a healer.

She had replied only that the greatest good she has ever known has been found in the guileless innocence of children, their capacity for unquestioning faith.

But that's why adults have to protect them! he had protested. That guileless innocence also invites the greatest evils . . . but it was to no avail . . . she only looked sweetly, if sadly, back into his obdurate rationality.

Then he grimly smiles in realization that even such sadness is only a silly and frail habit that soon he will shed, along with all the other silly and frail habits that cluster through agencies of trillions of molecules into Jason Hood. That man in the tan windbreaker and white plastic hat will step down from that dilapidated wagon of death and rip him out of all those peculiar molecular habits. He, they all, will be freed into vaguely anesthetic warmth of the sun! He thinks he hears someone giggle, realizes that it is himself and then as suddenly snarls at the prospective cold futility, the entropic pointlessness behind it all, "Let the bastard die! We're all going to, anyhow! Get it over with!"

Then he ceases caring. Greatest good? Greatest evil? Where? Awhere . . . he fails into a vacuous brightness blanked across by his closed eyelids. In his mind's blankness is surcease, terminal refuge to be pushed out into infinity by his refusal to care about anything any longer. He listens and watches and waits and gradually eases into a rhythmically surging contentment. What's one gray goose, more or less? In the sleepy warmth of sunlight he hears the rear door scrape open and feels the pavement bounce as Chloe jumps down beside him. He keeps his eyes squeezed shut, awaiting the eruption of death. Then Grasshopper's voice squeals in ridicule, "Look at that boy! Lying asleep and letting Ants walk all over him.

Jason keeps his eyes obstinately closed as from the hearse he hears a distant voice, trapped within its own frail and silly habits of concern and anguish, anxiously asking, "Grady? Grady? How you doing, bro'?"

At that moment the voice seems neither sad, nor apprehensive, nor grateful. He cannot recall what any of such moments feel like. Then Chloe is shaking him, rousing him from his reveried rendezvous with death. When finally Jason opens his eyes to meet hers, he finds them glowing in their own fabulous reverie, "Jason? Jason! He did it! Grasshopper healed him!"

"That's nice," Jason drowses, closing his eyes again to keep in the contentment and fend off remote distractions. Finally she lets go of his head and shoulders, releasing him back gently against the pavement. He cautiously opens one eye to find that ants are indeed cutting a trail across his chest and stomach. Now and again they nose about on his belly then keep going. He opens the other eye and becomes lost in the frail and silly insistence with which they move in only one direction.

~ 12 ~

Awakening to daylight, Kristin at first thinks that she is back in her Manhattan apartment, rousing from an intense nightmare. She falls back against an abundance of pillows. Sunlight streams through windows into brilliant trapezoids of illuminated oak flooring. Obviously, she must grapple with her situation, anticipate Cornrath's schemes and plan her escape—but not before she somehow avenges the people out on the freeway and especially those that have died there. She fantasizes that there might be an opportunity to kill him . . . perhaps poison or a knife.

Her resolve fails to form any idea of what opportunity might present itself and she gradually admits her own reluctance to indulge wishful vigilante revenge. From somewhere in her college days she recalls Gandhi's injunction, "Any good achieved by violence is temporary, but the evil of violent means always endures." But gazing absently into a blazing platter of light on the floor she affirms that she will not hesitate to exploit any chance to bring Cornrath's machinations to light.

The vow spreads like warm honey along her spine and a smile draws slowly across her lips. It is a secret she will keep from Eric Cornrath, no matter what may transpire. It is a hidden purpose to cloth the naked vulnerability she first felt when Reese Hatlo and his henchmen were unleashed onto unsuspecting prey. How deeply does Cornrath's complicity reach?

Determined to find some means of escape from house arrest, Kristin selects from the wardrobe a clean pair of slacks and a blouse that fit rather well beneath a simple corduroy and leather jacket. She marvels that even a pair of walking shoes fit her feet perfectly. As she dons the clothes her mind weaves a web of intrigue in which she imagines herself wreaking a subterfuge that foils Eric Cornrath. It sustains her as no hope of mere escape could—to turn the bizarre absurdity of what has transpired into a triumphant act of revenge wherein every detail will, in retrospect, seem predestined: Eric Cornrath reduced to trembling in the shadow of his own connivance and masquerade.

Kristin is invigorated by sudden resolve. She unlocks the bedroom door and strides along the carpeted hallway, down the stairs and into the lower house. She finds herself to be alone.

Her solitude is disappointing in that she cannot begin immediately to discover means of undoing Cornrath. She has imagined finding him somewhere downstairs, leading him into conversation and quickly homing in on an unforeseen Achilles heel. Instead, she finds only lack of any signs of habitation that would indicate he has been there at all. She wonders if he even uses the house. If not, then where does he live? The Colonel may have other accommodations.

Suddenly she is ravenously hungry. Plots of retribution are set aside in favor of a new objective. Both the refrigerator and pantry are fairly well stocked with canned provisions, bread and sandwich meat. She breakfasts on a bagel, cream cheese and jam, tuna and milk, then looks out the back door. A sudden desire to leave estimates the distance from the back door to the distant woods and she wonders how quickly drones might be dispatched to intercept her fleeing figure before it disappears into the trees. Too unlikely, she decides, recalling the stunning electroshock and disabling gases that hit the Berserkers. Shrugging slightly, as if some request has actually been denied, she begins a more thorough exploration of the house.

It is extremely spacious; new rooms seem to pop up at every quarter, both upstairs and down. When she has familiarized herself with the general layout and decided that she does not care for the desultory mix of contemporary and Colonial furnishings, she switches on teleview in what appears to be a game room, hoping to get some news of the national situation.

She is astonished that only game shows, soap operas, old movies, talk shows and a few weary situation comedies are available. Cable channels devoted exclusively to news are missing from the lineup. Turning to one of the talk shows, she hopes someone might refer to the chaos she has seen in New York and imagines for other cities. In astonishment she realizes that the show is several years old. Its guests are prodded into blithe confessions of weaknesses for chocolates, sex, money and vanity or ushered through cheery discussions of trivialities regarding their sex lives or gossip about other celebrated peccadilloes, heavily interspersed in commercial pitches of kitchen cleaners, sporty autos, virility pharmaceuticals and scented douches.

In disgust Kristin flicks a button on the remote control causing the wall screen to go black. A small radio she finds in a drawer is even less informative. It drums and hums old punk-rock and jetso amid a roaring blather of commercials. Suddenly it becomes apparent that she cannot get news of the outside world, only pointless shreds of useless-tainment. She throws herself onto a large leather sofa, the radio still raunchily punctuating the silence. The past several weeks of her life suddenly seem to be a feebly imagined concoction floating in an absurd quirk of memory.

She looks about the large room, her eyes slowly roam what seems a hollow buffoonery compounded out of gleaming chrome and supple leather upholstery. Autumnal golds and browns are woven through the carpet, a sprightly colorful print winds about a wing-backed chair. Even her own well scrubbed skin seems misleading. The walls display hoards of framed and kitschy Americana. If the past has any validity at all she should be tired and grimy and disenfranchised along with everyone else. She has been extracted from meanness, but into what? Calumny? Depredation? Treachery?

It seems that she recalls being happy, even while feeling disgruntled, back with the cluster of people that gathered about and moved with Dooley's Grasshopper. Now, absurdly, her present circumstances seem like staging to shoot another commercial, as if she has been thrown back to an earlier time when everything was available in unfolding scripted non-sequiturs. Her hand works absently at arranging her hair as she tries to recover her bearings.

She arises and walks quickly down the hall, through the kitchen and to the rear door. In the distance she sees Cornrath and Eddie Minnow walking slowly toward the house, engaged in lively dispute. For a moment she cannot recall just what it was she had planned. When she remembers it seems too futile, too silly, too impossible in these absurd circumstances. The sense of her own vulnerability returns like a crystalline fracture rendered only in dream fragments, wishes and implausible scheming.

Kristin turns from the door and heads back into the house, hoping to avoid Cornrath's presence. The kitchen offers no refuge, nor does the dining room, nor any of the several large rooms in the rear that seemed once intended as den and playroom, which now have become dump sites for all sorts of items and boxes bearing stenciled military acronyms, BATCOM 321 . . . ANT 207.

Wandering through the house, she distracts herself by imagining who has lived in these rooms and lolled about on its furniture. The antique building obviously has been restored, a new-old vogue exerts itself in every detail. She imagines that she might have lived as country gentry and wonders wistfully if such a tribe will ever return to those storied graces.

She thinks she hears him leave the house, but as she muses about the furnishings and drifts through rooms she comes near the living room and hears conversation. Just as she is about to go upstairs, Eddie Minnow rushes out of the room and stops, surprised to find her standing in the hallway by the stairs.

"Jesus! You startled me," he fusses at her.

"Don't mind me, I'm just going upstairs."

"Kris," Cornrath appears at the door, "I thought I heard you. Please, come back. We need to talk."

"No," Kristin says evenly, "we don't need to talk, we need to agree that you'll stop holding me against my will."

"Well," that's what I want to talk with you about," Cornrath urges, "come in. Let's work this out."

Reluctantly Kristin moves to the sofa; her image still gazes up petulantly from an array of old magazine covers. She flips them face down on the coffee table to rebuke that past and tosses a couple of other magazines on top as if to fully bury it.

"Is she always like this?" Eddie asks from behind.

"I really can't say," Cornrath replies as he sits at the other end of the sofa.

She gets up and moves to the chrome and black leather chair.

"Kris is offended right now, Eddie," Cornrath observes, "I think she'll come to appreciate her remarkably good fortune, though. I can be very patient. And she thinks I didn't do her a favor by bringing her here," Cornrath chuckles, shaking his head slowly, incredulous that she can remain so stubborn

Eddie shrugs and asks Cornrath, "So what are you going to do with her now?"

"I'm not sure," Cornrath lights a cigarette and stares into its cloud of smoke. He reaches over to pull an ashtray closer. "But it's possible that she may have a change of heart and seize this marvelous opportunity that has fallen right in front of her."

Kristin thinks silently how mechanical they sound, as if reading for a play about which neither cares. She is puzzled by their abrupt shift into sham histrionics. Then she realizes, this is a custom PR spot, produced for her benefit, the old ploy of two people discussing some sponsor's concerns.

"A marvelous opportunity?" Eddie echoes idly.

"Yes. Imagine. Kris was one among millions with no place to go, nothing to do but try to stretch today out into tomorrow, no probable future except in privation and misery. Now suddenly her prospects have changed into ample options. She is upset by the suddenness, unexpectedness and even messiness, as one might expect. Confused by a situation suddenly altered. Naturally she feels threatened and hemmed in by circumstances beyond her control."

"If this is for my benefit," Kristin snaps angrily, "you can save it!"

Cornrath ignores her outburst and continues in the same monotone delivery, his eyes meeting hers squarely. She cannot discern whether their expression conveys detached humor or some subtly bizarre malevolence. ". . . it's the classic dilemma of one ensnared by sudden and unearned enrichment. She has been lifted above thousands, even millions, among whom she has commiserated. The bond of desperation they shared has been broken and displaced, at least for her. She doesn't understand why she is singled out for special treatment. So she feels guilty. She doesn't understand that there is nothing to understand. Haunted by images of the women with whom she was terrorized, she can't accept her self-inflicted guilt, and so blames it all on her rescuer—me."

"Stop it!" Kristin shouts hoarsely," Just stop it! I'm not listening to this any more!" She arises and moves toward the door as Cornrath continues his soliloquy, his eyes tracking her."

"She will punish herself for a while by denying this new situation. She will think of it as a prison defying her desire for freedom. But truly, freedom lies in merely her accepting the fact that, for her, life has changed. She has been elected by fate to join the privileged. Why struggle against fate?"

As Kristin approaches the staircase, she is startled to hear the tenor of Cornrath's voice suddenly shift and begin echoing throughout the house. The place is wired for omni-surround sound! Cornrath's voice pervades every corner, rounds every turn, saturates every room. Even with her hands over her ears she still can hear him.

"Yes, Eddie, that is the marvel of this opportunity . . . freedom lies in acceptance. To realize that her guilt is senseless, that all accusations arise only from personal envy, that she has an advantage others out on the freeway would gladly kill to make their own. We must be patient . . . give her time to discover and accept that option, as ultimately she must."

Kristin runs back into the living room where Cornrath now stands by the bar, his hand pressing one of a row of buttons on a wall panel. He turns casually to watch her as he continues his mock dialogue. The sound reverberates and echoes throughout the house. "And we can help her. Such an abrupt transition is not an easy accomplishment. Even now she is filled with horror as she suddenly realizes that she cannot escape my voice and suspects that at night when she is asleep I can hear her cry out to the women named Agnes and Justine and that she is no longer secure within the privacy of her self.

"Yes, Eddie, the mind of Kristin Banner can't always be on guard and little by little it will seep out at night into the walls and ears of this marvelous old colonial house where it will mingle with mine and she will awaken one day and find that she has already made the transition. She has become one of us. See, already she carries herself with the pride and bearing of true aristocracy!"

Cornrath releases the button and closes the panel to lock it with a key that he returns to his pocket. He still watches her but in a demeanor of detached boredom, as if this has grown all too familiar. Eddie sips a drink, an eyebrow raised slightly, obviously glad to be finished with a routine that he finds somewhat silly.

Kristin leans back against the door jamb, staring back and forth at the two of them, trying to make some sense of their charade. To even have known Agnes's or Justine's names, Cornrath must have heard her cry out in her sleep. Is the place also wired with cameras? Has he also watched while she lay in bed vaguely plotting fantasies of revenge? Can he deduce her thoughts by simply watching her?

Then she remembers a moment from long ago: Eric Cornrath bragged to Bill that given enough time and access, anyone could be persuaded of anything.

She turns to flee the room, unable to bear their looking at her as if she were quarry to be pushed, prodded and turned this way and that. Stumbling up the stairs she reaches the bedroom and slams shut the door, but her vulnerability does not dissipate. She still feels their eyes, peeping from hidden vantage. It seems they are listening to her breathe. A recollection of Reese's eyes watching her breasts heave in terror flashes through her mind and she falls back upon the bed to smother a frantic scream into the pillow.

Downstairs Cornrath hears her muffled outcry delivered by a speaker on the wall near the panel of buttons. He nods grinning to Eddie and says, "There . . . score one for the lions."

Kristin glances from despondent wall to taunting window, treacherous in their pretensions of holding back the beyond while they permit the ravaging of her privacy—all along she has been on display. She knows now that an invisible and surreptitious trickling has sapped her confident independence. Worse, she is mired in trickery, deceit and hidden traps. This pseudo-place, inseparable from Eric Cornrath, ends in no particular locale or condition. It feeds from his intrigues and is decked out to accord with his intentions of the moment.

In the loss of certain boundary she seems to have given up her own place to be, to have become contingent upon circumstances beyond her control or even knowing. She feels stripped of any degree of privacy. Her muffled scream reveals a gasp of panicked desolation, an abrupt and menacing loss of soul.

No privacy! Who might she be, if not herself? Where once there was separation, now it seems that the world has quickly exploded through some lacuna that once permitted her to be a who and driven her back against a fretfully distressing and strangely alien what into which she is now on the verge of dissolving.

A senseless question tries to open into vacuous departure of her being, by some collapsing inner platform of her own self. What might it be if not itself? Foul taste and loathsome textures inhabit a formless question that horrifies her. It seems that she, or it, or somebody is standing outside, beyond these walls, looking back where her schematic-stricken eyes tell her such fey haunting is impossible. A muffled scream already has signified that pulsating inner walls about her are on the verge of failing—that she will see on their far side only her own self, staring back in wide-eyed and deranged disbelief. A stream of her own images on countless magazine covers and television commercials has suggested a clue to such hideous travesty of persona. Now she has to find out—and find out quickly—who is inside Kristin Banner?

In prelude of an eruption fizzing in ominous effervescence, she sees in her mind's eye or hears in her mind's ear explosive recollection sounding like a latch dropping into place, the hook on a screen door sliding crisply to clink against its mating metal eye, its spring-twanging tensions released. Curiously, for a moment, all she can remember is someone once saying to her in gravelly indignation, "You be Peter Dooley—I'm busy."

That single recollection draws itself unawares across her distraught lips to relax them into a smile reminiscent of those often seen musing above still empty boxes of working crossword puzzles: a thirteen-letter word of meaning? But what can it possibly matter?

And then it is over.

Eric Cornrath does not hear the giggle suffusing the radiant burst of private enlightenment that follows shortly after her muffled scream. Otherwise he would not have bothered to return to the locked wall panel, insert and turn a key as if cranking some hidden engine, flip back its walnut door, press a button and begin speaking. His voice is practiced and slides expertly among tones and semitones, threading urges and demiurges.

Upstairs, Kristin has moved from where she initially cowered back against the wall and now lies propped up on pillows piled at the head of the bed, thinking idly of Grasshopper lives. Cornrath watches her through a hidden telecamera as he speaks through mellow folds of persuasion. She gazes back at the wall, not minding if it falls or holds—she is too busy to concern herself with petty folly.

She hears, but in distant and wry amusement at the fool Eric Cornrath has revealed himself to be. Somehow she has stepped beyond the contained person he is addressing into an opening region he can never reach. It is strangely calm and peaceful, a realm of invigorating suspension that simply awakens opportunity for enigmatic smile to transmute into broader joy than Eric Cornrath can imagine or acknowledge.

She knows that his eyes watch, but takes comfort that he cannot see, for he cannot understand. She has moved beyond his grasp. She has been restored to mindfulness of the depth, clarity and sanctity of her own being. She listens through idle reminiscences of Peter Dooley putting on Grasshopper makeup that slowly become reveries of Grasshopper taking off his Peter Dooley disguise.

"Kristin . . . I want you to understand precisely what I am doing and what is happening to you. I've found that it is a more effective path to transpersonation." Cornrath pauses, then continues in the same practiced voice, "You may be displeased with me. But I assure you, that's only a misconstruction of opinion. I have spent my entire life changing opinions. I know that an opinion simply culminates a set of experiences. Change either the nature or the circumstances of those experiences and opinion alters. I will transform your opinions, of me, of you, and of the world about us.

"Gradually preconscious rhythmic suggestions subtly transform the hierarchy of forgotten and repressed associations that culminate in your current proclivities. You are a bright person, but you may not know about recent techniques that go far beyond crudities like subliminal perception. I want you to understand fully so that you can attempt to resist. Only then will you appreciate the transformation through which I will have brought you. Only then will you be willing and able to fully join me . . . ah, you have no idea of the delights that await.

"Moments of conscious attention have perceptual thresholds. I am saturating your environment with subsonic rhythmically encoded images and auditory stimuli that penetrate far below those thresholds. You will be largely unaware of them, yet they gradually accumulate in effect. By combining these within resonant frequency information derived by cybernetic sensors from your own body's living processes, it is possible to deliver absolutely precise inductions of information that slowly, like spring breaking across a snow bank, melt away the old to flower into new Kristin's awakening.

"Until the government-imposed transition into austerity, we were unable to use much of the research into such advanced persuasion techniques. But they have proven very germane to the national war against terror. And like any true professional, I have always brought my work home with me. The point is, these techniques are very effective. As a matter of fact, you are wearing some of the wardrobe left here by a recent test proselyte . . . only she wasn't a very good test since she didn't think I was disgusting to begin with. You do. I want to take advantage of that firmly held opinion, even encourage you to hang on to it. It won't hurt. It isn't even wrong—what will you have lost? Nothing! It will simply be as if you gradually acquire a different legacy of experiences in your life, experiences that reconfigure into opinions more amenable to our circumstances. I will make you happy. I prefer to think of such revision of your psychic structures as a fortunate reshuffling for all concerned. It is the ultimate morality to bring harmony where there was discord and amity to displace disaffection—to resurrect peace without depriving anyone of anything. And that is the pinnacled quest for all practitioners of public relations!

"Such efforts now can resolve errant conflicts that arise through unruly social interactions. When pacification is restored everyone benefits; everyone is happy seeing apparent necessity in which their lives unfold. Every consumer matched with a vendor; every market, a supplier; every artist, an audience. But I need your help. When we find Dooley, I'll need you to help me explore the persuasive implications of the technology he bears.

"Resist. You are very strong, so I know that you'll resist even without my prompting. But eventually you must understand, even laugh at, your misunderstanding of what is at stake. How can options that benefit everyone be wrong? Still, I do need to test and evolve these techniques, so, believe me, your resistance can only help us improve these measures. Believe the counter case as hard as you can: Eric Cornrath is disgusting. You'll be amazed at your slowly infused progression toward transformation, though you may be unaware of it as it is happening. Hang on! Keep the faith, Kristin. I am coming."

His voice dissolves into silence. Kristin still lies back against the pillows thinking idly about the ostensible proselyte whose clothes she now wears. What became of her predecessor? Had that person truly succumbed to Cornrath's suasion? And if she had, does she still use her same name, or did she garner a new Cornrath cult moniker? Does she feel dirty?

Kristin toys with the possibility of telling Cornrath that her opinion of him already has evaporated, even before his little lecture. She no longer finds him disgusting. She sees his pathetically confused blindness and obsessive inertia, which now seem trivial against the span of new realizations. If she were to inform his pathological apathy, he would misconstrue it as an opinion and fall into another paroxysm of flipping switches and constructing recordings to play back as she slept and God-knows-what-other-playthings-he-fools-around-with. He would fail to grasp that however he might tamper with the underpinnings of her judgment, he cannot alter a known truth. Her known truth. She smiles, then chuckles happily as she decides not to burden him with it.

Downstairs Eddie Minnow watches her image on the monitor and turns to ask his gloating companion, "Hey, Eric. Why is this one giggling?"

~ 13 ~

"What's going on, Hec?" Grady groggily inquires of his brother. The wounded man, still unsteady on his feet, leans wearily against the rear of the hearse, his conflicted eyes attempting to coalesce scattered observations and memories into some meaningful grasp of what has happened to him.

"I'll be damned if I know!" Hector Abbot shakes his head in dismay as he peers back at his brother, looking for signs of distress, "I thought you were gone, for sure!"

Grady's face erupts into a broad grin as he tentatively touches the still-knitting wound with his forefinger, "Just goes to show what a tough ol' bird I am," he cackles.

Hector, still incredulously shaking his head, says, "All I know is that guy over there got a-hold of you like something crazy! It was the damnedest thing I've ever seen in my whole life! He was beating on you and spitting and licking and just going on something crazy, like a dog working on a wound." Hector's voice quavers slightly as he inclines his head toward Grasshopper who now stands over near the side of the road, removing his blood-soaked shirt in dismissive gestures expressing his distaste for sticky, now crusting sensation of clotted and drying blood.

Jason, still prone on the broken pavement, watches them all through a tiny crevice in his tightly squinted eyelids. Distraught in a world suddenly grown insolent in absurdity and total abandonment of necessity, he now is pushed by rank curiosity to open a tiny peephole in his defenses against chaos. Chloe has given up her attempt to rouse him. Even the ants marching across his chest have disappeared. Lying on the pavement he listens and watches through a faintly gleaming slit of revelation, like some sea urchin ready to recoil suddenly back into its closet of private concerns.

The older brother suddenly realizes that he is among strangers. He enthusiastically extends his hand to grasp Chloe's as he exclaims, "I don't know what your friend is, but we'll both be obliged to him the rest of our lives!" He gestures at his brother, "This here's Grady. I'm Hector. We're both Abbots. The Abbot boys, folks call us."

Chloe takes the proffered hand and clicks off names of her companions, "I'm Chloe, Chloe Jernigan. That's Jason on the ground, Jason Hood. And he's Grasshopper." She gestures over toward Grasshopper who now is removing his blood-soaked pants, keeping only the iLink strapped across his hips.

The two brothers stare at him, periodically looking at each other quizzically, then in some tacit joint decision, ignore his disrobing. "I don't remember much after some guy came at me with a pistol," Grady recalls in a dazed voice, "I do remember being somewhere and Clara saying somethin' one time, and that's about it."

"Clara and I loaded you in the heap. I figured the only chance was to get you to an Army medic. You started bleeding again and we couldn't stop it. Then the transmission locked up and I stopped and asked these folks to help drive while I took care of you. Then that guy, Grasshopper, climbed in the back . . . and I don't know what happened then, except all hell broke loose and he was licking and rubbing on you and singing a crazy song about a gray goose, spitting and carrying on like a mad man. I was just about to knock the crap out of him or something when the engine blew up . . . and there you were . . . sitting up, looking at me like it was breakfast time!" Hector hazily summarizes events as he looks toward Grasshopper, obviously at a loss for an appropriate conceptual niche in which to store the tale for later telling.

"He really is a faith healer, then," Hector turns to Chloe, "just like you said. His voice is in awe. Jason wants to interrupt, to object that Grasshopper is a lunatic fraud, that he cannot possibly heal anyone. Unfortunately for his intent, he is recoiled into his would-be urchin shell, waiting for a large, unfamiliar and intimidating ontological creature to pass on by, one which he has never before even imagined, one that brings in its wake a contradiction to his most preciously certain tenets. He does not dare open up at this moment lest he be consumed by some newly revealed absurd proviso of a universe he is being forced to start all over again trying to explain to himself.

Now Chloe, for all of her zealous defense of Grasshopper, also seems at a loss to explain how a man with a bullet hole in his chest can step back from the precipice of death to idly finger the orifice through which his life has just tried to escape. She nods an uncertain confirmation of Grasshopper's powers, then looks again at Grady's blood matted chest. He grins groggily back at her, like a drunk getting ready to launch another round.

"I don't know what he is or what he has," Grady sways, slightly giddy, "but I feel really strange, and I guess he did it!"

"Strange? How?" Hector inquires in renewed concern.

"Like I'm coming a-fire!" Grady yelps, his shoulders twitch abruptly, "Like pure corn whiskey running in my veins! Like the Holy Rapture raising me up."

He shouts as his hips join in rhythmically spasmodic seizures of his shoulders, "Like if I don't take off and fly, I'll burst inside and out" Holy Jesus, Hec! It's crawling all through me! It's alive!"

"What is?" Hector pleads for an answer, fearful that his brother's reclamation from death will prove to be some transient fraud. He glances darkly at Grasshopper, then demands of Chloe, "What's he done to him?"

"I don't know!" Chloe implores, then she too quivers, "but I'm feeling a little strange myself."

"Hec!"

"What?"

"Not only do I feel strange, but I feel GOOD!"

Grady slowly backs away from his brother and begins to laugh rakishly as he twists and turns in ecstatic abandon. Hector edges toward him, hands outstretched as he begs for caution, "Take it easy, Grady! You'll start bleeding again."

"Too late, now . . . I'm off!" Grady shrieks an ear-rending whoop and prances across shattered pavement, down a gully washed out of the derelict roadway next to the abandoned hearse, past Grasshopper who now stands stark naked, smeared in the ecstatic dancer's blood, watching his giddily choreographed plumes of enthusiasm. Grady's shoulders heave to and fro as if trying to shake loose a weight or attempting to find wings. Grasshopper answers in cackling, echoing jubilation and the two men dance as if taunting the sky's refusal to accept them. Insistently, they keep hurling themselves upward in rhythmic frolic damped only by their own gravidly ponderous bodies.

Now Chloe joins their gamboling across the ditchbank, all shedding clothes in a pell mell paroxysm of laughing and pulsing squeals of exuberant energy.

"Whatever it is, its kind of fun!" Hector lurches his stolid frame into a bounding mean of Chloe's grace and Grasshopper's frenetic tirades. True to his recoiled urchin self, Jason watches from the slit that marshals his reserve and keeps his regard for the world from slipping away into their tribal ecstasies.

"I'm alive! Hallelujah, Lord! I'm alive!" Grady cries out in jubilation.

"Oh, Lord, the tongues of flame are dancing!" Hector yells as he tears away his jacket and shirt in one ripping flourish of his arms. Chloe is naked. Her sleek and rippling fairness play a counter theme to three caroming mock warriors. She twirls and swims like sunlight across the brush. They follow her beyond the pale of Jason's vision. He hears them whooping and railing the day in their madness. He is glad that he is not like them. He is glad that the sanity of the world has found respite in his bosom, at least. He holds it tight and refuses to let go, keeping it safe from their tirade. He cannot let their failures invade the last desperate hold of all that for centuries has held the world together in a semblance of reasonable commonsense.

Then suddenly, they are back and there is no stopping it. They dance a circle of madness about Jason and goad his eyes open by chanting, "Hang on! Hang on! Hang on!" He later remembers Hector caterwauling an attempt at old Celtic airs turned Christian hymn. And then it begins to come over him, too. In fading dismay he realizes that this body to which he has for so long laid claim now arises of its own accord to assert some hidden carnal privilege. He is hoisted up on someone's pulsing feet. Looking down he sees that they wear his very own shoes. A heat flares through his entrails and sweeps upward to cloud his eyes in a searing compound of sunlight and landscape as suddenly he feels those shoulders merge with his in an electric invigoration that threatens to rip him bodily from the world. If only the ants would come back and weigh me down, he implores even as he plummets off into a shrieking and naked melee of wanton abandon.

Then they are running through a woodland of jubilation, yelling up to startled birds and squirrels to come down and join them, sweeping like dervishes among tall timbers until arrived at a creek into which they plunge to dissolve and wash away their sheathes of blood, sweat and earth. They frolic in frothing, bracing waters turned turbulent in the heat of their glistening newness.

Then finally, as the moment exhausts its flailing delights, Grasshopper suddenly stands bolt upright and his companions all fall back to earth, discharged from epiphany's height as heavily pulsing engines of three helicopters pass overhead. Looking up they see the dark carriers glide like sharks above the trees. Across their drab green bellies, emblazoned in white block letters, is the acronym ANT.

"ANT!" Grasshopper screams in outrageous joy. "You're back!"

Sloshing up the creek bank, Grasshopper scrambles frantically up and flies away in pursuit, exultantly calling back to his cohorts, "It's Ant! He's back!"

Without hesitation they are impelled into his wake, only vaguely appreciating the magnificence of such an entity to so abruptly command their presence. They pursue Grasshopper's frantic plunge through the woods as he calls out to the fugitive ships skimming rapidly off into the distance,

"Wait up, Ant! I'm coming!"

Back near the Lebanon Command Center, about four miles from the main entrance, at the wooded edge of a glade, Lieutenant Nathan Pomeroy tries to reconcile what he has been told with the scene unfolding out in the open field. His orders are to disrupt a high-level transaction among criminal traffickers of drugs, humans and stolen property and to arrest its participants. Heavily armed squads of his sentinel platoon have been dispersed about the periphery of the clearing, awaiting his orders to move in as soon as there are actionable signs of a transaction.

In the distance a convoy of vehicles, comprising three limousines and several heavy vans, has formed up to receive men in business suits debarking from three Air National Transport helicopters that have just landed. He is confounded by what appears to be a teleview video crew. What criminal organization makes video recordings of its operations?

He is informed that the flanks of his deployed platoon have spotted what seems to be some homegrown militia unit, clothed in a mix of military and biker garb, secreted in woods at the far side of the clearing. Before he has time to interpret the tactical situation automatic gunfire and rocket grenades from the distant forest edge begin relentlessly to rake the scattered clusters of suited men out by the vehicles, some of whom immediately produce automatic weapons and return fire while others knock their companions to the ground or drag them behind shielding vehicles. Pomeroy orders both wings of his deployed squads to advance through the wooded periphery and suppress the unknown group's fire. As he moves to his left to gain better position for dealing with this new contingency Lieutenant Pomeroy is instantly killed by an errant bullet that shatters his neck at the base of his skull. The melee continues without him.

~ 14 ~

Kristin awakens from a nap into a mood of rather resilient tranquility. As usual, she has the house to herself, but somehow its atmosphere seems curiously still. Muted noises have been emanating from a locked room next to her own; perhaps it houses sensors and feedback equipment with which Cornrath fancies he is manipulating her. The entire venture has become quite laughable; it isn't working—she won't let it work.

But now the room is silent. Maybe he has altered the stimulus program. She places her ear against the wall that separates her bedroom from the equipment room. Not a sound; not even the occasional faint clicks she has come to associate with her own movements in the room. Probably a glitch in the equipment, she decides. It pleases her to note yet another flaw in the facade of mock competence by which Cornrath flatters his own prowess.

To deter his elaborate scheme for altering her feelings she has plugged her ears with toilet tissue. Kristin takes great satisfaction in this simple ploy, convinced that she has effectively reserved the privacy of her own drowsing mind.

At first, curiosity caused her to lie awake, acutely listening and attempting to decipher sounds supposedly pumping surreptitiously into her unguarded awareness. She was disappointed that only occasionally could she detect anything at all and even then the sounds were not anything like the melodramatic Svengali voice she had imagined, hypnotically impelling obedience. Rather, scant noises have comprised a subtle repertoire, sometimes strangely similar to recordings of whales, long and rumbling, or hauntingly whispered oboe-like passages. At other times there have been briefly squeaking whispers, as a window shade pulled slightly by a shifting breeze or lips puffing quickly at a pipe. She could never be quite certain that she had indeed heard them; they seem more akin to remembering a subtle echo than actually hearing specific sound.

She rapidly lost interest in deciphering Cornrath's schemes. Stuffing her ears with tissue, she began looking for means of escape. Outside, she has seen no monitor drones and has begun to think that perhaps she should test them, perhaps by trespassing the yard boundary then hurrying back inside to safety from electrical shocks or disabling gases. Thus far, she has been content to just stroll about, her muscles savoring exercise as she lets her mind wander in and out of various fancies, without committing to specific priorities. The crisp coolness of twilight has come to bode a gathering onset of winter. Frost already has already begun to tinge the hills in hints of earthen hues of brown, gold and red.

As she wanders about the grounds Kristin peruses events of the past several days. Eric Cornrath has been confirmed in her estimate to be an immensely pretentious fraud. No observation or intuition provides a hint of any redeeming quality in the man. He has been shown a fool for disregarding his own limitations. He seems aware of no bounds to any wish, however fantastic. Had there appeared any moral dimension in his machinations, she might have regarded him merely as evil. Her esteem, however, can weigh nothing more profound than merely adolescent—no, childish—disaffection. Annoying and dangerous, yes, but ultimately trivial in its prospects. Silly.

She finds that her thoughts of him are no longer nested in immediate hatred of all that he stands for. Cornrath's problem, she has concluded, stems from his being some kind of wounded child. She catches herself regretting that he has squandered very real talents in comic book fantasies. Damn, she wonders, am I trying to understand his motives? . . . what's next, forgiving him? Might his behavioral programs be having some transmogrifying effect after all? Then, dismissing such idle reflections, she turns toward the pool.

She stretches out on a chaise lounge to enjoy the remaindered warmth of September sunlight. Her eyes relax into drowsiness and time drifts in and out of discontinuous musings until she is roused by a woman's voice, "Kristin Banner . . . what are you doing here?"

Startled, Kristin looks up to a slight silhouette framed in dazzling sky. Blinking, she turns her head slightly, her hand shading her eyes. Then she realizes, "Stacy? Is that you?"

Kristin leaps to her feet; the two friends embrace amid girlish squeals of delight.

"My, God, I never expected to find you here," Stacy pulls back, looking closely as if to confirm that their encounter is not a fantasy, "Why aren't you in Harrisburg or wherever you were so determined to get to?"

Kristin pours out a summary account of her being waylaid, kidnapped, rescued and kidnapped again. "But what about you? How . . . what . . . why are you here?"

"Well, you know how insistent Jim was about going to settle his problems with the law. Well, once we got all the documents signed and property released he took me outside in the hall and told me we're finished . . . that he's fed up with my ignoring him and not taking care of business first. That I'm on my own."

"What did he mean?" Kristin seems astonished.

"Who knows? He can be such a spoiled little piss-ant, a real bastard at the drop of anybody's hat. Who knows what he's up to . . . anyhow, he just walked off and left me standing there in the courthouse, outside the magistrate's office."

"What did you do?"

"Well, that's just it. I didn't do anything. I was standing there trying to get a sense of what might come next when a man came up to me and started trying too hard. I wasn't in the mood for pickup lines, but, being suddenly stranded and without any notion of what to do next, I did think maybe lunch was a good idea. So we went across the street for something to eat. That's when I learned that my would-be knight is regional director, or something, of FERAL, that his name is Eddie Minnow and that he thinks maybe I'd like to go with him to ride in a surveillance skypod out over the freeway. And that was when I saw posters of you and Peter plastered all over the place; I called and told them you were bicycling toward Harrisburg."

"You got picked up by Eddie Minnow?" Kristin says in disbelief.

"Forget diamonds, honey—a man's hopes are a girl's best friend!" Stacy winks, "besides, there I was, stuck in the middle of nowhere with no prospects. So we went up in the skypod and it's all true. He is regional director, or something, of FER-

AL and we got up there, drifting along, all up and down, over the freeway. Then sometime around early evening the strangest thing happened."

"What?"

"The most bizarre feeling came over me . . . there I was, standing looking out the port down on a huge throng of people below, on the highway, telling Eddie what it was like being down there . . . when suddenly it was like I could hear Grasshopper. . . Peter. Or, I didn't hear him as much as I thought of him, but like I knew what he was thinking . . . you know how when you're with someone and you know what they're going to say? Well, it was sort of like that, but a thousand times stronger. More definite . . . like I was actually thinking with him, myself. And it was like he and I—we—were saying my name, but yelling it like on this little chain I wear, 'ec-Stacy', over and over again. It was so weird. Then I felt this rush throughout my whole body, like I was on fire and suddenly I got lightheaded. Then I fainted.

"When I came to we were already headed back to base and Eddie offered for me to stay with him, but I put him off. Something about him tells me to watch out. So I asked around and found a place to stay overnight. The next day he came by and offered me a little job as his courier—'gofer' is more like it. So now I run around delivering and picking up stuff in one of his little coasters. This morning I had to bring a bunch of papers out here to his house. I just happened to look out the window and saw you over here. Or at least I thought it was you, so I came over to see. And here we are." Stacy moans happily and again embraces Kristin, "I'm so glad to see you! But what do you mean you've been kidnapped again?"

Kristin tells her how Cornrath's initial insistence that she be available to assist with debriefing and decommissioning Peter Dooley has rapidly escalated into some bizarre control-freak brainwashing fantasy of his in which she is held under house-arrest, hemmed in by monitor drones and patrolling sentries.

"Honey, I think he's already gotten into your head," Stacy marvels through an appraising frown, "there are no sentries out here. The base is back that way, about five miles. I'll bet there are no monitor drones, either. Eddie says they both live out here so they can get away from all that government claptrap. Your Eric Cornrath is just playing games with you!"

Kristin flushes as she confronts evidence of her own gullibility, "No sentries or drones?"

"No, not for miles. Besides, even if there are, you can come with me—I've got a passkey. It tells drones to ignore us." She holds up the plastic fob of a key ring holding several plastic security passes. "Eddie says it comes with the job . . . it lets me move across monitor boundaries back at the base."

"But where will we go? If I leave Eric will just track me down again."

Stacy shakes her head as her eyes narrow in diagnosis, "I've got to get you out of here, honey. Come on, we're going to Harrisburg. It's only twenty-five miles from here."

In half an hour Kristin has gathered a change of clothes, travel kit, energy bars and two canteens of water for her break to freedom. Not until the 'borrowed' coaster glides silently far beyond sight of Cornrath's residence does she dare nibble at a new prospect of liberty restored.

"The navscreen shows this road going to Harrisburg," Stacy says, "The main road for vehicle traffic is up toward 81, so if we can get there on this one there's less chance of running into any hassles with monitors or sentries."

"Sounds good to me," Kristin exults, "I just want to get away from here. Let's go."

As they travel west the road quickly deteriorates. Pavement broken in recurring cycles of summer heat and winter freeze hammer the coaster wheels, often reducing their progress to a lurching crawl. Erosion has taken chunks of shoulder and parts of a lane wherever the roadway depends upon artificially terraced support.

Eventually, as late afternoon shadows cast long across the broken pavement they arrive at an impasse. Blocking the only usable section of the road, someone has abandoned a derelict hearse and scattered stained articles of clothing on the ground about it. When Stacy attempts to turn the coaster around its rear wheels dislodge a fractured clump of pavement and fall into an awkwardly gapped niche that hold it fast. Stacy sighs, shaking her head in dismay, "Honey, it looks like we either walk back or spend the night here and figure what to do tomorrow."

"Tomorrow sounds good to me."

~ 15 ~

Jason has begun to feel very silly. Gradually his head has cleared into disdain for the whooping enthusiasms of his companions. His skin has been whipped and scraped by branches and brambles. He doubts that even indigenous Americans have ever really run very far through a forest. He imagines early natives judiciously picking prudent paths through sharp and rough things sticking up and out among woodland greenery but is not cheered by the image.

Jason is especially chagrined at having seen Chloe repeatedly kissing Grasshopper, full on the mouth during their ecstatically skyward hopping. She had kissed him long and deep . . . Jason puts it out of his mind, except to recall that she had joyously proclaimed that she too was "on fire!" Now he follows disconsolately in the wake of her chasing after Grasshopper's reckless pursuit of three helicopters. The Abbot brothers seem also to have regained some level of prudence, beginning to regret so much exposed flesh as they meander more expertly than he through the screening undergrowth.

Eventually a couple of distant explosions followed by intense bursts of gunfire stops the Abbot brothers and Jason. Grasshopper chases on, unchecked, after the now unseen helicopters, which have descended farther ahead. Chloe follows blindly after Grasshopper. Jason decides he cannot let her go on alone and sets off after her, despite every instinct warning caution. The two brothers follow after him, seemingly game for almost anything. Jason suspects that neither of them really wants to confront the fact that they are running naked through the woods with three strangers. So long as they refuse to really think about it, they can just keep going until somehow the embarrassment manages to trickle out of their compromised circumstances.

Sounds of gunfire are unmistakable and very disconcerting. It sounds like a pitched battle, with explosions, shots and screams. Glancing at Hec and Grady Jason tries to call Chloe back but she does not even slow or glance back; she chases after Grasshopper as if the new commotion might be a lot of fun, too. She obviously is still drunk in this . . . this, whatever she has led him into. Jason wonders if everyone resents epiphany. He suddenly becomes very angry, at her persistent gullibility, at Grasshopper's imbecilic flights, at Hec for not restraining them all, at Grady for not dying like he should—and at himself, especially, for knowing neither where they are nor what to do.

Muttering a rosary of epithets he stalks after her. She chases Grasshopper. The Abbot brothers, for lack of any better sense, follow after them all. As they cross an old road, long past its most recent repairs, they see a collection of gleaming motorbikes parked in undergrowth well off a distant roadway. Draped across the seats of several machines are vests emblazoned with vivid depiction of snarling wolf beneath the sanguinary legend BERSERKERS.

"What the hell?" Hec pushes Grady's arm and gestures toward the motorbikes, "What're they up to now?"

"Berserkers? No good, for sure."

In the distant melee, two helicopters have tried to fly out of the clearing, but have been destroyed by rocket-propelled grenades. President Ira Crown and a Secret Service agent, Sam Horowitz, have escaped into the woods where they cower secluded in a thicket. They are startled to hear someone yelling and calling through the woods. Thinking it to be renegade soldiers shouting "Hunt!" they scramble deeper into the woods, away from the scene of mayhem. In actuality it is Grasshopper calling after the fleeing pair, "Hey, Ant!"

When finally the shooting ceases, they pause, gasping for breath; Ira Crown screams vows of revenge as he vents terror and rage against the brush, ripping off branches and banging them about trees,

"My mandate . . . I have a mandate."

"There was army down there," Sam growls, "I saw uniforms."

"They might be impostors," Ira Crown speculates aloud, then snarls, "No—you're probably right! Army units! They don't want me as Commander in Chief!"

An abrupt noise of rustling leaves causes Sam to suddenly shift as he tries to wheel from a crouch and bring up the muzzle of his automatic pistol. He freezes as he sees the muzzle of another automatic weapon pointed directly at President Crown, held by a hand and arm reaching around from behind a large tree twenty paces away.

"Hunh--uh," warns a raspy, deep throated growl, "Drop it. Hands on your head."

Stunned that he could have allowed someone to get the drop on him, Sam obeys, pleading with his eyes for President

Crown to somehow maneuver his body out of the direct line of fire. The President does not move; he seems to have accepted that he has been displaced as Commander in Chief.

A stocky, thick-featured man emerges from behind the tree, dressed in military camouflage shirt pulled haphazardly over a black BERSERKER T-shirt, shirt tail flapping about faded blue jeans. On his head he wears a military field cap. His face is disfigured by an ugly gash beneath his left eye and heavily scabbed lips, as if he has been hit with a brick across the mouth. When he grins triumphantly, Sam sees that two teeth are missing.

"Okay, now—on the ground, face down. Both of you." Georgy advances slowly to rest the muzzle of his weapon near Sam's lower back while his left hand removes a service pistol from a holster on his right side, beneath his jacket, and then pats along the agent's legs for additional concealed hardware. He similarly searches the prostrate President then growls, "Okay, real slow, now . . . on your feet."

"Who are you?" Sam hisses, watching President Crown at the periphery of his vision. He longs for a momentary distraction to let him knock the muzzle aside. His charge, POTUS, merely stands quietly, his hands resting on his head, his eyes studying the scabrous and swollen lip of their captor.

"Never mind," Georgy snarls, stepping backward in anticipation of Sam's wary gaze, "Who are you—what's this all about?" He glances back and forth between his two captives. Sam is puzzled by the question. How can anyone party to such an assault not know its purpose? He races through a range of hypotheses, searching for some clue until President Crown speaks up in a tone often used to remind unruly subordinates of his and their respective stations, "I am the President of the United States of America!"

Their captor tries to smile at a joke, but the scab seems to prevent anything beyond pained sneering of thickened lips.

Then he looks more closely at the bald man before him and suddenly seems to recognize the face. "Damn! You are! So that's it. Get rid of you and us too! All at once. But they didn't get me . . . now the question is what's up next?"

"What are you talking about?" President Crown demands in impatient officiousness that often asserts itself in moments of executive distress. Sam winces as the gun's muzzle swings menacingly toward his charge who has absently lowered his hands to reflect the dilemma of one trying both to be President and captive. "Oh, sorry," President Crown apologizes and throws his arms back up again.

"I'm talking about a double-cross. Bring us in to do a job on you, then sending in troops to take us out! Scum-sucking bastards."

Georgy's voice trails off into puzzlement over what to do next as Sam and Ira Crown are both catapulted out of the gravity of their circumstance by an apparition appearing in the distance, through the trees. Their captor grows suspicious as he sees them exchange quizzical glances.

"Don't try anything," Georgy snaps, covering the two men closely, trying to interpret their transformed facial expressions. He watches them briefly, trying to discern some con, but moves too slowly as Grasshopper yelps and dashes forward, the iLink flopping avidly about his groin,

"Don't shoot! Don't shoot!"

Sam lunges to seize the barrel of the machine pistol and wrest it from the grasp of their distracted captor who, as he wheels to confront the intrusion, is totally flummoxed by the approach of a naked man with ample locks of bright silver hair.

President Crown's gasp mixes in undulating paroxysms of abject fear to induce a fit of uncontrollable hiccups. Grasshopper ignores the two men wrestling for control of the machine pistol and approaches the hiccuping President,

"Ant? You're not Ant. Where's Ant? Did he come this way?" Grasshopper chatters, leaning forward, attempting to peer into the President's tearful eyes as he doubles over with the churning seizures.

"Hic--ugh?" President Crown tries to respond, "H-h-ant?"

The two men wrestling about for control of the disputed weapon are evenly matched, although Sam seems at first glance to have advantage of size. The other's agility compensates for his only slightly smaller frame. Sam is hard pressed to hold his own as they roll and wallow among the brush, grunting and cursing while Sam desperately pleads for someone to help him. Neither Grasshopper nor Ira Crown seem aware of his plight nor of any contribution they might offer. One appears helpless, beset by contortions of his abdominal diaphragm. The other shouts like a hapless tourist relying upon insistent loudness to overcome linguistic difficulties to wrest information from a recalcitrant native.

"What? Speak up!" Grasshopper commands as he leans forward to hear what erupts from the heaving distress of Ira Crown's frame.

"Hic--umph!"

"Hic--umph?"

Again Ira Crown tries to squeeze out to this naked interloper his own identity, "Presinantcrowd--hicumph!"

"Present Ant crowd," Grasshopper translates aloud, squinting his eyes with intensity of tracking the vowel sounds, "yes, go on."

"Hic--urkch!"

"He's here, then?" Grasshopper extracts hopefully.

"Hic--uph!"

Grasshopper finally notices the two men struggling for control of the weapon, but dismisses them when neither is found to fit the object of his search. He is disappointed to find no other candidate save the gurgling man beside him and he certainly does not qualify, either.

Georgy, unable to keep control of the automatic pistol has managed to thrust its muzzle deeply into the dirt, to prevent its being fired, then lets go and dashes off into the forest. Sam has lost his balance as his opponent suddenly stops struggling. He catches himself, looks at the barrel now uselessly clogged with dirt, then throws it aside as he rushes for his own weapons lying nearby. He wheels about to find that Georgy is now completely out of sight. He starts to chase after him, then thinks better of it as he growls at Grasshopper,

"Who the hell are you?"

"Grasshopper."

"Hic--urmph!" says President Crown. Sam watches the two of them momentarily, decides there is no immediate danger from one so obviously without weapons and sighs as he lowers his pistol.

"Damn! How could I let him get away?" Sam chastises himself.

"That wasn't Ant," Grasshopper says to relieve his concerns.

"Ant?" Sam repeats, suddenly suspicious, "Grasshopper? Those sound like code words." He looks closely at Grasshopper's iLink codpiece, "Just what are you doing out here . . . near naked?"

"Looking for Ant. He just flew over."

"Flying ants?" Sam frowns toward President Crown for guidance, but gains only an apologetic smile that says, "Hic-cuph!"

Sam peers at Grasshopper warily. He does not like the bizarre appearance of a man wearing only a bulky codpiece named Grasshopper searching for some kind of flying ant, especially mixed in with an assassination attempt on the President of the United States. "Why are you . . . undressed . . . like that?" Sam demands.

"I took off my clothes," Grasshopper replies, disdaining to explain the obvious.

"But why?"

Grasshopper turns to the bald man whose hiccups now seemed to be subsiding, "I'm looking for Ant."

"Quiet!" Sam orders under his breath, "someone's coming."

"It's Chloe and the others," Grasshopper reassures the agent who now has weapons in both hands.

"What others?" Sam reaches out to push President Crown suddenly into the protective zone between himself and a nearby oak tree.

"Four Ants," Grasshopper snaps, annoyed that Banner Ant has once again eluded him. "Over here," he yells out in the direction of the thrashing footsteps.

Sam's jaw drops as an additional four totally nude people suddenly emerge from the scrub brush. "It's a nudist colony!" he exclaims in disbelief.

Ira Crown has begun chortling in heaving convulsions that threaten to overwhelm and dissipate his hiccups, "Oh, dear God! What next?" he giggles hysterically.

"There he is!" Chloe shouts, thrilled that Grasshopper has finally ended the chase but apprehensive at his being in the company of two men, one of whom is armed.

Sam seems unable to decide what to do with either the pistol in his left hand or the machine gun in his right. He stares in embarrassment at Chloe, then at Grasshopper, as Ira Crown guffaws uncontrollably, tears streaming down his cheeks.

"They don't seemed to be armed, either, Sam," the President howls at the agent's discomfiture.

"Who're they?" Chloe asks Grasshopper.

"Nobody," Grasshopper replies, disappointed and indifferent to the two men's identities.

"Nobody," Chloe repeats in disappointment as Hec, Grady and Jason moved between her and the armed man.

"Did you get hit out there?" Sam asks as he notices the bullet hole, only recently begun knitting, in Grady's chest.

"That's Grady," Grasshopper intones melodically, "he was all leaky. But the blood was all dried until I spit on it to get it back in. He nearly died."

"I should think so," Sam whistles in appreciation, then asks of Grady, "Don't you need to lie down?"

"Yeah, I do," Grady admits wearily.

"Where are your clothes?" Sam asks them all, "Why are you out here like this. And who are you?"

"And that's his brother, Hector," Grasshopper continues with introductions as if shepherding a social gathering. "This is Chloe. And he is Jason." Jason is irked by the inflection of disdain in which Grasshopper includes his name.

"Oh, me no," Ira Crown sighs luxuriantly as he is released from recurring spasms of hiccups and laughter, "what is happening to me?"

"This," says Sam, as he recalls his duties, "is Ira Crown, the President of the United States!"

No one moves. None take him seriously. They inspect the rumpled bald man in his mid-fifties lifting himself from matted leaves where he has been wallowing helplessly in compounded tears, terror and titillation.

"You're kidding," Jason is skeptical, but the man's face does seem awfully familiar.

"I am not kidding," Sam retorts hotly, reaching for his identification. "See!" he extends a wallet with photo ID and badge, "Secret Service, Sam Horowitz."

"Hey," Jason announces his astonishment to the others, "this is real."

"I think you should show a little respect and cover yourselves," Sam adds disapprovingly, scowling especially at Chloe.

"With what?" she shoots back one of her "mind your own business" glances at him.

"Well, here," Sam replaces the ID case in his pants pocket and removes his jacket which he holds out to Chloe.

She sighs in resignation, "Okay, if it makes you so nervous."

"Oh, Sam," the bald man says, "what the hell difference does it make?"

Jason still has not accepted that he is indeed in the presence of the President. For one thing, the circumstances are very inauspicious and this rumpled, cringing man seems bereft of any trappings of power.

"Well, sir—your office. I just don't think people should stand before the President stark naked," Sam blusters.

"Turn about is fair play," Ira Crown muses darkly, "I've been held naked before the people by the press ever since I started campaigning."

"Sir," Sam changes the subject abruptly, attempting to take charge of the situation, "I don't think we can afford to stay here. They'll be looking for you. They can't let you escape now. And that guy knows where we are."

"You're right, of course, Sam," President Crown regains his composure and manages to brush off most of the leaves from his suit. "Lead the way." Looking back at the unclad four, he chuckles again as he says, "I'm so glad you came along, though. Until I saw you I thought I was going crazy." He turns to the agent, "Which way, Sam?"

"Why don't you come with us? To my house?" Hector suggests.

"How're we going to get there?" Grady interrupts, obviously not feeling very well after such exertion. He leans slumped back against the tree.

"I guess we'll walk," Hector says, "can you make it?"

"No way," Grady confesses, "I spent all my dollars just gettin' here!"

"Well, I'll carry you," Hector says decisively, "you can't stay here." He leans down to trace the remaining edges of the still open bullet wound with his finger. "Beats anything I've ever seen," he says, shaking his head incredulously.

"What happened to him?" Sam asks, finally able to indulge his curiosity.

"He got shot."

"Well, that's obvious," Sam retorts sarcastically, "so how's he up and walking around? That's what I want to know."

"Grasshopper healed him," Chloe sings out enthusiastically.

"Yeah, sure," Sam snorts.

"Well, he did," Jason is astounded to hear himself supporting Chloe's dubious assertion. Sam stares at him blankly for a moment, then unconvinced, decides that under the urgency of the moment, it is not worth pursuing.

"Let's go," Sam commands, "Which way?"

"That way," Grady points after glancing about to gain his bearings. Then he looks abstractedly at Ira Crown, "Are you really the President?"

"Yes, I am."

"Of the U. S. of A.?"

"Yes, that's right."

"My God, Grady," Hector exclaims, suddenly convinced, "It's true. He is!"

"Goddamn!" Grady exults, weakly, but inspired at the prospect, "ain't Clara gonna have a fit!"

"Ain't she, though!" Hector cackles. "Can't you just see the look on her face when we come walking up the driveway—naked as jaybirds and with the President of the U. S. of A. and a loony man besides. Hell," Hector shakes with laughter, "she ain't gonna talk to us for a month!"

The two brothers howl in shared delight. Hec gasps for breath as he pauses to say to President Crown,

"Let me warn you, though—Clara didn't vote for you!" Somehow that recollection renews their jubilation; they laugh and josh as the others follow along. Sam is disturbed by mention of a loony man. He looks over each newcomer in turn, then seems satisfied that the remark is directed toward Grasshopper.

"A lot of people are not voting for me lately," President Crown sighs in weary resignation.

Jason is convinced. This man is the President. But he is disappointed, also. If such a man has been reduced to chasing around the forest with Grasshopper, then the times have indeed become hopeless.

"Come on," Hector urges eagerly as he helps Grady to his feet," I can't wait to see her face!"

As they thread their way among the trees, behind are shouts in the distance. Quickening their pace, Sam and Jason take turns helping carry Grady. He seems drained of energy and grown heavy with drowsiness. Jason realizes that he has grown cold and aching and totally without any shred of purpose. Worse, he cannot dispel the despised image of Chloe kissing . . . he blames it all on Grasshopper.

~ 16 ~

Clara Abbot again drags her rocking chair away from the edge of the front porch. Unevenness along one of its rockers makes the chair track gradually to the right. Despite, or perhaps because of, Hector's puttering, it stubbornly seeks the porch edge. She even tries not rocking, but forgets and inevitably a nervous pitch and yaw resumes, driving her slowly toward the edge.

It has been a fretful day. All morning Clara has distracted herself from worry by washing, cooking, feeding chickens and hogs and inflicting hundreds of readjustments upon things already accomplished. By afternoon, worry has eclipsed practical efficiency. She listens and watches the winding driveway that leads from the road up to the house and pours her apprehensions about Grady into the anxious metronome of her sidewinding chair.

Now uneasiness wrestles exhaustion. Her last reserves of energy have been sapped in the hours since her husband dashed off with his brother in the old hearse that the two of them have kept around for hauling and tinkering—mostly tinkering. She had earlier that day helped Hector move Grady to the hearse when something in his chest had opened and his heart seemed determined to transfer his vital substance to the bedclothes.

Clara dozes in and out of fitful dreams, her lean yet sturdy body greedily reclaims its due for sleep lost since Grady was first brought home, his life trickling out his chest and back. Her eyes snap open to find Hellbender, a mix of Beagle and Blue Tick Coonhound, watching sympathetically, looking up at her with his great brown, all-comprehending eyes.

"You're a good dog, aren't you, Hellbender?"

His tail thumps agreeably against wood planks as again she gauges the rocker's distance from the porch edge. Clara smiles, leans forward to scratch just behind his ears and then repositions herself at the start of another grudging trail toward the precipice.

Her ears remain alert for any sounds of the decrepit hearse. If Grady still lives, it is a miracle. She declines to consider that even now he may be dead. Silently she prays that the Good Lord might allow at least one more miracle before calling the numbers of the earth and then, in the midst of her prayer, catches herself and apologizes for the vanity of a sinner impudently instructing Holy Spirit.

She is angry with both Hector and Grady. Ever since they got caught up in that militia she has known that no good would come of chasing over hills and gullies in pretend military exercises. Her fretting over both Hector and Grady's welfare has kept it penned inside. Now her checked anger storms and assails the stalls into which she has consigned it, vainly fuming over their folly at being drawn into trouble that is none of their concern in the first place. Besides, what can a handful of farmers and used-to-be soldiers with shotguns and deer rifles do against a misbegotten horde of criminals anyhow?

Leave that motorbike scum to the Sheriff, she had pleaded. Look to yourselves. It's the Latter Days. The Good Lord can be expected to return any day now to claim His Own. Hector and Grady would best attend to their own affairs, look to the readiness of their souls and await the rightful conclusion of the world. He soon will return, she is convinced. The Messiah will appear when no one expects and in disguise, as He Himself warns in the Scriptures—like a thief in the night. Now is the time. It is fixed in the Word.

Through doleful eyes Hellbender watches Clara's visage strive from frown to hopeful vision, fretfully knit its brow and then lapse back again into scowl. His muzzle rests patiently on outstretched paws, awaiting any sign of the advent of dinner.

We know no more about it than he does, Clara smiles sadly down at the hound, pitying his animal ignorance. She only wishes that Hector and Grady might share her beatific resignation to the woes of the world, to the preordained weeping and wailing and gnashing of teeth that must lead and attend the Day of Judgment. She did not want them going off, risking getting killed in a dispute with Mammon's vile legions, leaving her here on earth, alone, to await the triumphal return of the Redeemer.

Behind eyelids drowsing in the chilled approach of evening she dimly marvels as hosts of Beasts and Angels strive above a pit fastened by adamantine chains. Somewhere beyond, Hector urges on the old hearse he and Grady long ago found and laboriously rejuvenated with all kinds of makeshift parts. They even converted it to run on biodiesel fuel they concoct from discarded vegetable oils. Above its wafting fumes that smell variously of peanuts or french fries she fancies angelic voices chorusing *Are You Washed In The Blood?* over Grady, prone in the back, the crimson flow of his life streaming out the bullet's wake.

Thieves had robbed and beaten old man Windsell. Hector and Grady had joined an upstart vigilante posse made up of their militia buddies, given to chasing about the woods all night in jovial would-be blood sport, fitfully revisiting martial enthusiasms of their younger years. They had departed the previous day, bound in some clammy oath of vigilante justice only to return the next morning with Grady barely holding onto a borrowed mule. Men and their guns! Clara had raged and fretted aloud as she inspected and cleaned his wound. She thanked God that the bullet had passed all the way through, apparently missing his vitals.

She had cleaned and bound the raw opening, gotten Grady into bed and then tended him until Hector got back, astride their own mule Jake, from returning the borrowed animal. Once he was again within earshot she railed against men and their greed and games and godlessness that has done this, not only to Grady but God-only-knows-how-many-others!

"For Christ's sake, just let it go, Clara," Hector had snarled when he finally had had enough. "You don't know anything about it, so just let it go."

Hec's swearing usually pushes her over the brink into silent sulks but this time she would not be so easily shushed. It was stupid for Grady to die because the two wanted to chase through the woods pretending they are the law, when they are not. God is Law; vengeance is His, she reminded Hector bitterly. It was a silly, childish, sinfully stupid thing to have happened and she would not shut up no matter how often he took the Lord's name in vain.

Then after a night without much change, Grady's wound had suddenly opened up again. The suddenness in which the bleeding reasserted had worried her. She feared it to be punishment for the wickedness of vanity. Would she and Hector be left with only the memory of his amiably beaming face?

They had put him in the hearse as the quickest way to get him to medical care. In haste Hector had wrenched the rickety gearshift too hard into low as he gunned the engine and something had frozen down inside its metallic labyrinth of a transmission that refused to release into higher gears. Roaring malevolently, belching fragrant plumes of scorched french fry smoke, the hearse had crept out of the yard, down the path to the road as Clara chased alongside, venting her anguish through the wail of the engine, Hector acting as though he had already left her far behind.

Hellbender now rumbles in his low alert voice and Clara is instantly awake. Hector, at least, should be back by now, but Hellbender would not have growled at him. Remembering what happened to old man Windsell, she arises warily and reaches inside the door for an automatic twelve-gauge shotgun. It stays there all the time now, loaded with double-ought buckshot.

Peering through the trees, she tries to force her vision beyond the path's drop onto the road below. Apparitions tease her eyes. She glances at the still pitching rocker as if to reassure herself that she is not still dozing there in prolific dreams. A group of seven figures plod slowly into view: four men are totally naked, a young woman wears only a jacket that is too large for her and two men are dressed in rumpled business clothes. Even more astonishing is the dawning realization that two of the naked men are her husband and brother-in-law! Hector and another naked man are helping Grady, also in the buff, walk up the hill! Hector apparently has not yet seen her; he is calling boisterously, "Cla-ra!"

She does not answer. For the moment her throat and its capacity for making sounds are vagrant. She gasps, feels her cheeks flush in embarrassed confusion, drops the shotgun to clatter against floor boards and flees into the house. As the group draws closer, Hector yells again, "Cla-ra!"

Then she is at the screen door, refusing to step outside as she regales him, "Hector Abbot! Don't you come in this house naked like that! Grady! Do you want to catch pneumonia, too? You fools! Scaring the wits out of me! You get some clothes on, all of you—that goes for your friends, too!" she fumes through the screen door.

Hector guffaws happily at her diatribe. Unintimidated, he leans forward slightly to see if he can catch her peeking through the screen wire. "Clara," he protests, "how do you expect us to find any clothes out here?"

"I'll throw you some," she barks, still invisible on the other side of the door. "You get some more clothes on that girl, too! Don't any of you set a foot in my house 'til you're covered!"

"Well, toss'em on out here," Hector roars in delight. He grins triumphantly toward Ira Crown and Sam, both fidgeting nervously in a social predicament with no ready protocol. Hector taps Grady lightly on the shoulder to urge his attention as he raises the other hand in silence. In a mockingly punctilious voice he calls out, "And Clara?"

"What?" she yells impatiently back from deeper within the house.

"After you throw us some clothes . . . I've got a surprise for you!"

Hector and Grady scream in painfully profound joy, falling back into the dirt, oblivious to their companions. The two brothers howl as if this might be the culmination of a lifework devoted to confounding the poor woman. Hellbender dashes down off the porch and dances about the duo, trying to reckon just how this new game works. The screen door bursts open and the air blossoms with clothing of every description as Clara furiously hurls out two armloads of garments.

Grasshopper plucks a faded pair of overalls from the air, searches among a pile of shirts lying near the two brothers, who have been reduced to panting and chortling, and finds a red plaid one suiting his taste. He quickly dons the garments and sits on the porch steps to pat and rub the hound's head. Jason watches him nuzzle Hellbender, who responds by licking Grasshopper's face. Grasshopper enthusiastically reciprocates, avidly licking the dog's muzzle and flailing tongue.

"Good grief," Jason points out the spectacle of mutually slurping tongues to Chloe, who seems to find it amusing.

"Aw, that's sweet."

"Kissing a dog?" Jason grimaces in revulsion, "God, he's licking its tongue!"

At the door Clara peeks out, satisfied that everyone is indeed decently covered before storming out to scold the two authors of her discomfiture.

"Hector Abbot! What have you been up to? And Grady—you were near death. Now I find you parading about in the cold, stark naked with strange men . . . and . . . and . . . a half-dressed woman to boot!"

Hector and Grady glance sheepishly about as they accept that it is time to wind down their prank. Hector mounts the steps, his voice placating, "Aw, Clara, you'd never believe it anyhow and it was just too good a chance to let pass. Now cool down some and I'll introduce you to everybody. That's Chloe, Jason, Sam, Grasshopper," he pauses for the final flourish, "and the President of the United States, Ira Crown.

Hector points with his index finger as he calls out each name, expectantly awaiting her response after the final one. Clara stares at each new face fixedly for a moment before moving on to the next. She seems unable to get beyond Grasshopper sitting on the ground below, tussling energetically with the dog, oblivious to any dictates of etiquette.

"Grasshopper?"

"Yes, yes," Hector impatiently prods her toward recognition of his true scoop, "and President of the United States of America, Ira Crown!" He jostles her elbow, hoping to crank out a little respect for the prominence of their guest.

"How'd he get a name like that?" Clara asks, still looking distractedly at Grasshopper.

"Ira Crown?" Hector is puzzled, "I guess like everybody else. He just got it." The politician nods that indeed his christening had not been unusual in any respect.

"No. Grasshopper," Clara replies, staving off the hurdle of recognizing that the Ira Crown standing in their front yard might be the same Ira Crown for whom she had not voted. Hector scowls impatiently at her reluctance to look to the next face.

"Hell, Clara," he gestures impatiently, "that man right there is the President of the United States. That is President Ira Crown!"

Her eyes follow his finger; slowly they grow very round in dismay as she recognizes a widely publicized visage, "Oh, my Heavens . . . it is! It's him! What's he doing here?"

Hector cannot determine if she is offended or simply baffled. He answers from within his own consternation, "I asked everybody to dinner. And they'll be staying the night. I thought you'd be pleased."

Clara moans pitifully; her voice falls into a raking whisper as too many challenges mount and merge into one terrible maelstrom of things needing attention, "But Grady, what about Grady? He is . . . was dying. And the President is here . . . Hec, what am I going to do for supper . . . oh, these assuredly are the Latter Days!"

She collapses heavily against her husband, who now is at a loss what should be done. Hector has never seen Clara faint. He has not even conceived that it might be possible. Instantly he regrets the boisterous foolery that has fostered her collapse.

"Take her in the house," Grady suggests ruefully, "we shouldn't have sprung it all on her like that." He staggers slowly up the steps and manages to hold open the screen door as Hector carries Clara inside. She moans fretfully, her eyes squeeze tightly as her head rolls side to side in protest. "Come on inside," Grady motions to the bedraggled group still dawdling in the yard, his voice confident in the resilience of his sister-in-law, "she'll be okay. She's the original iron lady."

As they troop into the hallway, Sam takes control of the castoff shotgun. He notices a telephone on a small table under the staircase. "Perhaps we should notify headquarters. They could have a squad down here pretty quick."

"Let's not rush into any notifications," Ira Crown counsels gravely, "there are some I would prefer to keep in the dark for the moment. At least until I have a better idea of precisely what is going on and who is responsible." His voice is bleak as he considers their plight, "I'll contact Duane Odenheimer directly. Just now, though, I want to sit down, rest and think this through."

"Certainly, sir," Sam assents as he scans cautiously about, glancing balefully into each room, his automatic pistol held at the ready, the other arm cradling the shotgun.

"Relax, Sam," President Crown urges.

"I can't afford to. And if I may say so, neither can you, sir."

"Of course, Sam. I'm sorry. You never forget your duty. But I don't think these people are any cause for concern."

"Maybe, but still . . . ," Sam reluctantly lowers both weapons to his side and follows the group into the kitchen where Hector commences reviving Clara by splashing water on her forehead. She sputters, regains control of her feet and surges into exasperation, apparently for lack of any handier emotion.

"Hector—get away! I'm all right. I can't believe any of this. Just leave me alone!"

"Mrs. Abbot," President Crown speaks up, "if we will be any inconvenience, we can seek other accommodations."

"You're no bother," Clara snaps, "I'm married to the tribulation of my life. Anything else ain't worth fretting over!" She glares at Hector, who sheepishly acknowledges the merit of her accusation.

"Come here, Grady. I want to look at your wound." For a moment Clara glares at Grady, too, as if to exact some toll for her flustered bewilderment. Then her expression softens into one of concern with even a hint of humor added to entice him around the table to where she reaches out her arms, waiting.

Despite his exhaustion, Grady shambles around the table to lean against the sink next to her and point toward where Grasshopper squats by the fireplace, frolicking in communion with Hellbender.

"Clara," says Grady in awe, "it beats anything that ever happened to me. He did it! He healed me!"

Clara's eyes pinch into a focused wariness directed momentarily toward Grasshopper, then spray about the room to scrutinize everyone else, looking for any hint of another unfolding prank.

Ira Crown and Sam are both puzzled by Grady's remark. Chloe happily nods her confirmation lest any find the claim dubious. Jason merely stands and waits; during the trek to the house he has reviewed events in his mind, mulling them repeatedly. His initial disbelief has emerged into an inkling that this is no question of healing, but of credulous misperception. Grady still exhibits a nasty bullet wound that might at any moment open up again. His energy is depleted and he seems near collapse. Actually, all that has happened is that he has stopped bleeding so profusely. His vitality is nothing more than a case of healthy youth with a great deal of luck.

Clara, too, seems skeptical. She ignores the claim, but in a moment of rote social manners says to everyone, "Well, now, all of you just make yourselves at home. As soon as I tend to Grady and get him to bed, we'll set about making some dinner."

Grady settles into a white painted wooden chair that has been reinforced with iron wire struts. His fingers fumble at his shirt buttons until Clara brushes away his clumsy hands and deftly opens its front. She peers closely at the wound now exhibiting a solidly formed scab fringed by lightly lustrous new skin. A moment's reflection fights through resurgent emotions to observe, "Well, I thought you were a dead one for sure."

Hector again kindles his high spirits, "I wish you could've seen it, Clara. It was something! I picked up these kids on the road to help me with Grady—he was dying for sure! Then little Chloe here said that ol' Grasshopper could help. I didn't know what else to do so I said okay and he hopped in the back and started the damnedest healing you'd ever imagine! The way he was beatin' on Grady's chest, and spittin' and lickin' him and singing that crazy song about a dead goose, I thought I was going to have to kill him right there before he did Grady in for sure.

"Then the transmission fell out—or the engine blew up or something terrible—and when the smoke cleared away, ol' Grady was propped up on one elbow wanting to know what's going on.

"Then we all just went a little crazy, I guess. We started tearing off our clothes and running and jumping on the ditchbank and throwing clumps of weeds up in the air in showers of dirt. Then we took off running through the woods. We went swimming and . . . ."

"You can skip the 'what else'," Clara's eyes incite shame as she glares at Chloe. Hector chuckles sheepishly and continues,

"Then Grasshopper takes off chasing these three helicopters that flew overhead, yelling crazy-like about ants. When we caught up with him he had stumbled into a whole mess of trouble with the President and Sam here who had just escaped an ambush by Berserkers and maybe even the army and, well," Hector's narrative prowess finally is exhausted, unable to construct any sensible blending of inchoate events, he simply whimpers optimistically, "here we are!"

Clara swings her head slowly side to side as she searches their faces for sanction of Hector's narrative. She looks at Grasshopper curiously for a moment, then more carefully at President Crown as if reassessing her political allegiances. Finally she says, as if to dismiss the entire matter, "Well, you're going to have to take pot luck for dinner."

"Is that all you've got to say?" Hector's disappointment at her cool reserve savages his spirits.

"What else can I say?" Clara asks quietly. "Plain folks don't hear something like this everyday, Hector. I will tell you one thing, though," her voice slowly gains urgency, "if Grady hadn't come back from the lap of death, if the President of the United States wasn't standing in my kitchen, I'd tell you to start packing and to get out of my sight until you grow up . . . running around, corrupting a young girl. Hector you ought to be ashamed of yourself! A full-grown man gallivanting naked all over the countryside with boys and girls! What if somebody had seen you?"

Her voice rises in pitch to gain a tenuous intensity; now it swoops back down into casual sonority as she assesses available options, "There's more here than I can take in all at once. Until I understand what mysterious ways the Lord is moving in, all I can say is what I know for sure . . . you're gonna have to take pot luck for dinner!"

"Oh, Clara," Hector recoils from her reproach. At a loss for some vindicating response he scratches his head and looks haplessly about. "Just pull up a chair, Mr. President," Hector sighs, "we don't stand on formality. Just find a spot anywhere."

"You said something about berserk . . . or something. What do you mean?" Ira Crown asks.

"Berserkers. They're renegades, outlaws. They've been in these parts since the troubles started. And they're getting worse. Nobody can seem to stop them. Either the law is unable to do anything or they've been bought or scared off. Anyhow, Grady and I saw Berserker motorbikes near where we ran into you and Sam."

"And that character who came at us was wearing a T-shirt under his field uniform, I think Berserker was printed on it," Sam reminds the President.

After a moment's reflection President Crown says, "There's no maxnet signal out here and our satellite phones were lost in the battle. Could I use your telephone? You'll be reimbursed, of course."

"It may be intercepted," Sam cautions.

"Odenheimer has set up a safe line, encrypted, just for me to use. I'll call that number and everything routed to him will be scrambled," Ira Crown explains, "besides, I doubt a remote rural line has been tapped, so if I keep the call short we should be okay. Besides, if that channel is compromised, then we've already lost the fight."

"Sure, help yourself," Hector laughs, pleased to be distracted from Clara's reproving ethic, "and don't worry about the charges."

"Don't give away money you don't have," Clara snaps as she lifts pots and pans from the drying rack in the double sink to the stove.

Chloe whispers something to Hector, who glances at Grasshopper, then goes to the pantry where he retrieves a bottle of corn syrup. Grasshopper takes the bottle, removes the cap and pours a stream of its contents into an opening atop the iLink.

"What's he doing?" Hector asks.

"Refueling the iLink," Chloe answers, "that thing he wears around his waist. It's a kind of computer. It can convert syrup to electricity."

"Well, I declare," Hector marvels despite having no clue what Chloe is saying.

"Can we help?" Jason interrupts, hoping to get on the good side of what appears to be Clara's formidable temperament as well as to avoid a digression into Grasshopper and his iLink, or even worse, invite his activating the limner.

"Yes, I suppose so," Clara answers tautly as she produces a pan of potatoes and two paring knives. "You can wash and peel these." She situates the paraphernalia by the sink, then pauses to stare abstractedly at Grasshopper as she mutters to herself, "Like a thief in the night."

As he screws the cap back on the syrup bottle and sets it aside Grasshopper smiles politely back in her direction to convey his innocence of any thievery. Clara returns to the stove, thinks for a moment, picks up another pot and instructs Chloe to go into the pantry and pick out several onions.

As Chloe opens the pantry door, Jason grimaces to see Grasshopper switch on the iLink. Hellbender has been lying patiently on the floor, but seems suddenly interested in new developments. In the dim kitchen light vague images swirl lightly about Grasshopper, swell out across the hound and then open into the entire room. Chloe now sways slightly to and fro, as if enchanted by a tune on some distant radio. Hellbender moves with her in an improbably shuffling waltz as she makes her way back to share Jason's work space where she commences to peel and slice onions into a large cast iron Dutch oven. Hellbender saunters back to Grasshopper's feet.

Clara watches Grasshopper closely. "What's that?" she asks.

He says only, "Happiness."

Grady has gone upstairs to get into bed. Hector has followed Sam and President Crown to the telephone in the hallway, whispering not to worry about the long distance charges.

Clara's head pivots back and forth in a barely perceptible indication that what she is thinking must be impossible. But then it reconsiders and moves up and down to signal plausibility emerging among all the events Hector has recounted, occurrences she has confirmed with her own eyes. Grady has been healed. Or at least seems to be healing. Or at least he is alive, which in itself seems miraculous. Tribulations vex the nation. The President of the United States has come to dinner. It is the Latter Days.

But just to be certain, Clara opts to test this curious being Hector has fetched into her kitchen, "What is the greatest good on earth?" she quizzes him, wheeling to witness any manifestation of otherworldliness.

"Are you asking me?" Grasshopper chirrups.

"Yes, you're the wonder."

"Oh," Grasshopper thinks for a moment, "the greatest good?" he repeats.

"Here on earth," Clara reminds him.

"Here on earth," Grasshopper repeats studiously, seemingly intrigued by her question. The iLink limner purrs lightly, golden images lavish hints of fruitful fields of wheat waving in the wind, dissolving through montaged layers where happy children play, sunrise beckons and salmon fight upstream of mountain waters. The kitchen blossoms into a fulsome golden glow.

"Yes, here on earth," Clara ignores the dynamically flickering displays, wondering that he seems so easily stumped.

Jason glances over at Chloe. Tears are streaming from her eyes, braving showered effusions of onion juices. He sees her lips form words as if she were reading, "un-ques-tion-ing faith."

At the same moment Grasshopper repeats slowly, "The greatest good here on earth is," then carefully enunciates, "un-ques-tion-ing faith." He smiles pleasure at the astuteness of the answer.

Jason frowns in disbelief, questioning his own witness. Clara glances sharply at Grasshopper, but she seems to like his answer. "And the greatest evil?" she adds expectantly.

"Oh, that's easy. Un-ques-tion-ing faith," Grasshopper parses the syllables with the same deliberate care and smiles proudly. The last answer has done so well that he tries it again.

"But that can't be," Clara protests. She turns and places a hand at the waist of her white apron. "If unquestioning faith is the greatest good, then certainly it can't be the greatest evil, too."

"It can't?" Grasshopper is crestfallen. The iLink limner paints the room in imaged reveries gleaned from old movie clips: spectacular grand cathedrals, mosques, synagogues, temples, monks, holy clerics disputing from varieties of ornate robes and predilections, crowds of condemned heretics aflame, torture dungeons dispensing agony, armored soldiers indiscriminately slaughtering peasants, coercing holiness, imposing salvation. "Then maybe the greatest evil is worrying about the greatest good?" Grasshopper suggests helpfully.

"That doesn't make any sense either," Clara snorts her contempt as she tries to disregard the limner's flickering, "if we never worry about the greatest good, how could we ever attempt it?"

Chloe tilts her head, giggles through onion tears and sniffs.

"But, then, neither would we have any idea of greatest evil," Grasshopper counters, "and therefore could never conspire with it."

"Oh," Clara knits her brows, unable to isolate the faulty link she suspects is lurking in a veiled reference to the Garden of Eden, the greatest good lost?

"Then what is the greatest power in our lives?" Clara tries another gambit, gleaned from a rich mental archive of sermons attended, dissected and absorbed.

"Stupidity. Intolerant ignorance," Grasshopper answers. Jason watches tears stream across Chloe's cheeks as she vigorously but silently mouths assent, affected by far more than onions.

"No!" Clara thunders, "the greatest power is love!"

"Oh," Grasshopper acts bewildered, "maybe it just seems that there are more intolerant ignorant than there are either loving or loved."

"The greatest power is love, God's love," Clara insists hotly.

"Of which most people are intolerantly ignorant," Grasshopper observes sadly.

Again Clara frowns and turns her back to him as she prepares to cut up a chicken carcass taken from the refrigerator. She seems determined to bring some private matter to resolution. Chicken carcass in her left hand, butcher knife in the right she wheels about, "If you're the Redeemer, seems to me you'd act like it. Are you the Redeemer?"

"Wait," Jason interrupts their foray toward the supernatural, "why can't 'good' just be natural, just the open expression of healthy minds?" Neither disputant pays any attention him. They blink in his direction, but ignore his question.

Grasshopper is genuinely puzzled, not so much by the question as by the way she gestures at him with a chicken carcass in one hand while the other threatens the air with a butcher knife.

"I'm Grasshopper," he offers to clear up any misunderstanding.

"Grasshopper?" Clara refuses to be dissuaded, "Don't you have a Christian name?"

"What's wrong with Grasshopper?"

"It's not normal," Clara answers curtly.

"Normal? What is?" Grasshopper snorts. Hellbender has perched himself by Grasshopper and now carefully monitors the sweeping flights of the chicken carcass in Clara's gesticulating left hand.

"Hector says you're a healer."

"Hector lets me be Grasshopper," he has grown impatient with her interrogation.

"You do heal, don't you?" Clara presses for an answer.

"No. I don't," Grasshopper counters. Chloe's mouth drops open; she turns and looks straight at Jason. Grasshopper ponders for a moment, then says, apparently straining to make sense of something troubling him, "At least I don't think so. At first it seemed so simple, but then got very complicated, at the same time. I don't know how to say it yet."

Hellbender nods either in assent or hoping to liberate the chicken from Clara's grasp.

"Did you heal Grady?"

"Nope, sorry."

"Then who did?" Clara fumes, frustrated by his evasions.

Grasshopper thinks for a very long time.

"I guess Grady did it himself."

"Grady couldn't heal a brick!" Clara snaps.

"Well, I don't know," Grasshopper thinks for a moment, "but if he did, I guess you'd be mortar-fied." He wiggles his eyebrows to signal that he is finished answering questions, then adds, nodding toward Chloe, who is blotting tears onto the sleeve of her blouse, "Maybe she knows."

A rush of jealousy surges up within Jason. The paring knife slips to bang against the side of the pot. "Sure . . . go to the source," he mutters balefully. Chloe seems exhausted. She looks wistfully back at him, eyes pleading for understanding.

"Have you ever seen him heal anybody else?" Clara demands of Chloe.

"Yes," Chloe sniffs, fighting back more tears, looking back into her hostess's flinty skepticism, "There was a little boy out on the freeway that no one could wake up. But Grasshopper did."

"See?" Clara challenges, "She says you do. Why don't you admit it?"

"Because," Grasshopper says, rising to his feet, "I don't want to be anyone's healer! You've already decided I'm something I'm not and now you want to change that into something else. And I'm not that, either!

"I'm Grasshopper. You've never seen me before and you'll never see the likes of me again! Everyone is different and unique. Just like you're you. I'm me. And I don't heal squat!" The golden wash of imagery across walls and furnishings shades into garishly disgruntled purplish grey with splatters of livid green. A blotch of scarlet darts about Clara's apron as if to tag her annoyance even while holding it at bay.

"But you do," Clara insists, "either by God or of Satan—and I want to know which before you sit down at my dinner table!"

"I'm Grasshopper! I feel like Grasshopper! I taste like Grasshopper!" He licks the back of his hand to signal his distinctive flavor. "I sound like Grasshopper. All because I am Grasshopper! Just let me be . . . me." Angered by her intimations of Divine Instrumentality, he opens his arms toward the entire world, "It's all just hoppening!"

"These are the Latter Days. We must heed the Return of the Redeemer!" Clara's steely voice measures patience, undaunted by his resistance to her effort to place him among her catalog of supernatural beings.

"I am not your Redeemer!" Grasshopper vows, "Maybe you're in your Latter Days! Me? I just got here!"

"Then you must be the Antichrist!" Clara trumpets a basic syllogism.

Exasperated, Grasshopper jumps up and down on the fireplace hearth. Hellbender prances about happily as the chicken strives to leap free of Clara's accusatory left hand. "So be it!" Grasshopper yells, "Okay! All right! Just for you. Your very own Ant egress!"

Clara draws back against the counter in horror at what seems to be a confession drawn from the very jaws of Hell itself.

"I knew it!" she hisses, somehow managing to work both triumph and failure into the timbre of her voice, "I knew it!"

She tosses the chicken carcass over into the sink. Hellbender slumps disconsolately while she directs a clammy left forefinger toward Grasshopper, shrieking, "Fiend—get thee hence!"

Grasshopper suddenly bursts into laughter and leans back against the fireplace mantle, ignoring his banishment from her kitchen.

"Who's a fiend, Clara?" Hector walks back into the kitchen.

"This . . . this . . . demon. This Lucifer. This Beelzebub. This Grasshopper you brought home," she rants, "he is the Antichrist."

"Oh, dear, God," Hector swears, "who got her started off on that?" He glares darkly at Grasshopper, who promptly switches off the iLink, as if it might be a source of mischief. Hector says to his spouse, "Look, Clara, it's been a long day and a lot has happened. None of us is in charge of everything, but one thing I am certain of, he is not—I repeat—not the Antichrist! I can assure you that the Antichrist is nowhere in the county!"

"Blasphemer!" Clara snarls back into his face, "How would you know anything about it? You haven't been to church since I got on you for making goo-goo eyes at Phoebe Spencer!"

"Leave Phoebe Spencer out of this," Hector slams his fist against the table.

"I will not stay in the same room with the Antichrist!" she declares.

"Then leave. These people are our guests and the very one you're accusing saved Grady's life."

"He says he didn't," Clara responds angrily, "If so, then he's a liar, too!"

"Well, I say he did. So just let it go! He's here with me and I won't have you acting like this, especially with the President as one of our guests!"

Clara turns her back upon them all and pouts in righteous silence. Hector looks back toward Jason, Chloe and Grasshopper as he pleads, "Please don talk with her about religion. She is the most scrupulously religious woman in the state. She reads the Bible morning, afternoon and night. I wouldn't be surprised if they haven't already named a whole bunch of mansions after her in Heaven! The Clara Abbot Reception Center, or something!"

"Hector, don't blaspheme!"

"See?" Shrugging his despair, he edges over to where Clara now pulls futilely at the chicken with her bare hands. His voice becomes gentle as he tries to soothe her, "Honey. Don't take on like this. Just let it go. Whatever happened, we owe these people a lot. Now just put it out of your mind," he cajoles.

Clara says nothing but eventually Hector seems to think that he either has gotten his point across or has given up ever getting it across. He turns back to the others and attempts to revive their earlier camaraderie. "I guess everybody has to take things the best way they can," he says cheerily, "Clara didn't mean anything by what she said."

"Speak for yourself, Hector!"

Ira Crown's voice suddenly erupts impatiently from down the hallway, apparently finding hurdles in completing his telephone call, "The Signal Corps is already operating the Alexandria exchange!" Ira Crown rants aloud to Sam, "What the hell is going on? I'm the first President in Exile!" he laments despondently.

Ira Crown's voice booms again down the hallway, "Duane? Ira Crown. Yes, really. By the way, why is the Signal Corps operating the Alexandria telephone exchange?

"Well, I wish someone would tell me when they plan to move the seat of government again. Yes, Richmond was getting as bad as Washington, but still.

"Duane, never mind—I'll find it wherever you put it—you just can't very well hide a thousand civil service workers.

"Yes, I know there are more than that, I just pulled a number out of the air.

"Look, Duane, this isn't why I called.

"I know it's important. That's another problem, there are too many pressing matters for anyone to handle. But we do the best we can, anyhow, right?

"I am not on vacation, Duane! Who told you that? You knew I was flying over to Lebanon Evacuation Command—that's what I'm calling about! They tried to kill me!

"Well, I'd hoped you could come up with more than a 'Gee, that's terrible!'

"Never mind! It's all right. Yes, I'm okay, for the moment.

"You're the only one I can trust to oversee what we have to do.

"No, I can't tell you where I am. Everybody's phone is tapped, Duane, you know that . . . yes, I'm routing this through the encrypted line. But I need to keep it short.

"No, I can't tell you where I am . . . the wrong people might be listening. I'm taking a chance even calling like this, so listen, I've got to wrap it up.

"Yes, that's right. Anyhow, here's what I want you to do. I'm afraid you won't get much sleep tonight. Call every agency chief in the executive branch and borrow every available agent, bookkeeper, janitor, I don't care what. Swear them in as Federal Marshals.

"No, you're wrong about that. You have the authority to do that when I give it to you. Yes, I'll send you a cover memo. Yes, Duane, you may regard this as official verbal authorization under the Emergency Powers Act!

"Now look, Duane—we're literally fighting for our lives now. No, that is not just a figure of speech. Yes, you, too. If there's a coup, you can imagine what will happen to the Attorney General. Yes, it is grim. If you'll just shut up and listen, I think we can gain the upper hand in this one. Fine.

"No. I'm not sure who's behind it—but I do know where to start looking. Have the Joint Chiefs relieve General Michaels and General Flach of duty. Place them and their entire staffs under house arrest and hold them incommunicado until we get this sorted out. Grill the hell out of them, somebody will break. No, don't tell anybody else in the Army. Go to the Navy if you need more muscle. Use shore patrol officers or Marines, use your judgment as to which is appropriate, but don't trust anybody in the Army just now.

"To hell with interdepartmental courtesy, Duane. Don't tell anybody. Not anybody! Do you understand?

"Duane. Duane. Forget that! This is different. Get those forces out to the Lebanon Command Post by dawn tomorrow and get some trained investigators into the operational files. Also, seize the communications records and start auditing who's been calling whom. And lock down Lebanon FERAL offices, too. Who knows, they may be involved, as well.

"Good! Yes, I'll come onto the Post tomorrow morning some time after you've secured everything.

"Oh, yes, Duane. Tell Frank that I'll need a new squad of Secret Service. I'm afraid it was a pretty bad massacre. Sam Horowitz and I managed to get away. I don't know about anyone else. Yes, there definitely were U.S. troops on the scene. There were some local hoodlums as well, but we saw unmistakable evidence of Army involvement.

"Yes. It is a bad sign. It was a bloody, horrible and despicable mess.

"Yes. Other than that, how are things?

"Well, you know what you have to do.

"Oh, yes. And Duane? Make sure that everybody knows what I look like. No, I'm not trying to be funny. Nobody seems to recognize me. It's embarrassing.

"Okay. You'll be there at dawn, then? Good. I knew I could count on you. Say hello to Ellen for me and give my love to your daughters. Yes, 'til tomorrow, then. Good night, Duane. And thanks for your loyalty and friendship."

President Crown's voice is morose as he murmurs upon heading back toward the kitchen, "Duane Odenheimer—Defender of the Constitution. It does give one pause."

~ 17 ~

"I envy you people," Ira Crown sighs, pushing his plate away to signify that he has finished his meal. "You must have a wonderful life here."

"It's all right for a day, I suppose," Clara wipes a hand with the corner of her apron. "Mostly just a lot of hard work, though."

The Chief Executive overlooks the haggard sharpness of her reply. Smiling empathetically for rigors of her unending toil, he glances about to gauge the expressions of his dinner companions, reddening only slightly at her rebuke as he searches for a new conversational thread.

Clara does not elaborate, though her eyes declare him a fool. Obviously finding little to celebrate, she resents his implication that somehow she has overlooked something.

"You boys do much hunting?" Ira Crown asks Hector.

"Yeah, actually we manage to get in a good bit," Hector says, grateful to escape a pall that has pervaded the entire meal. "Ol' Hellbender there noses up whatever you want to bag. Practically all you have to do is just think of it and he finds it."

Hector grins broadly. A glint in his eyes anticipates one of his favorite yarns. He glances at Grady who has rejoined the company for dinner, declaring that he feels a little sore still, but otherwise is okay.

"Last time we were out, Grady says, 'I sure would like to get some quail.' We walk about ten paces and Hellbender flushes out a covey. Grady picks off two of 'em—blam! Blam! Clean as you please.

"I say, 'that's all right. Birds are okay, but me, I want some rabbit.' Next thing you know, there's a rabbit running lickety-split across our path with Hellbender hot on his tail. I pick him off neat as spit."

Hector pauses to lean back in his chair, looking about the table as if he is about to divulge some cherished secret. He winks at Grady who suppresses a snicker. Each has an often-practiced role in the narrative and they each take great pains to draw out each flourish for greatest effect.

Hector chortles as he lowers his voice, becomes nonchalant, and pretends to concern himself with the bare facts of the story, "Then Grady says to Hellbender, 'Hey, boy, how 'bout some squirrel?'

"Lord! You'd a thought that dog understands the English language. He turned right around and treed one not more'n twenty yards away! Honest to God—I mean Betsy!" Hector looks apprehensively toward Clara's raised eyebrow. The momentum of the narrative carries him on through as Grady giggles expectantly in falsetto,

"And blam! Grady bags a squirrel scrambling for cover."

Glancing at his brother, Hector leans forward, letting his voice drop conspiratorially. His audience bows closer to follow his whisper, "Then Grady leans over to me and says down real low like this, like he don't want Hellbender, who's over by the tree guardin' the dead squirrel, to hear.

"'Hec,' he says," Hector bumps Grady affectionately on the shoulder with his fist, "'Hec, For Christ's sake, don't say nothin' about no grizzly bears 'cause I ain't got nothin' bigger'n number six birdshot!'"

The two brothers howl in unison; Grady tries to fend off the pain of wrenching guffaws. Everyone but Clara laughs at their narrative triumph. Even the dour faced Sam smiles as Ira Crown seems to contemplate a medal for the hound now bouncing and barking happily as if he never tires of hearing of his exploits.

"That's some dog," President Crown observes as he reaches out to pat the elusive animal who seems reluctant to be touched by hands that have shaken so many others. The dog returns to his station by Grasshopper's feet.

Jason has been eyeing Ira Crown during most of the meal, unsuccessfully restraining an urge to call him to account for scandals riddling his administration and widely rumored pandering to special interests. Thus far Hector has managed to stave off Jason's surly overtures by abruptly changing the topic or by requesting that some platter or bowl of food be convoyed down to him. Even President Crown notes Jason's impatience once he resorts to simply staring at the politician.

Sam watches the young man eye Ira Crown. Ignoring the agent's stolidly daring gaze, Jason announces, "I've never eaten a meal with a President before. Do you mind if I ask you some questions?"

Ira Crown's eyes are instantly wary, then quickly veil in practiced amiability. "It's my pleasure, I assure you," he turns slightly toward his petitioner, "no, I don't mind questions. So long as the rest of our friends aren't bored by them." Ignoring the tone of constructed affability Jason looks down at his plate, trying to recall the list of crucial issues he has so carefully rehearsed. Sam shifts to balefully watch him.

Suddenly at a loss that he is unable to recall specifics of his own discontents Jason merely asks, "What's going on? And what are you going to do about it?"

Sam seems to relax as President Crown bursts out laughing, "Well, that just about covers everything. It sounds a lot like the Washington Press Corps. Compact every concern into a single question!"

The question is a little broad, Jason admits to himself, but he silently waits for a response.

"Frankly—no one really knows," Ira Crown drops his voice into a confiding tenor. "What began as a global collapse of interacting financial bubbles has now devolved into stagnation, hobbled by our debt and aged infrastructure, exacerbated by widespread unemployment. Social unrest has steadily worsened until the only common factor seems to be universal discontent.

"There are widely varying opinions. Many say, and I believe it to be the case, that we are at the cusp of a new technological age, that old ways are collapsing before new technologies and their widespread use have matured. A fourth wave, if you will. You know, the successive waves of agriculture, then industry, followed by cybernetics and now nanotechnology conjoined with biotech. These waves induce severe dislocations at such times of transition. We're considering all our options and I can assure you that the entire government is addressing the full crisis with the total measure of commitment permitted by our current resources."

President Crown seem pleased with the craftsmanship of his response. He leans back, looking about the table to gauge its reception.

Jason jumps right into it, "In short, you don't know and you aren't doing anything—except trying to remain in power."

Ira Crown does not attempt to hide his sudden flush of anger.

Chloe gasps, "Jason, that's rude!"

"It may be," Jason retorts hotly, "but it's also the truth!" He looks about the table for any hints of agreement and is irked to find only Clara nodding in support. "Just look at where we are," Jason waves an arm wide, toward the whole world, "millions of people out wandering the highways. And what does the government do? It 'helps' by setting up FERAL, a corporation to hand out credit cards and franchise 'service providers'. Now what set of fools thought that up? For-profit public assistance?"

"It's just a way of tracking public money spent for assistance," Ira Crown protests, "by harnessing the power of markets to solve a problem. Besides, people don't have to pay for basic necessities they acquire with their assistance cards."

"Hell, for your information, out there on the highway, basic necessities always seem to be sold out . . . and all anyone seems to actually have available is non-basic stuff that you have to go in debt for just to survive!"

"Don't swear at my table," Clara warns, then adds, "but the boy is right. There ain't no government of the people, by the people and for the people anymore."

"Of course not!" Jason interjects, "It's government of the ignorant, by the corrupt, for the privileged!" His indignation resonates among his table mates, "And every election just makes it worse. Candidates make promises that are downplayed or forgotten by the winners who set about cutting their own private deals, for themselves and their buddies, regardless of the effect on the people who voted them into office. It's government by imposture! And who are the real players? The money people, of course. Bankers and brokers. And corporations. And it's been that way ever since the courts let corporations have the same legal standing as people. And it got worse when Woodrow Wilson turned the U.S. financial system over to the money wolves."

"Oh, that old humbug," Ira Crown dismisses threads of familiar argument in exasperation, "Maybe you just don't understand modern finance."

"I understand that the common good has been fractured and fragmented to be sold off in pieces, conveniently renamed markets, to the highest bidders shoving cash under the tables of government!" Jason explodes, "Yours. Ira Crown's government. You're creating a nation of debt serfs. You pretend it's democracy but actually it's a growing hidden feudalism of debt serfs."

"Young man," Ira Crown slams a fist against the table, upsetting a number of lighter dishes, "a great many people with far more experience than you have grappled with these problems. I will not sit here and listen to your upstart evaluation of my capacity to deal with them! If you are interested in serious discussion, which I doubt, then I will entertain any questions or suggestions you may have. Otherwise, if you cannot demonstrate common courtesy to me, then I must at least ask you to regard the office to which I was elected—by a respectable margin, I remind you—with some degree of deference!"

"I've kept my tongue because you're an important man and I guess you've got important worries," Clara speaks up in a voice that tends naturally toward scolding, "not to mention that you're a guest in my house . . . but the boy's right—even if he don't know how he's right.

Jason is surprised to hear Clara enter the fray; especially since she seems to take his side. He has gathered that she doesn't care for any of these unexpected guests Hector has inflicted on her.

"Alas, alas, that great city," she quotes, heavily intoning a biblical cadence, "that was clothed in fine linen, and purple, and scarlet, and decked with gold, and precious stones, and pearls! For in one hour so great a riches is come to nought. And every shipmaster, and all the company in ships, and sailors, and as many as trade by sea stood afar off, and cried when they saw the smoke of her burning, saying, What great city is like unto this great city?'" Clara abruptly pauses to emphasize what she feels to be an adequate synopsis with a quick nod of the head that implies 'so there'.

"What is that?" Chloe asks, her eyes baffled.

"That," sniffs Clara, "is *Revelation*, chapter eighteen, verses sixteen, seventeen and eighteen."

President Crown is uncertain as to the appropriate response for the Biblical citation and so sits simply staring at the woman. Sam is still watching Jason, daring the young man to foist any more distress on his boss.

"Clara," Hector says impatiently, "what do ships have to do with it?"

"Ain't you ever heard of the 'ship of state'?" Clara demands heatedly, "And ain't New York City—the richest city in the world—burning and fallen into ruin? And don't you know that John the Divine saw all this happening in his great vision over two thousand years ago?"

"Oh," Hector's timorous reply betrays long experience with her biblical reveries.

"John Divine?" Chloe, a devotee of children's television of olden times, but a little shaky on minor character actors, echoes helpfully, "Didn't he play in some old cowboy movies?"

"You mean Andy Devine?" Jason corrects, annoyed that religion has hijacked his opportunity to protest injustice.

"Saint John the Divine! The author of *Revelation*! Lord, who raised you children?" Clara exclaims in dismay.

"Don't you ever get even a little suspicious that your church traditions were instigated and promoted by kings and emperors just to help keep them in power? Look at Emperor Constantine and the Nicaean Council. Or King James I and his version of the Bible, concocted especially for the Church of England to help solve his political problems with Puritans. And as for the English Church, itself, it was established just so Henry VIII could divorce his Catholic queen and remarry again and again and again. Now does that sound like holy inspiration to you?" Jason exclaims, exasperated that she is sidetracking his challenge of Ira Crown's motives.

Clara looks blankly back at him, then pronounces, "God moves in mysterious ways."

"Mysterious ways? What's more mysterious than quantum entanglement, or gravity . . . or even consciousness itself?" Jason's voice falls into quiet persuasion, "Look up at the stars some night. What's more mysterious than that? Every day we swim in a mystery called light. What is it? Why is it? You don't have to whip up delusional conceits over some poor guy strung up two thousand years ago. Or portray blood-drenched agony that only needs a ball-gag to qualify as torture porn. Divinity is sufficient to itself . . . how can you or anyone else pretend you know what God wants or is like? Such mystery is beyond even that of light. If anything is beyond imagining, it is the Divine. So please, just stop trying to imagine it! Or telling us about it. We'll all be better off."

"It's all laid out for any to see who will," Clara gazes fixedly into Jason's dismissal of her grounds for complaint, "take care for your immortal soul, young man."

"So you say. And that may be very well and good for you," Jason dryly disdains such chary readings of his motives, "but I prefer to think that these are real problems with solutions that can be affected by government action—if government is adequately managed.

"Never mind for the moment who's responsible for the dead and dying out on the freeways. I want to know what plans are afoot to restore the living? You've tried to hide problems under blankets of public relations and election rhetoric. You lied and you knew you were lying!"

Sam leaps to his feet, briskly stretches his massive frame across the table and snatches Jason from his chair and halfway onto the platter of ham. A glass of water tips over into the pan of biscuits. Glowering down into the young man's face, the agent's eyes flash malice as he hisses through heavy breath, "That man," he points with his free hand, "is the President of the United States of America! You can't talk to him like that because I won't let you talk to him like that! Now close your sniveling little mouth, you upstart little wad of shit, or I'm going to start with your ankles and work up to your teeth!"

"Sam!" commands President Crown, "Put him down—that won't solve anything."

Reluctantly Sam drops Jason back into his seat, shakes his arms to straighten the crumpled jacket that has bunched up about his shoulders and raises an intimidating eyebrow for the pest's benefit. Jason writhes uncomfortably in aftermath of sudden fear, anger and embarrassment at being so easily manhandled.

"Unfortunately," President Crown sighs gloomily, there are people far more discontent with my efforts than our friend Jason here." He smiles graciously but Jason will have none of it as he glares back into Sam's taunting gaze.

"I'm going to check around the yard," Sam gets up abruptly, annoyed that he has not been allowed to pummel anyone into submission. He shoves aside his chair and reaches for his automatic pistol which he cradles in his arm like an infant. Then he lumbers down the hall and out the back door, which slams violently behind him.

"By all means," President Crown acknowledges with a nod. He frowns briefly, then calls after the agent, "Remember. We have to be at the Lebanon Command Post early tomorrow to meet Duane."

There is no reply. President Crown looks about, gauging expressions on the surrounding faces. "He usually is so composed . . . I don't know what. Oh, well, he feels, ah, responsible for what happened today. But he'll be okay."

Returning his attention to Jason, he attempts to leaven the moment with a note of humor as he continues, "I suppose it is useless for me to assure you that I do not as a matter of policy lie. In fact, so far as I know all of my so-called lying has been directed toward evading social dilemmas."

"Bullshit," Jason erupts, "you can't tell me that you didn't know about, about . . . the transition to austerity, for instance!"

"Well, certainly I knew about it. It was inevitable. The debt crisis is unmanageable, simply unmanageable," Ira Crown protests. "Would you rather have money that becomes totally worthless? That's what would have happened if we'd broadcast the true nature of the crisis."

"Then why didn't you exercise some leadership? Tell everyone what was going on and how we were going to deal with it?" Jason asks so loudly that Chloe jumps and Hellbender stands up. Grasshopper coaxes the dog back down by tickling him under the chin, watching the troubled exchange with a curious tick that tilts his head sideways, just like the hound's. President Crown looks patiently at Jason for a moment, glances about at the other faces, then replies in a voice tense with worry and depleted in weariness, "Because they would not have accepted it in those terms. It would not have worked. All the surveys showed."

"But it didn't work, anyhow—and mostly because once people find they've been lied to, they don't trust anything you or anyone else says!" Jason counters, "Who can blame them?"

"But it might have," President Crown sighs, "there was the chance. It almost worked."

"There was the chance that you could make people believe that things were continuing as they always had—but things only got worse! More corruption, hidden takeovers. Corporations and banks run everything now. And they're all transferring their headquarters to Rio or Tokyo or Beijing or who knows where? Anywhere but here. There's nothing left here. The carcass has all been sucked dry. Ordinary people have lost control. We're left to live like parasites in a theme park closing for the season. We've lost control because you and generations like you betrayed the public trust in a web of lies and deceit aimed at staying in power."

"It wasn't my doing! It started a long time ago!"

"But you didn't try to stop it either. You just kept escalating a fantasy world of squander and waste. Now it's all used up! Gone."

"But it was out of control. The whole bureaucracy is too fragmented, too unmanageable."

"Who stole it? Are you really surprised that there are people who want to kill you?"

"Jason!" Chloe is shocked that he can berate such celebrity.

President Crown slumps back into his chair, his face registering defeat, "No," he sighs, "not really. Now it appears that elements of the Army . . . maybe Morning Star Militia, maybe not. I don't know yet. I just never realized what was happening until it was too late, that's all."

"That's all?" Jason echoes sarcastically, "I just wish it were all."

In the brief interim of silence Chloe stares in bewilderment at him. Then she whispers, "Jason, I've never seen you like this."

"You've never seen me after wading through weeks of hell coming across the man who's chiefly responsible for it!" he mutters.

"You're not the only one who's been through pain and misery," she reminds him gently.

"I seem to be the only one who cares about what's happening. I seem to be the only one who cares about what lies ahead."

"That's not true," Chloe counsels softly, "But you are the only one determined to pin the blame on the most convenient target. But explanations without understanding can't solve this. they're just words. And harsh accusations like these just make people feel worse. You're not accomplishing anything here except making everyone feel awful and embarrassing these people who have been good to us."

"Don't worry about that," Clara snaps, "I didn't vote for anybody here."

"Grasshopper, make them stop," Chloe pleads.

Grasshopper snorts derisively, his eyebrows hoist above a vacant stare that dips and swings across their faces. "Ants don't know how to stop. They only change directions."

Another Ant tirade! Jason explodes, "Ants? Ants? Here's the Ant of Ants! The Presid-ANT" he yells suddenly, pointing at Ira Crown, "now what?"

"What are you talking about?" President Crown demands warily, suspicious that another conspiracy is afoot while his bodyguard is absent.

Grasshopper looks briefly at the politician, then explains, "Ants. I'm talking about Ants. They're everywhere, you know."

"What does he mean?" President Crown begs for someone to dispel his anxiety.

"Ant and Grasshopper," Chloe interjects, nodding toward the silver-haired man sitting back away from the table. "You remember, he played Grasshopper on teleview."

"Of course," President Crown slaps the table, "I knew I recognized that face. That's why you . . . Grasshopper. Dooley, isn't it?" He blathers through several layers of recollection competing at once.

"Yes, but he doesn't remember," Chloe interjects, "His amnesia."

"Amnesia? I don't have amnesia," Grasshopper retorts, "I have a very clear memory. I just don't remember what you want me to!"

President Crown seems to find some solace in no longer being the sole celebrity in the group. Additionally, he sees in Grasshopper someone who might deflect some of Jason's scorn for the derailed austerity plan. Ant and Grasshopper, after all, had been the symbolic beacon of the entire effort!

"Here is someone who had an intimate role in the public relations campaign you now find so abominable," President Crown gloats as if he has just discovered a wild card in his hand.

Jason frowns at so transparent a ploy. "He didn't conceive of it. He didn't approve of it. He was just a small part in a large campaign of deception."

"Oh, ho," President Crown trumpets, "a lot you know about the Presidency. So now it's a matter of being a part of it! Let me tell you, son. I play far less a part in anything than you can imagine. You would be hard pressed to find someone who can tell you more than just a small sampling of what goes on in any aspect of government. That includes me! My office may be responsible for what happens, but there is no way for a single person to know all that transpires and keep track of all options."

President Crown looks slowly about, then, satisfied that he has tapped into a workable vein of oratory, he continues, "Let me give you an idea what my part has been," he leans forward eagerly, pleased that somehow he has found a way back to the initiative, "My aides briefed me on various aspects of an unfolding situation long before it became apparent that a new crisis was approaching. I asked for a range of solutions and considered them all. I listened to consultants recommended as the best in their fields. And when I was finished listening, I told them what seemed obvious to me: we must undertake a transition to austerity.

"A week or so later several proposals—all very much alike, except for details—were presented to me. Included were estimates by several corroborating sources that some sort of interactively responsive public relations campaign would be advisable to stimulate public acceptance and cooperation. We discussed the pros and cons and eventually arrived at a strategy, using the latest media technology and techniques, that was implemented by aides who worked out the final details on everything, ranging from recommending legislation to the austerity public relations campaign you find so offensive."

Ira Crown watches Jason with gravely empathic eyes. He feels a sense of hopelessness and helplessness as he says quietly, "Do you realize what my part was? I sincerely chose among the best alternatives presented to me by the best minds I could find. I don't even recall any details. Your friend, Grasshopper, here has been closer to the actual events than I."

"But why all the secrecy?" Jason remains unconvinced.

"We're living on a globe with real, live enemies just waiting to exploit any weakness! The Chinese, Russians and Brazilians hold trillions of U.S. debt. All they have to do is pull the financial trigger! There'll be no war because it will be over before it's even launched!" President Crown is enjoying his opportunity at sarcasm and drives it expertly home.

"But you're still President," Jason insists, ignoring the implied rebuke of his naivete, "that makes you responsible!"

"Exactly," the politician agrees firmly, "That's what I bought into. I get paid to be responsible. Responsible for that and everything else—including the deaths of all the good people with me this morning. I'm responsible. But, Jason—that is not the same as being guilty."

Somehow Jason's unflinching quest for truth has failed into a deflated crusade. Searching through all his recollections of upheaval and flight he can find no plausible response.

After a few moments Ira Crown adds, "It's not an excuse to say that I didn't design the system. It's the way I found it. Ira Crown is a very small part of it, an elective scapegoat for things gone wrong. In the old days people would say the gods are angry and sacrifice some poor fool. Nowadays we say the polls are down and hire someone to sacrifice for when the populace grows angry—it seems the gods are powerless."

"President or not," Clara pivots at the sink where she is washing dishes, "I won't have you blaspheming in my kitchen!"

"I'm sorry, Clara," President Crown apologizes, "it's only a figure of speech."

"Those are the ones that get you into trouble," she snaps.

"How right you are," he chuckles, recalling a number of unfortunate similes he would like to recant. Returning to Jason, he smiles paternally, "So you see, mythically, at least, until there is a clear culprit, or even a more compelling scapegoat, we have a tray full of problems like you can never imagine."

Jason gives up. Somehow his accusing finger keeps being turned back against him. "I hadn't thought of it that way. I guess there is another side to it," he admits, "but still I think there must have been something you could have done. There must have been!"

"Maybe so," President Crown acknowledges, "I only wish I knew what I could have done differently."

"If you'd quit running around with the Antichrist, it'd be a good end to a bad business!" Clara observes in a voice thinned by dry impatience.

"Clara," Hector raises his voice to warn, "I said get off that. I mean it!"

"If you've got such faith in him," Clara retorts, addressing everyone as she points accusingly at Grasshopper, "why doesn't he change things? Make everything okay?"

"Clara!" Hector raises his voice another octave.

Jason notes that the iLink has been switched back on as Grasshopper looks back at Clara, grinning magnanimously as he says, "I know what's bothering her." He shifts Hellbender awkwardly under one arm, slowly stands and then walks part way around the table.

Jason looks over where Chloe and Grady are following his progress toward Clara, their heads pivot in unhurried unison. He tries to catch Chloe's eye but, along with Grady, she is wholly absorbed in the confrontation.

"What're you up to?" Clara demands of Grasshopper as she backs away from his prospect of occult perils.

Grasshopper smiles encouragement as he stands holding the dog, shifting him again to smooth the folds of skin gathered about Hellbender's shoulders and neck. Jason looks alternately from Grasshopper to Chloe. She, as well as Grady, seems caught up in and a part of his movements, even to silently moving lips, as if they were but shadows cast by a secret light.

"You're afraid of me—you think I heal people," Grasshopper says, "and that if I can heal them, I may hurt them, too."

"They say you healed Grady," Clara makes the comment sound like an accusation, narrowing her eyes as she tries to gauge his intentions.

"I didn't do anything you couldn't do," Grasshopper replies evenly.

"Hmmph!"

"I'll show you," he says, depositing the dog in a heap on the table.

"I don't allow the dog on the table," Hector objects, "Hellbender, get down."

Hellbender ignores the command and gazes dolefully up at Grasshopper. Clara watches through dread-filled eyes, apprehensive that the dog might burst into flame or unleash other sorts of demonic theatrics.

"Hellbender is going to tell you all about faith," Grasshopper announces, "aren't you, Hellbender?"

The dog looks morosely back up at Grasshopper as if he is about to reply, then sits down heavily to deal with a distracting flea. His right rear leg beats fiercely at his ear. Jason sees Chloe scratch sympathetically at her own ear, as does Grady.

"There!" Grasshopper shrieks triumphantly as he points to the still flailing hound's leg, "Did I make him do that?"

"No," Clara admits hesitantly.

"Did you?"

"No, of course not. He has fleas."

"Get off the table, Hellbender!" Hector repeats; he especially does not like the idea of fleas on the dinner table. Again Hellbender ignores the command, but does stop scratching to sit and await further developments.

"There you are," Grasshopper emphatically nods as he pats the dog gratefully.

"But nothing happened," Clara protests, "just a plain old dog scratching at an itch!"

"Exactly," Grasshopper agrees, smiling back into consternation gathered in her face, "It happens all the time, every day. But, Antwardly, you don't notice anything unusual in it. You ignore it."

"What are you doing to my dog?" Hector complains at being ignored, "Hellbender. Get down!"

Jason sees Chloe look quickly toward Hector, then back at Grasshopper, her lips moving in some soundless utterance.

Grady, in perfect synchrony, mirrors both her head and silent lip movements, seemingly mouthing, along with Grasshopper,

Ring around the rosey,

Pocket full of posies,

Ashes, ashes, ...

The dog looks woefully up into Grasshopper's face for guidance. Jason thinks he sees Grasshopper shake his head just before Hellbender sighs to reluctantly settle back onto his haunches. Chloe, too, slowly shakes her head, as does Grady. Then Grasshopper winks at the dog just as Hector reaches out to swat him, yelling, "Damn you, get down!"

This time it is Chloe who initiates, mirrored in perfect unison by Grady and Grasshopper, silently chanting,

All fall down.

Hellbender keels over, as if he has been bludgeoned. He lies motionless on the table, his eyes glazed and vacantly unresponsive.

Chloe and Grady simultaneously slump in their seats. Hector jumps to his feet, startled by the dog's collapse.

"What's going on?" Hector attempts to rouse the prostrate animal. He waves his hand in front of the hound's motionless muzzle and listens at his chest. "He's dead," he cries in dismay, "he's dead. Hellbender's dead!" He turns furiously on Grasshopper, "What'd you do to my dog?"

"You hit him," Grasshopper reminds him, standing on tiptoes to peer over Hector's shoulder at the prone animal.

"Not hard enough to kill him!" Hector rages, "By God and All Creation. See if I don't pound your face some, that was the best hunting dog I ever had!"

Grasshopper casts a doubtful look back toward his antagonist, then says to Clara who is trying to find some sign of life in the limp animal. "You'd best get him off the table, Clara,"

Grasshopper shrugs, "I wouldn't want to fall and squash him when my face gets pounded."

Clara reaches down to pick up the dog, recoiling slightly as her fingers touch the motionless form. Hellbender instantly springs to life! Wagging his tail and jumping about, he rejoices in a new trick and looks to be rewarded. Clara screams and drops him back onto the table where he catches and rights himself. Then, as if remembering Hector's injunction, the dog hops down to check about the floor for any stray morsels he may have missed earlier."

When Jason looks back, the chorusing duo, Chloe and Grady, have jumped to their feet to prance about, uproariously mimicking Grasshopper's shrieking parody and Hellbender's happy resurrection, "Clara healed him! Hallelujah, she's raised him from the dead!"

Grasshopper grabs Clara's hand and pumps it vigorously up and down as he congratulates her for a marvelous healing. Hector still glowers angrily as Grasshopper avidly congratulates the distressed and disoriented woman by enthusiastically pumping her hand. Hector reaches out, spins Grasshopper about, and hits him beside the head with his fist as he shouts, "That's for messing with my dog!"

Grasshopper bounces off the sink and lands on the floor. His cheek and ear bloom suddenly red then fade to faint lavender and gradually back to normal skin color. Hellbender suddenly is insane. Snarling viciously, he advances against Grasshopper's assailant. Hector is caught somewhere between astonishment and sudden fear. He howls, "Stop him! Get him away."

Grasshopper chuckles as he picks himself up from the floor and then clucks. Hellbender relaxes, glances about as if to apologize for any misunderstanding, and then walks away from Hector. "That," snorts Grasshopper, "is what happens when you mess with your dog's friend."

"This is crazy!" Hector moans in anguish.

"Unique, perhaps, but hardly crazy," Grasshopper dismisses any cause for consternation.

Chloe looks curiously back at Jason, frowning lightly as she tilts her head to one side as if to ask why he is watching her so closely. Then she and Grady exchange glances and giggle.

"You, you," Backing away from this silver haired apparition, luminous in the limner's wash of imagery, Clara's voice quavers as she searches for a name sufficient to momentous peril, "Lucifer!"

Flames jauntily dance in the limner's faint surround. Grasshopper, in unison with Grady and Chloe, push tongues through their lips to blow raspberries that bathe the kitchen in luxuriant sounds of flatulence, "#%&@*!"

"Fiend," Clara's tremulous finger targets her nemesis, "risen from the pits of Hell!" She catches up her apron as if it might be the surest talisman against the worst of evils and scurries from the room.

Ira Crown leans against the table, raising himself up on one elbow for a better view. He now sinks back into his chair as he looks dubiously from the still delirious duo over to Grasshopper and back again. "How did you do that?"

Grasshopper, Chloe and Grady pivot their heads toward the baffled executive and demonstrate with rousing tutorial raspberry, "#%&@*!"

"No, no. Believe me, I know about Bronx cheers. How did you get the dog to seem really dead?"

No one answers. Hector is trying to make amends with Hellbender, patting and rubbing him tentatively as he encourages, "I didn't hurt you, did I, boy? And you wouldn't bite me, would you?"

"Ha!" Grasshopper mutters defiantly as he switches off the iLink, "I bet he would! Hit me again."

Hector's face clouds in anger, "What've you done to my dog? He's mine, not yours."

Grasshopper snorts, "Hellbender makes up his own mind. He's liberated."

"How did you do that?" President Crown repeats, "How?" He is unaccustomed to waiting for responses to his questions. His voice has grown impatient and taut.

Grasshopper mocks him, "How did you do that? How did you do that? Keep your little Ant teeth off me," he snarls.

Jason dreads another manic episode erupting in the akimbo tension of Grasshopper's frame and widened flare of his eyes. He glances briefly at Chloe and Grady playing with Hellbender. They seem uninterested in Grasshopper's gathering tirade.

President Crown flushes with anger as he tries to think of something to restore a shattered protocol.

"Here, you do this," Grasshopper orders imaginary aides into action, "And you do that!" As Grasshopper's caricature grows more merciless President Crown squirms about, wishing for Sam's return.

"Why did you do that?" Grasshopper whines, then becomes flippant, "Don't do it again." His hands are a flurry of pointing fingers as he directs imaginary legions of subordinates, teasing and exhorting them through their paces. Into his executive pantomime he incorporates fleeting suggestions of construction foreman, lion tamer, football referee, incensed parent and used car salesman—all comically instructing some underling. Then apparent catastrophe arises; Grasshopper jumps back, whimpering, distraught in imaginary onslaughts, "You did that. I didn't do it. They made me do it! I don't know what is happening."

"That's enough!" President Crown bites the words in clenched teeth, his face tinges apoplectic purple as he springs to his feet.

Grasshopper dances a happy little jig, spreads his hands wide apart in flourished finale, then says to empty balcony seats, "Now I've done it."

Nothing has so amused Jason in a very long time. Suddenly he is howling, helpless in laughter. Through tears he watches President Crown attempt to control lips working futilely against silent air. The executive's hands form into tight fists, their skin drawn in a white tautness across his knuckles. His patch of baldness is a crimson tonsure above dark eyebrows oscillating up and down in counterpoint to still hopelessly pulsing lips.

Gradually the crisis to the Chief Executive's nervous system subsides. Jason manages to restrain his mirth. Ira Crown draws himself up slightly, glowering invective back at Grasshopper. In measured tones he says evenly, "You are a lunatic if you think I will stand for this." Pausing for a moment, he ruminates over some unspoken thought, then adds, "I have explained the intractable problems facing this nation. I don't have to take your abuse simply because you don't care for my executive style."

"Nearsighted," Grasshopper smirks "Ant."

"What do you mean by that?" President Crown demands.

"Weak little eyes only see the ground right in front."

"What the hell is this fool talking about?" the politician fumes as he looks about, "I give up! I don't understand any of you. I hope you all get elected someday!"

"He just gets this way sometimes," Chloe offers, trying to keep from giggling.

"It seems to be a very dangerous state," Ira Crown observes darkly, "he should be locked up."

"Maybe you could lock him up in the Oval Office," Jason snickers. "Oh, that's right, you've been locked out."

President Crown stares, "That is quite enough! You young people don't understand. You just don't have enough experience to understand."

"That's right!" Jason admits, "We don't understand. Maybe that's why we're all suspicious. That's why no one believes that things have to be this way. They don't. Corruption, greed, deception—what makes them necessary? Why don't you do something about that?"

Grasshopper, seated across the table from Jason, selects a slender snap bean from the now cold bowl and holds it in his lips like a limp cigarette. Puffing imaginary smoke with which he blows pretended smoke rings he says slyly to Jason, "People like you . . . it doesn't really work, you know."

Startled, Jason looks back at him, "Me? What doesn't work?"

"You want it all . . . everything . . . just for yourself. But you can't have it." Grasshopper presses his lips into a tight smile. Jason is irked that Chloe vaguely nods her head. He tries to grasp what is meant but finds no clue.

"So? What are you talking about?"

Laying out several rows of snap beans on the table, Grasshopper waves a hand over them, saying, "Words. You think you're in control when you explain things to yourself with words. But they are just sounds. They don't mean anything until someone hears them and absorbs them through the filter of their own experience. Everybody digests words in their own way.

"So. Words can't explain anything more than what people can understand. So. You really aren't in control of anything.

"But you think you are because you believe in your own words. They are your secret charms. Your own faith. You get upset when no one else follows you into your own gizmo-world. But self-serving words are really just a bad and silly habit."

Grasshopper flips the snap bean through the air to lightly tap Jason's forehead, "Snap out of it, Ant."

Jason wipes the splatter of bean juice from his forehead, "Don't do that."

"You already know," Grasshopper teases nonchalantly, "you really do know. You just don't accept it yet. Because you're obsessed with what you think you know, which really isn't very relevant to much at all.

"Because you want it all, everything. You think, therefore —you are, the reason for everything and everyone. Whatever you find you break up into little pieces and rub your greedy antennae all over it and hold it close to your peculiar Ant body. Then you take it back into your solitary hive and put it with all the other pieces of things you've found and broken into submission."

Grasshopper reaches into the bowl for another snap bean which, after taunting hesitation, he adroitly flips at Jason's forehead, adding as in amusement, "I know. I've seen you."

"Cut that out," Jason warns. Fleeting recollection of some long-ago and horrific déjà vu opens to reveal a glimpsed store of skewed dreams, angry suspicions and seemingly every disconcerting thought and feeling one might ever have. Now it opens again, yawning, inviting him to slip and dissolve into implausible depths.

"What do you really want?" Grasshopper hisses.

"To see, to know the world as it really is. Without pretense. Or delusion," Jason stammers, forcing a note of defiance as he stresses the word 'delusion'.

Grasshopper smirks. Narrowing one eye, he reaches for another snap bean. This one he holds suspended between the tips of his index fingers, pushing against its ends to make it bow up, then relax, bow up, then relax.

"Then here, don't forget these!" In a quickly fluid motion Grasshopper grabs and flings a handful of beans in Jason's face.

"Stop it!" Jason is livid. He wants to slam that insolent silver-wreathed face onto the table but his body is confused and listless. Worse, Chloe seems to find their confrontation amusing. She's giggling at him.

Jason takes a deep breath to control himself. This is worse than a mere demon; this is an unpredictable lunatic, capable of things beyond understanding or anticipation. And Chloe seems in sympathy with the onslaught. She laughs as Grasshopper screws up his face into a leering challenge. The snap bean he's been toying with is inserted between his lips. From a taunting grin revealing teeth gingerly holding the bean Grasshopper squeaks a tight whisper, "Only, I ain't goin'. Tell everybody else what you think, but leave me out of your flimsy daydreams."

"You crazy son of a bitch!" Jason howls in exasperation, "At least I know who I am!"

Grasshopper cackles gleefully, "Si, Meester Words-worth!"

Clara steps back through the doorway. She swallows, then intones ominously to them all, "You won't listen to me, but mark my words! He's a demon. All the demons of this earth will gather to raise him up and extend his dominion. And none shall renounce him. He carries the Number of the Beast!"

Jason's momentary distress has given President Crown time to regain a degree of composure, even amid the flinging of foodstuffs. He tries to resurrect the affable air that had allowed them to share the earlier part of their meal. "Fortunately or unfortunately for us, Grasshopper is the least of our worries," he observes in faint humor.

Clara will have none of it; she repeats her admonition, "Just remember when you're trying to find some way to tear him down. He's a demon and can only be destroyed in one way. All else will feed his power."

"And how might that be?" Ira Crown inquires tiredly.

"Denial. Denial will burn him and drive him back to where he came from."

Ira Crown looks thoughtfully at Grasshopper, contemplating Clara's portents. Jason snorts to himself that among the gullible a lunatic may indeed become king. In that moment he senses, to his own chagrin, that Ira Crown and he are too much alike. It is the most disheartening notion he has yet encountered. He senses that only Chloe and Grady—along with Grasshopper and Hellbender—seem unperplexed.

An idea emerges out of bemused reflection upon his seeming powerlessness to deflect the bond growing between Chloe and her bizarre mentor. He considers a way to fend off competition for Chloe's allegiance: hack into Grasshopper's iLink and tie its inference routines into his own AiPAL, thereby revealing this rival as a misbegotten maxnet marionette. But, first, he'll need to reclaim the pack containing his equipment that he left in the hearse.

~ 18 ~

Just before dawn Sam still has not returned and President Crown worries in dilemma of whether to wait for the agent or to proceed on to the Lebanon Command Post. "I have to be there when Odenheimer arrives with the news teams." He quickly adds, "And the investigators. We can't let this opportunity slip by. Damn! Where is that Sam?"

His complaining grows tiresome. A very early breakfast has been prepared for his convenience. He seems to expect catering as if everyone else is a member of his staff. Finally, in exasperation Jason asks,

"Well, why didn't you just have Duane come by and pick you up?"

Ira Crown looks back at Jason as if he is a dunce for not understanding the obvious. "Because then I'd have to tell him where I am," he answers obliquely.

Not comprehending, Jason continues to stare blankly, still awaiting an explanation.

"At this point I don't know who to trust!" Ira Crown adds in a pang of anguish, "I'm playing it by ear."

"Isn't there anything we can do?" Hector asks, "I mean, can't we help you get to the Command Post?"

Sighing deeply, he says, "I suppose so. We should leave now. There's no time to waste."

"Grady? 'You feel like comin' with us?" Hector asks his brother.

"I guess so," Grady answered. "I'm still kind of sore but maybe the walk will do me good."

"It'll kill you!" Clara objects tersely. She scowls at Hector for even putting the idea in his head.

"Well, not with Grasshopper along," Grady forces a note of cheeriness.

"Is it possible to get there any more quickly than by the route we took yesterday?" Ira Crown asks.

"Not really," Hector shakes his head, then thinks for a moment and says, "Maybe we can move a little faster if we hitch that squirrelly mule up to the junk wagon."

"Yeah," Grady brightens, "then we can all ride!"

"Jake?" Clara laughs contemptuously, "that mule's likely to take you the other way just for the fun of it."

"You just gotta know how to handle him," Hector assures Ira Crown that his plan will work, "he'll get us there."

"Well, we'd better get to it," Jason tries to hurry them along, "let's all go."

"I'm staying here," Clara announces stubbornly.

"Clara, what if people come lookin' for the President?" Hector pleads, "You might get hurt."

"Well, they ain't gonna find him if he ain't here and I ain't gonna do'em any good," she observes pragmatically. "But I ain't gonna go traipsing all over the countryside with the Antichrist for anybody, not even the President."

"Let's go," Hector ushers everyone from the kitchen, "she ain't comin'. I can tell by the look in her eye." He picks up two rifles, leaving the shotgun for Clara, and gives one to Grady. Ira Crown seems not to mind that he is unarmed. He has long been accustomed to delegated personal security.

They follow Hector out to the back yard and down a path to a small ramshackle barn under which is parked an old gray-weathered wooden wagon. Two worn but still sturdy rubber truck tires have been mounted on it's single axle. Hector and Grady check it over, musing aloud that they have not used it since the hearse has been available. They agree that it will serve to haul everyone to Lebanon and beyond, if necessary.

Hector manages to harness Jake the mule while Grady supervises Chloe and Jason in clearing junk out of the wagon and laying out harness and reins in accordance with his instructions. Neither has ever seen such a contraption, except in movies. They are thrilled at a novel ride to new adventure.

Jake has other plans, however. He balks at being placed into harness. But a combination of curses, threats and cajoling finally manages to get the mule rigged.

Finally Jake is hitched and seemingly reconciled to a morning's work. Hector climbs onto the faded green board seat mounted on the wagon front and clucks the mule forward with a light flick of the lines, "Giddy-up."

After a moment of reconsidering his options, Jake moves reluctantly forward a few steps, drawing the wagon from under the shed. Hector urges them all up into the creaking contraption. First Chloe, then she and Jason help Grady. Finally President Crown clambers clumsily onboard.

"Let's go!" he says, perilously infected with a spirit of adventure.

"Shouldn't we call for Sam?" Chloe asks, obviously distressed that he is being abandoned.

"No," President Crown advises, "I don't want to wait. Besides, we don't want to call attention to ourselves."

Hector flicks the lines again and urges the recalcitrant mule forward. They pass by the house where Clara watches from the rear porch. She calls back toward their waving hands,

"Hector Abbot! You take care that Grady don't get shot again!"

"What about me, Clara?" Hector asks.

Jason thinks he sees Clara flicker a hint of a smile as she says, "Well, you're supposed to have some sense. If you get shot, you deserve it!"

Hector grins broadly, "I love you too, Clara. Giddy-up, Jake!" he orders and pulls the reins to turn Jake down the driveway.

"If Sam turns up, send him to the Command Post," Ira Crown calls back to her. Clara nods and watches them bounce down the rutted driveway and out of sight.

After brief consideration, Hector and Grady decide that an old fire path cut through the forest will be the best route since the paved road is in such disrepair that the wagon might have been shaken apart. Besides, the path through the forest will come out on the road near where they had abandoned the hearse on the previous day and would save at least an hour. Ira Crown agrees that it sounds like the preferred route.

As the morning opens into bright clarity they travel at a good pace along a barely discernible trail through a forest of hickory, elm and oak. At times the wagon barely manages to squeeze between young saplings that have grown up in the neglected path, but generally it proves to be a pleasant journey.

They talk of many things, mostly trying to avoid any kind of unpleasant confrontations. Grasshopper does not say much; he either plays with Hellbender or sits in a kind of childlike reverie looking at features of the woodland scene. Occasionally Hector throws a brooding gaze upon the camaraderie growing between Grasshopper and Hellbender, but even he seems eventually to accept that he has been displaced in the hound's fealty.

After an hour they arrive at the broken slab road and soon come into view of the abandoned hearse where it had been left sitting awkwardly angled in the roadway.

"There she is," Grady exclaims as they round a curve. "Hey, there's something else, too. Somebody left a little coaster there." Unconsciously he rubs his chest where a bandage covers a still tender wound. He stares curiously at Grasshopper, then dismisses some fleeting notion as he asks his brother, "Do you think we can get the hearse running again, Hec?"

"I don't know," Hector says, "the main bearing's probably shot to hell. If so, it dumped all the oil in the crankcase. There's no telling what happened to the valves and lifters when that went. I wonder who that coaster belongs to."

"I don't know," Grady sighs dolefully, saddened by the loss of one of their favorite pastimes as he tries to get a better look at the small vehicle sitting awkwardly part off the pavement behind the hearse.

He climbs down from the wagon and walks behind the hearse as the wagon draws nearer. "I almost died in there!" he proclaims in awe. Throwing open the door, his body freezes in rigid ambivalence. His eyes widen while his hands lift in slow surrender above his head. Someone is in the hearse!

Hector motions Jason down off the wagon as he thrusts the reins into Chloe's hands. He grabs a rifle and hops down to move alongside the old vehicle, trying to peep through the interior curtains that screen the side windows. For a moment Jason is immobilized in confusion, then in gathering fear as he realizes that Grady is being threatened someone who is armed. He, too, grabs the other rifle and hops down off the wagon. Feeling painfully awkward, he creeps along the side of the hearse, near Hector, not knowing what to expect nor what to do with this unfamiliar firearm.

Grady finds his tongue and urges in an oddly quavering voice, "Hold your fire! It's two women. One has a gun! She's got a pistol pointed square at my chest. Just take it easy, everybody."

Grady lowers his voice to plead with someone inside the hearse, "That goes for you too ma'm. Please," He gulps, then tries to feature his best boyish grin, "I'm just getting over one gunshot, ma'm. Please don't shoot me again!"

In the brief interim of taut stillness, Grasshopper has climbed down and meandered over behind Grady to peer over his shoulder into the dim interior. "Just everybody hold on," he scolds.

Hector yells out, "Hey, you in there. Throw out your gun and come out. You're outnumbered. We won't hurt you."

"I said hold on!" Grasshopper snaps impatiently. Squinting his eyes he looks past Grady's ear, "It's only Ant's Sister." Then he squeals delightedly, "And ecStacy," His voice quickly escalates into a whoop of surprised joy.

"Peter!" a woman's voice exults. "Grasshopper!" another trills happily as Kristin and Stacy together scramble out of the hearse, knocking a dumbfounded Grady to one side as they enfold Grasshopper in their arms.

"It's you! It's really you!" Kristin exclaims through tears of grateful wonder. "I thought I'd never see you again!"

"Who's she?" Hector approaches cautiously from around the side of the hearse. His apprehension abates at the sight of Kristin showering kisses on Grasshopper, who now is both distraught in the entanglement of her arms and raining kisses and flummoxed by her overture of delight in encountering him.

"This," Grasshopper extracts himself from Kristin's avid embrace, "is Ant's Sister, Kristin. And this is ecStacy!" Dooley reaches out to emulate Kristin's enthusiasm by grabbing Stacy and trying to shower her with kisses.

"Oh!" Kristin squeals breathlessly, "I thought you were gone forever and ever. So much has happened!"

"Hello, sugar," Stacy pulls back from his ardent overture, then as she gets him to calm down, puts her left arm around Grasshopper's waist. Her right hand still holds a .380 caliber automatic, dangling forgotten at her side.

Kristin takes a deep breath, briefly glances at the others and commences an outpouring of her recent experiences, "Eric Cornrath. Remember?" she inquires, "No, I guess not. Anyhow, somehow he got to be a Colonel and he's built a little warlord empire out here. He wouldn't let me leave. He's really crazy. I once said you were, but really, he's seriously pathological! Anyhow, thanks to Stacy, I finally got away, though, and now here you are. Oh, it's so wonderful to see you again."

As the onlookers stand in the verbal downpour of her cascading joy and relief, she pauses for a few deep breaths. Ira Crown catches a name from the list of suspects compiled in recent investigations into complaints about the Lebanon Evacuation Command. He asks, "Cornrath? Did you say Colonel Cornrath?" His voice is urgent, "Did I hear you correctly?"

Kristin frowns, "Yes, do you know him?"

"Not yet," Ira Crown said in contempt, "but I do expect to meet him shortly."

"Oh," Kristin's voice turns icy, "and who are you?" She glances darkly at Grasshopper, disapproving his falling into bad company.

"This is the President. Of the United States," Jason offers, "Ira Crown."

Kristin stares dumbfounded at Jason for a moment, as if he has spoken in some foreign tongue. Then, as she looks more closely at the President, recognition ripples across her face. "You really are," she flusters, "What are you doing out here? With Peter? I mean, Grasshopper?"

Stacy, too, is startled. She lets go of Grasshopper's waist, "Why, Mr. President, aren't you way off course? Where are all your people?"

"That is a very long story," President Crown smiles ruefully, "At any rate we are now on our way to the Lebanon Command Post where I expect to place a number of people, including your Colonel Cornrath, under arrest until we get to the bottom of a very long list of charges and complaints. If you have any information regarding his enterprises, you may be an invaluable witness in his prosecution."

"Well, whatever he's been up to, you can be sure that Eddie Minnow is in the thick of it too!" Stacy declares, recalling the long chat they've had after spending an uncomfortable and chilly night trying to sleep in the hearse.

"Arrest him?" Kristin repeats warily, "And then what? If I may be so blunt."

"Certainly, my dear," Ira Crown laughed with pleasure at her obvious dislike of the renegade Colonel, "frankly, we anticipate a number of courts martial, as well as civilian trials. Considering tales I've heard, frankly, I think maybe someone needs to be shot."

Kristin laughs, "Frankly," she giggles, "so do I!"

They are startled by a sudden voice erupting from the brush.

"Hold it! Nobody move!"

The group standing near Ira Crown spin about to see Sam standing at the edge of the road, breathing deeply in recovery from a marathon run, his automatic pistol trained upon Stacy. "Okay, lady . . . drop the gun, now!"

Stacy gazes back at him for a moment, then remembering she still holds a weapon in her hand, exclaims in fretful surprise, "Okay. Okay. Wait a minute, will you?"

"It's all right, Sam," Ira Crown interposes himself between the agent and the young woman. He reaches out and removes the pistol from Stacy's hand. "She's with us," he assures the wary agent, "And these young women appear to have very interesting information regarding activities at the command center." He asks curiously, "Where have you been?"

Sam moves onto the roadway, lowering his weapon as he gradually accepts that there is no danger. He begins sheepishly, then quickly grows irate. "I fell asleep out in the hayloft, way back from the house. I must have just missed you . . . but it's good that I did. That crazy woman, Clara!" He shoots an accusatory glance at Hector as if somehow it's his fault. "She called the cable news hotline. There are news crews all over the farm and scouring the countryside looking for you."

"Clara! Goddamn it!" Hector howls in dismay at his irrepressible spouse, "Why'd you do that?"

"Why would she do that?" Ira Crown laments the prospect of adverse press coverage transmogrifying this opportunity for a PR coup into another punishing episode of plummeting polls.

"She was standing on the front porch talking to cameras and microphones," Sam explains, "All I could hear without getting too close—I didn't want to be spotted—was that the President of the United States is in league with the Antichrist. That they are riding around the countryside in a mule drawn wagon in furtherance of some elaborate conspiracy against Saint John the Divine."

"Oh, Lord," Grady moans, "she's really off on a bender, now. She won't come back down to earth until she speaks in tongues."

"This is terrible," President Crown groans, "I just want a nice clear cut instance of military justice!" He sighs furtively, "They're going to turn this inside and out and back again, bit by painful bit, rumors floating alongside outright fabrication. Oh, Lord, I've never been on supermarket tabloids before. And with the Antichrist!" He moans in abject helplessness at so devious a turn of fortune, "They won't care that Cornrath and Flach and whoever else is in this are brought to justice. No-o-o! That's too slow and boring. They'll jump all over this nutty Antichrist thing. Oh, Christ, what am I going to do!"

Ira Crown mourns the abrupt souring of his public relations triumph. He rants, fumes and regrets his inability to get quickly to the command post and salvage whatever might be left of chances of restoring the electorate's faith in his leadership.

The President is angry now, and also concerned that he may suffer further punishment of plunging poll numbers. His prosecutorial training resurfaces as he avidly grills Kristin and Stacy for whatever shreds of evidence, or even just sensational tidbits with which to distract the press from his own gathering misfortunes.

Kristin does not know many names, but her narrative provides some clues and confirmation of some evidence already in hand.

"Slavers and drug traffickers! God knows what international connections are in this! Xanadu. That's good. We can put that out, and maybe push Saint John the Divine off to the side. Harry Flach! That pompous bastard," Ira Crown crows, "I knew he had to be involved. I'm going to enjoy watching him twist in the wind!"

"There was a general talking to Cornrath on a video segment I happened to see while he scanned through some videos."

"A general? Which general? Was it Harry Flach?"

"I don't know any of them. This one had four stars on his uniform, though."

"Victor E. Michaels?" Sam and the President both exclaim in chorused dismay.

"Dear, God," Ira Crown complains to the clouds, "if so, this is worse than I ever imagined. Is there any loyalty left anywhere?" He urges Hector, "Let's get going!"

Ira Crown frets and tries to urge everyone back to the wagon as in the distance they all hear the whine of an approaching engine.

"That may be a news crew," Sam speculates, they've been looking for you all morning. Probably picked us up on satellite."

"Good!" Ira Crown voice quickly turns confident, "Maybe we can render this to our advantage." To his companions he says in taut expectation, "You boys take your guns and hide in the woods over there. They'll recognize me if it's a news team. If not, you'll have the advantage in case it turns out to be, well, hostile forces. In any event, we must commandeer that vehicle. This is an emergency. I have to get to the Lebanon Command Post as soon as possible! Try not to injure any newsmen, though. Please! This is bad enough, already," the Chief Executive implores, distraught at the prospect of another PR debacle.

Grady, Sam, Hector and Jason jump quickly to seclude themselves in undergrowth near the edge of the road. Chloe, Stacy and Kristin work at trying to keep Jake positioned. The mule is growing increasingly skittish as the whining engine nears. Grasshopper and Hellbender cavort about the rear of the hearse as if this is still their recreational period. President Crown, to no avail, tries to persuade them either to calm down or get out of sight.

From their secluded vantage point Jason and his three companions watch an all-terrain vehicle maneuver easily along the rutted and broken road. A steproller! A multipurpose transport and camera platform in great favor among larger news agencies, it is capable of either walking or rolling smoothly across practically any type of terrain. The spider-limbed vehicle slows as it approaches the hearse, coaster and wagon blocking the roadway. Someone exults above the engine's whine, "Hey, it's Crown!"

Four men jump out. One already has a pancam pointed at the stranded Chief Executive as another launches a barrage of questions while pointing a small microphone. A third is fiddling with a miniature satellite transmitter. The fourth stands by and tunes equipment to lock on to the distant receiver. The door is ornamented with a corporate logo, *ConRelCo*, painted in yellow and orange with green blandishments.

Jason recognizes Al Statler, the news anchor who had left WNYX to join ConRelCo not long after Jason had been hired as a technician for his morning broadcasts. Jason suddenly feels his loyalties divide. Here he is preparing to subvert his former colleagues in their news gathering.

President Crown announces loudly above a clamor of questions and exclamations, "Gentlemen, this is a national emergency. By authority of the Constitution and my Office as Chief Enforcement Officer of the United States, I am commandeering this vehicle. You may contact the General Accounting Office at your convenience for any restitution due you."

"Ha!" one man yelps, "you can't do that, dude. We're on assignment. Besides," he adds hopefully, "we'll take you anywhere you want to go!"

"I'm truly sorry, but that won't do," President Crown signals with his hand. The hidden four step out from the brush, weapons leveled at the four hapless newsmen, of whom three immediately step back, throwing their arms into the air. The cameraman, in marvelously cool aplomb, swings his lens toward the armed men and then pans the entire scene of his colleagues being held at gunpoint.

"Get that damned oprock!" President Crown shouts. Sam leaps toward the cameraman but he manages to dart into the brush and disappear.

"Never mind!" President Crown calls the agent back in anguished anticipation of the evening news, "Come on! We have to go. Now!"

He climbs into the back of the steproller canopy as Sam occupies the pilot's seat. A hasty conference decides that Hector should accompany them to the Command Post since neither is familiar with the countryside. Grady can drive the mule and wagon, bringing everyone else, news team included, if they wish to come.

Yelling at each other to be careful, the two brothers part, vowing to meet up at the Command Post. Having established the outlines of a plan, Ira Crown is anxious to go. The engine surges as the vehicle lurches and then begins smoothly striding across uneven pavement, down across a ditchbank and then back toward Lebanon. As he watches it skim effortlessly across the rough terrain, Grady asks in wonder, "How does it do that?"

Jason says, "The suspension system consists of articulated legs, each of which has a track-wheel assembly at the end, controlled by sensors that raise or lower each wheel to compensate for the terrain. Just like when any six-legged critter walks. The wheel assemblies have treaded tracks that can be locked to act like flexible feet where it's rough or roll over smooth ground. That's why they're called 'steprollers.' "

Two of the three stranded newsmen swear and curse Crown's executive piracy. The fourth, a younger man, merely stands to the side awaiting instructions. They had been so close to breaking a plum story over the competition and now likely will miss the entire drama playing out at Lebanon Command Center.

Pacing back and forth to work off their outrage, the newsmen snipe at each other as they ponder a focus of blame. The fugitive cameraman soon returns from the woods and joins the diatribe. The two journalists congratulate the videographer on managing to keep the optical crystal that stores video depicting confiscation of their steproller. At least that will play well for a couple of days. "Nice going, Charley. That'll be a good bit."

Charley beams in accomplishment, "You'd best get somebody here to pick us up. And we should upload this segment before somebody else comes along and takes it!"

"Right," Al agrees as he removes a button mike from the side of the box shaped transmitter on his belt. "ATV-4 calling base. Come in, base."

After a brief pause, the box responds in a thin voice, "ATV-base, come in, four.

"Mel?"

"Yeah."

"This is Al. We found Crown, all right. But get this. He took our stepper. He claims it's an emergency and that he's commandeering it. He and a Secret Service agent took off toward the Lebanon Command Post. Anyhow, this road is useless for anything but a steproller. Do you have any air transport?"

"Aw, naw, you let him take the stepper?"

"We didn't have any choice, Mel. He was with an agent and some farmers . . . they had guns. What were we supposed to do?"

"Okay. Okay. AirThree just refueled and went back up. I'll send him your way. He can only take three of you, though. You still on the road you called in from before? Over."

"Yeah."

"Okay, I see you now. Well, don't go anywhere. Equipment okay?"

"Yeah, fine. We're just stuck in the middle of nowhere. Hold on, though. Charley's got a nice bit for you. A shot of Crown and his cronies in an armed take over.

"Great!"

"Are you ready to receive?"

"Count down from five and proceed."

"Just a minute," Al replies impatiently as he watches Charley plug a cable from his camera into the belt transmitter. "Coming in on Channel A—sound copies quad. Minus five . . . four . . . three . . . two . . . one . . . copy."

Charley flicks a switch on the camera and listens as it emits a low hum.

"Tell AirThree that we're on the road not far from Lebanon. He'll see an old black station wagon or something, a little blue coaster and a mule drawn wagon. Tell him to get on it. Over and out."

"Roger. Will do."

"Good grief," Al exclaims in exasperation.

The ConRelCo team has thus far ignored the strangers milling about nearby as if repaying their complicity in stranding the crew far from any viable news action. They still manage, however, to keep a wary eye on the strangers' efforts to usher Grasshopper and Hellbender back toward the wagon. Grady has managed to dissuade Jake from roaming down into the ditch and possibly overturning the wagon. Now he is urging the animal to pull the awkwardly angled conveyance out of its precarious position.

Having uploaded the clip, Al now turns his attention to the individuals milling about the wagon, considering what to do next. Having been around a few veteran news teams, Jason is impressed and forgets that he is still holding a rifle. He feels sort of like one of the news team.

"Who the hell are you guys?" Al inquires as he cautiously eyes the rifles.

"Grasshopper and friends," Chloe announces as she, Stacy and Kristin come over to join Jason.

"Grasshopper?" Al repeats, "Oh, yeah. That batty woman kept saying something about plagues of grasshoppers and the Antichrist." Al motions Charley into action. The camera swings about as the lens zooms back smoothly to include them all.

"What is Grasshopper?" the newsman asks idly.

"That's him," Chloe points to where Grasshopper and Hellbender are inspecting garments formerly cast off on the day before, still lying about the hearse. Al studies him for a moment, then apparently finds little of interest to connect with Clara's proclamations.

"What's Crown up to?"

"He's going to the Lebanon Command Post," Kristin explains. "Its a nest of renegades. He's going to arrest the criminals."

Al chuckles, "He's going to leave us without much of a government, then." The other three newsmen laugh appreciatively.

"Do you know anything about the assassination attempt yesterday?" Al asks. He absently watches Grasshopper and Hellbender snuffling about blood-encrusted garments on the ground behind the hearse. Grasshopper appears to be brushing up on his olfactory skills as he tries to emulate the beagle's interrogatory sniffing." What is he doing?"

"I think he's brushing up on how to smell," Chloe laughs, twitching her own nose as she breathes deeply, "and from a pro, too."

"We were nearby when it happened," Jason answers his earlier question but Al seems to have lost interest in political upheavals as he watches dog and man compete in ferreting out information from bloodstained apparel, as well as from the pants legs of the newcomers.

"Whose clothes are those?" the youngest of the four newsmen asks.

"Ours," Chloe's unassuming air is somewhat abashed at the obvious incredulity with which the four newsmen are beginning to regard them all. "Yesterday, Grasshopper healed Grady's gunshot wound. That's where the blood came from, on the clothes . . . then he kept the President from getting shot." Chloe tries to remind them of more newsworthy matters.

"Oh, yeah?" Al moves toward Grasshopper, "What do you know about the attempt on President Crown's life?" He pokes the mike out toward Grasshopper.

Grasshopper looks tentatively at the mike, then sniffs it for olfactory gossip he can share with Hellbender. The newsman draws back, "I don't think we'll get much from him," he growls, "he's a bit loopy."

"Leave him alone," Kristin approaches Al, "he's not himself. He doesn't know who he is. It happened back at . . . oh, never mind. If you want to know about what's wrong with the Harrisburg Evacuation Route, start looking into Colonel Cornrath's activities."

"And don't forget Eddie Minnow." Stacy chimes in, "He's the regional director of FERAL, or something."

"Really?" Al appraises the two women as if suspecting them to be a little daffy as well. "And how do you come to know about them?"

Briefly Kristin and Stacy summarize their experiences on the freeway. Kristin tells of being captured and held against her will by Eric Cornrath at the Lebanon Command Post. Gradually Al's expression transitions from remote wariness to profound interest. Presently, their accounts of mischief, malfeasance and maliciousness is interrupted by the churning approach of AirThree.

Al breaks off the interview to hastily confer with his two colleagues as the younger casually waves the helicopter down to a suitable landing site near the road. Al returns to speak with the group, accompanied by the youngest member of his team.

"Here's the deal I'm offering. This is Rico Entrada. He'll go with you over to Lebanon. Then, later today, once we get a few things sorted out, Rico will get you squared away in some convenient accommodations. ConRelCo will put you up somewhere comfortable where you can clean up, get some food and rest, if that appeals to you."

"Clean up and rest!" chimes a gleeful chorus.

"Yeah!" Al laughs, "And I want to talk to each of you, at length and on camera, about what's happened. We can work a deal so you'll get something out of it, especially for any eyewitness angles you can provide on the incidents you've told me about. Okay?"

"Wonderful!" Kristin and Jason jointly accept the proposition. Neither has worked in weeks. Suddenly it seems they have stumbled on real collaborative opportunity.

"Too bad about him, though" Al shouts above the raucous helicopter engine, as he gestures toward Grasshopper, "I don't think we'll get much there . . . looney as can be."

"Don't say it like that," Kristin protests hotly, "he can't help it. It's just been too much for him."

"He's not crazy," Chloe defends her friend, reminding Kristin where her true loyalties should be, "he's just different than most. He healed Grady, didn't he? And he'd been shot!"

"Healed who?" Al asks, waving at the helicopter pilot to hold on and to throttle down the engine.

"Grady. Right there," Chloe points to where Grady stands adjusting the mule's harness, trying to keep him from bolting away from the noise of the helicopter.

"It can't have been much of a wound," Al is skeptical.

"Actually, it was," Jason speaks up, "a nine millimeter bullet went all the way through his chest. He was in pretty bad shape. I thought he was as good as dead."

Al looks at Jason, then acts as if something is tugging at his memory, "You look familiar. Do I know you?"

"Not really," Jason is flattered at being recognized, "But I did work briefly at WNYX when you were there, before you joined ConRelCo. I'm Jason Hood. I worked the boards for your seven o'clock broadcast."

"I thought so," Al exclaims as he extends his hand, smiling and nodding as he recalls the times Jason used to sit behind banks of monitors, massaging the signal for broadcast.

"And he looks familiar, too," Al points toward Grasshopper, "what's with him?"

"Ant and Grasshopper. Remember? PR spots the government ran. That so-called austerity campaign?"

"Of course! That Grasshopper. Look, Jason, we have to go. But I'm counting on you and Rico to keep everyone together. We need to go deeper into this. Something tells me you guys may be at the center of the story of the century. Maybe we can even find a place for you at ConRelCo, if you're interested."

"Great!" Jason is thrilled at the prospect.

"I don't know about this healing thing, but you guys seem to know a lot about Ira Crown and what's been going on over at Lebanon Command, not to mention conditions along the freeway. We'll sit down tonight and begin to sort it out. How does that sound?"

"Fantastic!" Jason yells as the engine again screams to remind Al that news is happening elsewhere. Al puts a phone plug in his ear and listens, apparently to exhortations from the impatient pilot. Then he lifts the mike to his lips as he gestures back toward the pilot, saying, "I'm coming right now. Listen. This may be important. It may be nothing. Relay a request, right now. Tell Mel to get a forensic team out here and go over this old hearse and the FERAL coaster. Get lots of video and just keep it handy. I don't know why yet, just a hunch. Just tell him I say it's important to do it."

"Yeah," he repeats, after listening for a moment, "A forensic team. There's blood all over the place. I want them to reconstruct what happened."

"It's Grady's blood," Chloe offers.

"All of it?" Al looks at the brown encrusted carpet and garments scattered about, then swings around to look carefully at Grady still persuading the mule to hold still.

"Yes," Jason confirms, shrugging to indicate that he is equally at a loss.

"Okay, we'll talk later," Al reminds them. He motions Rico to take over in his stead and reminds him that no one is to be allowed to talk to the Grasshopper group before he does. Then he waves farewell to them all and trots off toward the waiting aircraft.

It rises slowly from the ground, clears the tree line and then hurls itself against the sky, eager to overtake events at the Lebanon Command Post.

In the sudden quiet the group looks about to contemplate their options and then begin climbing into the wagon. Chloe and Jason both retrieve their forsaken packs as Jason points out to a disconsolate Grady that likely Clara has improved everyone's circumstances by inciting the ConRelCo news team to chase after Ira Crown. Grady seems to find some comfort in that thought which soon brightens into his usual good-natured grin.

As the wagon lurches forward toward Lebanon and moves onto the roadway, Rico flashes a broad smile of white teeth glinting against his swarthy features. He stretches out his arms and says, in expansive good humor, "So, you guys are from New York City!"

~ 19 ~

Al and Charley find the Lebanon Command Post effectively sealed off. The sudden appearance of Attorney General Duane Odenheimer accompanied by hundreds of marshals and Treasury agents along with assorted teams of military police had caught the base by surprise. No shots were fired. New sentries have been posted and these are taking no chances. Not even the most carefully wrought persuasion evades the order of the day: No entry.

The journalists' exasperation at being unable to penetrate the security wall is eased in noting that other competing news teams have also been effectively excluded. Despite their tardy arrival upon the scene, no organization has gleaned more than has the ConRelCo team. Their exasperated and voracious peers find common complaint expressed in muttered imprecations about the few and generally vague news releases distributed at the main gate. Circulating among his rivals Al quickly appreciates how invaluable are the eyewitnesses he has squirreled away with Rico back in their field operations center at Lancaster.

The steproller has been reclaimed, despite some unpleasantness with sentries at first refusing to recognize their warrant. Several telephone calls and repeated inspection of credentials by various officials eventually pried loose the vehicle. In it the Grasshopper band were conveyed to the ConRelCo base camp operating out of a former motel in Lancaster where Al and Charley have managed to throw together an interview set, equipped with portable field equipment and various available lighting devices rigged together to emulate actual studio conditions.

Charles du Lac, an Algerian of French extraction had begun working field cameras for the Paris Bureau of ConRelCo twelve years earlier. He and Al first served together in the rubble of Beirut and then later in covering the food riots across North Africa. Of converse temperaments, each respects and trusts the other's professional judgment and treasures the deep friendship that has grown between them. Their rapport and individual skills have marked them as ConRelCo's hottest news team, a reputation that practically guarantees automatic endorsement of any project they propose in tandem.

In all but camera work, Charley is a slob. Proud of his proletarian heritage, he is determined to make good as a latter day peasant and not be proselytized into unnecessary social affiliations or graces.

Al Statler, in contrast, is a gentleman dandy of the trade. Enormously erudite, suave in manner and with a reputation for generosity balanced by keen insight into the motives compelling people, Al is widely accepted as one of the most respected and innovative of narrative journalists.

In the time since they have entrusted the Grasshopper group's welfare to their production assistant, Rico, Al has outlined a project that he envisions as being tandem to the main narrative of Ira Crown's struggles. Waste, fraud and incompetence in government as well as burgeoning insurrection and treason among the military are all encapsulated in the interbraiding narratives of the Grasshopper saga. He has broached an idea with their bureau chief, Melrose Bartholomew.

"A journal piece about a whacko named Grasshopper?" Mel repeats incredulously.

"Yeah. These kids have been traveling with him, I want to include their recollections. They were with Crown just after the attempt on his life and they've all been down on the evacuation route for several weeks."

"What about the one that says the Colonel kept her prisoner?"

"Kristin? Kristin Banner. She's the sister of the guy that used to play Ant. You remember those bits that used to pop up, the Ant and Grasshopper energy-ecology campaign?"

"Oh, yeah, that Grasshopper." Mel fits two of the pieces together. Now if Al could only gain his approval to play with all the other pieces as he sorts out the incipient pattern Al senses among them.

"But the hook is with the Banner woman," Mel objects, "if her story holds up, she's the one with the artillery to put away this Cornrath guy and anyone he's been working with. And that might even get Crown reeelected, if he can focus enough public attention and disaffection onto renegade generals."

"I know that she seems like the centerpiece, Mel," Al pleads, "but so far, that's a pretty sketchy case. Lots of talk and circumstance, not much evidence. I have a hunch about this Grasshopper slant. All I need is your permission to put it together."

"Al, Al, Al," Mel holds up his hand to cut short entreaties, "Look-it, we're way over budget. So far this entire expedition has generated a lot of color shots and human interest stuff. Just can it. The home office in Rio wants news, not literature! Get me the grit on Ira Crown and what's been happening along the Harrisburg Evacuation route, especially problems at the Command Center. The story here is mutiny. We're not going to waste any resources or time on Grasshopper human-interest fluff. Anything you put out on that score had better be your own."

"My own?" Al narrows his eyes.

"Yeah. ConRelCo is not interested. Period."

"Okay. You've got a deal. I'll lease the equipment and processing. You can take it out of my bonuses. The rights to the Grasshopper story belong to us, then."

"Us?"

"To Charley and me."

"Who else?" Mel laughs, agreeing as he shakes his head at such fluff, "Just see that the Ira Crown story doesn't suffer."

"It won't," Al snaps, "And Mel . . . I want it in writing."

"What?"

"This agreement."

"That hurts, Al."

"Well, I'm sorry. Business is . . . well, you know. I don't want some ConRelCo fig to come along later and pressure you into a lapse of memory."

"All right, okay," Mel snarls at being reminded of occasions on which he has given in to pressure from above. He rumbles about in his desk for a printed pad of copyright waiver forms.

Al and Charley are meticulous in extracting accounts of the Grasshopper clan. At first they interrogate each person separately. Jason supposes they plan to look for any inconsistencies that might betray their ostensible witnesses as either credulous or deceitful. Grasshopper, of course, stubbornly refuses to participate in any fashion, declining to sit still and tattle to Ants. The newsmen give up and turn to Jason, having already questioned Kristin, Grady, Stacy and Chloe at length.

The prospect of employment with ConRelCo has led Jason to resurrect a lately decrepit journalistic credo of skepticism and practicality. Despite earlier lapses on his part, it now seems obvious to him that Chloe's fatuous tales of Grasshopper's exploits on the freeway, later with Grady and then finally with Ira Crown desperately need a degree of circumspect grounding, which he tries to provide.

He ponders whether to mention the peculiar communications channel he has discovered to be afforded via Grasshopper's iLink. While locked in Lloyd Hershey's basement, in desperation he had used his maxnet euPhone to urge Grasshopper to help Chloe. Imprisoned unawares, he had quickly assembled abusive images gleaned among the torture trappings of that hateful dungeon, combined them with old clips from his favorite *Rocky and Bullwinkle* cartoon episodes hoping to convey the urgency of the moment across the tenuous maxnet connection to Grasshopper. That had been key to their escaping apparent abuse, enslavement, or worse. But Jason decides against complicating an already bizarre tale with details he has not explored and cannot yet fully explain nor appreciate.

"I didn't know Peter Dooley," Jason says, "I don't really even know Grasshopper. He's not accessible right now. At least, not to me. Chloe is close to him. And he apparently has a thing with Stacy. He does relates to Kristin, sometimes. Even Rico is able to deal with him. But he generally ignores me. Says I'm just an ordinary ant, which is fine with me.

"I agree that Grady was badly wounded, all right. A nine millimeter bullet had gone through his chest and out the back. Hector and Clara had doctored him for a bit when it apparently opened up again. We came across Hector hauling Grady in an old hearse, trying to get him to a medic at the Lebanon Post.

"It seemed certain that Grady was dying. At Chloe's urging Grasshopper started splashing about in his blood and singing a crazy song about an old gray goose. Then the engine blew out on the hearse—I was driving—and when the smoke cleared the bleeding had stopped and Grady seemed to be a little improved. Actually, a lot improved. A kind of hysteria came over us. All of us. We went dashing all through the woods. It seems like a dream now. Then Grasshopper went chasing after helicopters and stumbled across President Crown, just after he escaped the first attack, but as somebody—Hector says a Berserker, you know, the motorcycle gang—was about to shoot him.

"Then, later, at the Abbot house Grasshopper pulled off some kind of trick with that dog that's with him, Hellbender. He made Clara, Hec's wife—she's real religious!—think she had healed him, the dog. Or, anyhow, he acted like she had, to make some kind of obscure point—it's nothing really. He's just a guy that's lost his memory of who he is, or something like that. There have been some strange things, curious coincidences, I'd say. But no, I can't say I've seen any miracles."

In a few hours they have interviewed, reviewed and re-interviewed their subjects, individually and as a group. Al wants to devote every spare minute to the enterprise. Already he detects their restlessness with boredom of being cooped up in a couple of little rooms while the newsmen ask questions and adjust camera angles.

Grasshopper, especially, is becoming cantankerous. He has begun spitting whenever Al or Charley come near him, muttering that "no Ants are going to rub their antennae on him!" He especially hates the camera. All are astonished that somehow he allows Rico to play with Hellbender and after a while even seems to grudgingly accept that Hellbender might have found another friend.

Al has sent Charley out to shoot more footage on the Abbot farm and around the hearse. The forensic team has reported that all the blood in the hearse came from the same individual. They estimate that the volume of blood loss certainly was debilitating and likely fatal.

During lunch Al relates the arrangements under which ConRelCo will archive and distribute the video essay. An oprock copy will be set into the ConRelCo data lattice, accessible to subscribers via cybernets such as maxnet. Every time someone or some institution makes use of the piece, their account will be credited with their shares of any royalty payment. Even quotes of quotes will be automatically credited to their accounts, building up over the lifetime of their copyrights. Over the years he and Charley have managed to build a rather substantial side income from such royalties as well as an enormous estate of linked information to bequeath to their children.

From its inception as a data switching network, *Consolidated Relay Corporation* has grown into a global entity comprising many enterprises generating, using and storing information. Satellite links provide subscribers all over the world with immediate access to any archive material on demand, whether text, video, sound, or any other form, even StimuVision, the epicurean zenith of the online pornography industry. Whether for an entertainment conglomerate needing production footage or neural scans for StimuVision or for a gourmet seeking novel combinations of foodstuffs, ConRelCo provides access to everything that has ever been thought, created or suspected by anyone caring to store it away in the archive. Sneeze and ConRelCo probably makes a micro-eDime from selling copies of it for sound effects and medical research.

The New York evacuation assignment has been an unusual endeavor for Al and Charley. A couple of previous ventures have proven unprofitable for both themselves and ConRelCo.

They both have agreed to represent the company, along with a full production crew, as a means of evening up credit/debit columns of their quarterly ledger sheets. Furthermore, a successful engagement gathering news from turmoil threatening to overwhelm the States will enable them to reeestablish their clout with the head office in Rio.

Al recalls his surprise and dismay that he was so little moved by scenes of catastrophe smoldering below upon first circling above the devastation of New York. Similar forays over other sites around the world have long ago inured him to the thinness of the veneer of civilization. Such destruction, he observes haplessly, merely reveals how shallow is the esteem humans have for themselves and their fellows.

Now he wants to relate what has happened to the city and one of its citizens, Peter Dooley. Everyone seems reluctant to accept that this disaster is a consequence of patterns established decades ago in relentless fouling of the globe and unconscionable squandering of resources. Now people seek a magic pill, a miracle cure, a way to quickly discover that conditions are not truly as dire as they might have seemed. He has found his title in Chloe's telling of how her grandmother had a name for people who could kiss a hurt and make it well. He is going to call the essay *The Makewell Kisser*, a lunatic at large in a world craving miracles.

After jotting down a few notes Al searches for Charley. He knows what he wants to do. A few more interviews and shots, a review of the forensic team's report from the hearse. Maybe Rico can persuade Grasshopper to tolerate the camera without evoking torrents of spittle and flatulent raspberries.

Al was intrigued that Rico persuaded Grasshopper to return to the motel and made a mental note to put in a good word for the assistant. But now, as Al attends to production details, Grasshopper has become unmanageable.

Rico tells Al that out on the road Grasshopper had said that he was not waiting for anybody to come back, especially Ants and their box-eye antennae.

"But we can't go anywhere," Kristin had argued, "we promised to wait here, to go back to town with Rico."

"I don't like towns. I'm leaving."

"But you've lived your entire life in a city!" Kristin had fumed.

"Grasshopper in a city! You don't know much about me, do you?"

"Look, just this once," Kristin had sighed wearily, "you might like it."

"I'm certain that I won't. They're Ant hives. Everyone marches about and waits in lines . . . it's horrible!"

"But we promised," Chloe also begged, anxious to return to dispossessed comforts, "they want to interview us."

"Vanity!" Grasshopper spat the word.

"And we can have real food and showers and beds," Kristin dreamed aloud.

"Idleness! Gluttony!"

"But Al says we can negotiate a deal," Jason interjected, "He says it'll be 'worth our while.'"

"Greed!" Grasshopper snarled, "Haven't you learned anything?" He had stomped over to the edge of the road, Rico relates, Hellbender following along to plop down beside his buddy now sprawled disconsolately out across a ditchbank, idly plucking blades of grass to toss at the dog.

"There's something you don't realize, man," Rico had remarked to the incorrigible.

"What's that?" Grasshopper placidly flipped another fluff of dry grass toward Hellbender.

"Your friends, they're not like you."

"Of course not," Grasshopper had recoiled at prospect, "cities, indeed!"

"But in other ways, too."

Rico had squatted down to stroke Hellbender's head; the hound seemed to find the touch gratifying, his eyes closed in pleasure.

"Other ways?" Grasshopper seemed unconcerned at Hellbender's new liaison.

"Yeah. They been walking all over the countryside with you for many days now. They can't do that all the time like you can. They gotta have some city life occasionally. Clean sheets, hot food. Maybe some StimuVision, you know." Rico's resonant voice had urged tolerance for mortal failures.

"Who needs it?"

"They do," Rico insisted, "they gotta see what's going on, make sure there's something left of the world they know. See some nice neat buildings, some straight lines. All that—and more. It's important."

"Not to me."

"Maybe not, but it is for them."

"Let them go, then."

"But man," Rico laughed pleasantly, "I been here just a little while and already I can see that they need you."

Grasshopper had seemed surprised; he carefully inspected Rico, then asked, "Why do they need me?"

"They depend on you, man. They really got no place to go. They got no reason to do anything. They stay with you because you don't seem to need any place or anybody. That's how they can get along in this craziness."

"Well, I don't need them," Grasshopper sulked.

"That's where you're wrong," Rico had laughed easily, "Look, let me explain . . . is this your dog?" Rico gathered folds of skin at the back of the hound's neck.

"He's his own dog," Grasshopper avowed seriously, "I don't own him, he doesn't own me. We just go the same way."

"I can see that, too."

"Oh, don't waste your time arguing with him," Jason had growled to Rico, "he's stubborn. Just plain stubborn. And it's always at the wrong time." Already he could see his opportunity with ConRelCo evaporating in the flare of Grasshopper's refusal to return with them. No Grasshopper, no Chloe. No Chloe, no me. More wandering around, living off of chance.

"He's not stubborn, man," Rico said, placating, "He just don't agree with you."

"Phooey," Jason had stalked to the other side of the road and sprawled out in the weeds to wait for whatever blows Grasshopper would deal his hopes. He wanted to work for ConRelCo. He wanted to make a deal for comfort and a way to feel like he was accomplishing something. He wanted Grasshopper to wake up and be a person like the rest of them.

Rico had continued conversing with Grasshopper. Jason did not understand why Grasshopper talked with Rico and not with him. Now as he sulked he persuaded himself that didn't really care.

"You been wandering all over the countryside. I see you looking around between things for a place that will fit you. I see that alone you could just as well be a tree or a rock or maybe just a puff of wind. That bothers you. It confuses you."

"I'm Grasshopper!"

"Sure, you are," Rico chuckled affably, "And you hop from this to that, from here to there. You're never content. At least with your friends you can listen in on their hopes, even while you laugh at them for being the way they are."

"Shut up!" Grasshopper hissed sullenly, "I don't think I like you."

"That's because I know you," Rico said quietly, "I've seen you before. Where I grew up there are many like you, at least a little bit. Enough for me to know what frightens you."

"Nothing frightens me."

"Ah! No thing. No thing frightens you. But nothing, you're scared of nothing, man. It terrifies you and makes you hang around outside cities even while you look down on everything they represent. And all the time you know you can't live there. Because you don't know how."

Grasshopper did not respond for a while. Because of the sudden eclipse of conversation Jason thought he had gotten up and walked away. Turning his head slightly he peeped through closed eyelids. Grasshopper still sat on the other side of the road, looking at Rico and watching flies swarm. He rested for a long time, then said, "All right. I'll be with them for a while longer."

"Good!" Rico exclaimed, "It is good." He stood and straightened legs grown stiff. Walking over to me he asked, "You okay, Jason?"

"Yeah," Jason sighed, after a moment adding, "How do you know how to talk to him?"

"I can usually feel people out," Rico laughed, "I'm a pretty good bullshitter."

"It's more than that," Jason mused, "you said that where you grew up there were many like him. What did you mean?"

"I grew up with no family, no home, in Costa Rica," Rico replied, "and it's true. There were many like him, including myself: a child who is everybody's orphan."

Rico had paused, saddened by some recollection. Then he sighed and said, "You mentioned something about his amnesia. I watched him fiddling with those ants while you talked to Al. His lips were working through a snit of some kind. He fussed at the ants and warned them of Latter Days. Then he started a little tirade about vanity and greed and gluttony. His voice was shrill like that woman preaching back at the Abbot farm—Clara Abbot. It all fit together somehow. I think he's afraid. And he keeps working through things bothering him. There's nothing dependable inside him and his way to hold anything at all is by pretending, imagining. He becomes the thing he fears, or hates or loves. And then he gets angry because it's not his and he jumps to something else. Just like the Grasshopper he says he is."

"You're right about one thing," Kristin had joined in. "Sometimes he is just a lost child—sad and small. Then at others he can become vicious. There were times out on the freeway that I thought he would kill someone, especially at first. I was afraid of him. Then he became a little more calm, when people started responding to him, some even acting like he was their leader. It seemed to guide him somehow. He handed out the food and made sure that everyone got their share, like a little kid playing grownup!"

"That's why he'll come with us," Rico added as he watched the little man across the pavement, "he can't stand not being."

"What do you mean?" Jason was puzzled by the curious ontological proposition.

"You are the way he knows who he is. You let him be Grasshopper."

"But I don't want him to be Grasshopper," Kristin's voice had trailed off into a silent speculation as she watched Hellbender endure handfuls of dry grass being heaped onto his back.

"Oh, Peter," she lamented, "come back, please come back," she moaned.

"Maybe he'll snap out of it," Rico had smiled encouragement.

~ 20 ~

Jason had known it would happen. His chance to reconstruct a career is shattered as Grasshopper erupts from the bathroom, shrieking, "It's a parade! A parade!"

Before anyone can respond he has sprinted through the room, out the door and into gathering darkness. Hellbender barks in excitement and bolts after him.

"Hey, wait!" Rico yells, jumping up to chase after him.

"Wait for us!" Chloe squeals, "I want to see the parade!"

"Don't let him go off on his own," Kristin cries out of concern that he might yet disappear again. She, Grady and Stacy grab up coats and dart out behind Chloe.

"Damn! Damn! Damn!" Jason fumes at Rico for allowing Grasshopper such liberty. He too pulls on his coat that has been provided earlier in the day as they all shopped in a local mall at ConRelCo's expense.

Even then Grasshopper had refused to give up his overalls and plaid shirt, insisting that they are Grady's and Grady wants them back. Chloe did manage to persuade him to take a pair of yellow socks and running shoes. Jason now debates briefly whether he should stay or follow after. Growling at the unfairness of fate he reluctantly follows after the others, yelling back to Rico as if it were all his fault, "Well, come on. Can't miss a parade, can we?"

The companions pause above an embankment falling precipitously from the motel to the sidewalk and street below, hearing nothing even remotely like the jubilant furor of brass bands and whistles typical of parades. Instead they watch a throng of shadowy figures moved inexorably toward past, in no apparent order, but marching ineluctably up the avenue.

"That's no parade," Kristin gasps, "that's a mob!"

"You're right," Jason agrees, watching sullen figures approach, carrying assorted handmade signs, banners and flashlights flashing across them as they move past. An angry rumble of shouts and curses precedes the procession through an intersection just beyond the motel entrance.

"It is a parade!" Grasshopper insists. A police car briefly appears at the intersection, blue lights flashing ominously. Grasshopper yells, "See, they even have an escort!"

"It's a mob, you twit!" Jason rages at him above the sudden metallic whine of the siren, "and I don't like the look of it!"

"You don't like the look of anything," Grasshopper retorts as he slides down the embankment onto the sidewalk. Hellbender dolefully choruses with piercing sirens of two other patrol cars that now arrive on the scene. Grasshopper strides in step as he barges into the relentlessly surging crowd.

"Pete! Grasshopper! Don't go," Kristin pleads. He gives no indication that he even hears her.

"Let's go, too," Chloe says, her face flushed with excitement. Without waiting for anyone's answer she hops down onto the sidewalk.

"Gee, I don't know about this," Rico complains, "at least let me go get a camera."

"There isn't time," Kristin announces as she jumps down after Chloe, "come on. You can't work all the time! Grady, Stacy, are you coming."

"Hell, we'll make it a parade," Grady laughs as he and Stacy follow after Kristin and Chloe.

"You don't know Al," Rico mutters aloud. Then he, too, commits to follow after his charges. Shaking his head in dismay, Jason slides down the embankment and tries to keep up with his scattered companions.

The crowd chooses to ignore repeated police orders to disperse. Outnumbered officers can do little more than watch angry fists and listen to obscene taunts. Jason is able to make out a few of the signs in the dim and fleeting spots of illumination from occasional flashlights. They bear emblazoned phrases: WE WANT JOBS! WE'RE PEOPLE—NOT SCRAP! MONEY CAN'T FIGHT—WE CAN—AND WILL!

He may have been shaking his head from side to side in dismay because a grizzled and shabby old man looks him over, appraising the new clothes he is wearing and snarls, "If you don't like it, get the hell out of here, asshole!" The sign he is carrying swings forcefully against Jason's head. Pain explodes across his scalp and down into his eyes as suddenly he reels and falls, trying to evade an onslaught of trampling feet. Chloe shrieks in anger and fright as she runs back to support him. The old man grins proudly and looks about to see if his comrades have witnessed his ardor. All ignore him, sweeping his heroics and Chloe's plaintive protests along the avenue and past police cars poised in anticipation.

Only Grasshopper seems happy. He struts at the vanguard like a drum major. Hellbender trots patiently near his precariously stepping heels, moving with a slight sideward gait to avoid those extravagantly happy prancing feet.

The mob heads through dismal streets and disheveled avenues for a mile or two. During that time more police cars skirt the mob, scanning it with searchlights and intoning ignored orders over loudspeakers.

A field of gravel and pavement dotted with warehouses, their fenced yards filled with steel girders jutting askew in clusters like giant heaps of discarded popsicle sticks, proves to be the crowd's destination. The rusted wire yard fences have recently been reinforced and festooned in spiky braids of stainless razor wire. Above the gate a corroded sign proclaims admission to *Allied Steel Yard #4*. In its shadow, cast by a half moon now brilliant in the fulfillment of night, stands a platoon of armed and uniformed security guards.

"Oh, God," Kristin gulps as they move unwittingly into the field of fire of two machine guns stationed just inside the gate. Along the fence on either side stands more private guards in various combinations of professional and improvised riot gear, their hands clasping large carbon-fiber staves tipped with metal prods. Others are festooned in belts of draped canisters of what probably are anesthetic gases recently introduced to control mobs in major urban areas. All the guards have masks at the ready, attached to their belts. A few apparently anticipate immanent use of debilitating gases, their masks are suspended just below their chins. Both sides are expecting trouble. Jason has seen the glint of moonlight on concealed firearms carried by many in the mob.

Then, amazingly, the guards at the gates move aside and let the crowd pass through, as if by prearrangement. Some of the men milling forward nod in gruff recognition to a few guards. Jason feels a degree of relief that at least some of these people know each other. Still he is eager to leave. "Let's get out of here," he whispers hoarsely to Chloe. Kristin, Stacy and Grady and Rico agree by quickly nodding their heads.

"Where's Grasshopper?" Chloe stands up on her tiptoes, trying to locate him in the crowd.

"Look," Jason howls in an amalgam of urgings toward flight, "he'll just have to look after himself. You're not his mother. Let's get out of here!"

But already it is too late, even had she assented. They are carried along by the coursing host that pours through a large opening into an immense warehouse. A table sits in the middle of the space, three men in business suits fidget nervously as the crowd approaches. About the perimeter of the warehouse, more armed and uniformed guards stand alertly at parade rest. Four empty folding chairs have been placed on one side of the table, opposite the three men already seated.

"I'm scared," Kristin whispers to no one in particular.

"Me, too," Jason confesses, rubbing the knot already forming on his head. His fingers testify that the sign's blow has drawn blood.

"Let me see," Chloe pulls his head down to her eye level, "oh, Jason, I know that hurts. I'm sorry."

"You just had to follow him, didn't you?"

"I didn't know," she says sorrowfully, "I thought it might be exciting after being cooped up all day."

"It may be too exciting," Jason warns, "already I'm almost thrilled to death!" His exasperation takes over and forces its way out in hoarse throaty whispers. "I don't know why you can't use some common sense. Just because he's crazy doesn't mean we have to tag along in his escapades. This is stupid!"

"All right, Jason," Chloe retorts hotly, "you've made your point!"

"We gotta get out of here, man," Rico advises as he rejoins them. He has been looking for Grasshopper among the mass of bodies thronging about. "Everybody but us is here ready to kill somebody. This is going to be war." He grimaces anxiously, trying to grin.

"We can't leave Grasshopper," Chloe is obdurate, "He might get hurt."

"Dammit, Chloe!" Jason exclaims, "We all might get hurt!"

A young guy standing nearby overhears his protest. He smiles grimly as he furtively displays the muzzle of a sawed-off automatic shotgun, "It won't hurt long. This is gonna be over real soon," nodding to emphasize the firmness of his resolve. Then he returns his gaze to the distant table where one of the seated executives singles out a face in the crowd and begins speaking loudly. The crowd noise hushes in anticipation of what might be forthcoming.

"Mr. Waldo," the voice is ironic, "I see you've come prepared to support your proposals. Perhaps a committee of several hundred can accomplish what six cannot." The executive smiles grimly; the man he has addressed steps forward, flanked by two grim-featured companions.

"I hoped it wouldn't come to this, Stevens—you know that. But we don't have any way to turn. There's no unemployment, the government doesn't give a damn, the national union ignores all the locals. For us there's only this company—we're gonna be damned sure you don't ignore us!"

Mr. Waldo pauses to let the threat settle upon the men at the table, then continues, "Some of us have worked here for thirty, forty years. That counts for something . . . loyal work . . . good work. You can't just dump us like so much sludge. We're not just part of your inventory!"

The crowd rumbles assent. Stevens looks back at into the assembled discontent and sighs heavily. His voice seems sincere, but undaunted by looming threats. "I know. We've been through this again and again. But you just won't hear. We have to close this plant. Investment firms have unloaded us like so much scrap paper. The company right now is little more than a letterhead on useless stationery. There is, for all practical purposes, no company. Unless we can operate on fond memories and pay your salaries in faint hope, this yard will close."

"Very poetic," Waldo snarls, "but unless we can work out some kind of agreement, there won't be anything to close down. This yard won't be here."

The two men glared stolidly at each other.

"Take it to court," Stevens suggests humorlessly.

"Ha! With what? For what? How long would that take? It'd be years just coming to a hearing."

Stevens throws up his hands in helplessness.

"Then I guess we have nothing to discuss. We can keep a few men on as watchmen and a few others to dismantle and ship out equipment and scrap, but that's all. Everyone else is laid off. Please. Save everyone a lot of unnecessary grief. Go home. I intend to do my job."

"So do I," Waldo straightens up, "we can dismantle this whole plant tonight. For free!"

The mob bristles. Metallic clinks can be heard in the ranks. Guards shift their stance and move to test the heft of the metal capped staves in their hands.

In the ponderous silence Mr. Waldo is distracted by a sound of ticky-ticky-ticky moving across the concrete behind him. He turns to see a large doleful-eyed hound slowly moving toward his feet, claws ticketing against the concrete floor.

Hellbender looks up at Waldo, plops his haunches heavily onto the floor, rears back his head, closes his eyes and emits a mournfully lingering yelp that grows into an earsplitting, prolonged wail. The dog's forlorn moan seems to echo the sentiments of everyone present.

Grasshopper strolls sheepishly forward. Occasional side glances apologize to the crowd for the interruption. He stops by Hellbender's side, grins tentatively while he waits for the hound's wailing lament to cease. He glances along the panorama of disconcerted expressions and shrugs slightly to convey his mortification at such blatant outpouring of woe.

Eventually he's had enough and with a single elaborate flick of his right index and middle fingers, he snips off the noise in mid wail to stop cleanly, as if an invisible, plaintive thread has suddenly been clipped above the hound's head. Hellbender does not move, but sits poised, head still thrown back and eyes tightly closed as if he has not yet completed his song and awaits any opportunity to fully vent his anguish.

Grasshopper flips on the limner as he then tucks his thumbs into the bib of his overalls to take the measure of his audience. He shakes back the long snarls of his hair, scratches at a growth of beard along his chin and beams back at the gathering who have fallen entranced by this luminously swirling apparition clad ridiculously in old plaid shirt, baggy overalls and bright new tennis shoes over brilliant yellow socks.

"He's a good dog," Grasshopper drawls pleasantly as he summons his composure. Extending his arm above the hound, he jerks the invisible thread and Hellbender continues from where he had left off. Again Grasshopper seems to tire of the din and again cuts it off with a flourish of imaginary scissors. As he slowly cleaves the fabric of sound, Hellbender's voice fades through stepped levels as shearing fingers work deftly above his head, miming chunks of audio falling away while Grasshopper avidly chops at the air.

"I finally learned how to cut off his racket," Grasshopper gloats as he finishes snipping. Again the hound falls silent, but keeps his head poised and eyes tightly closed. Someone snickers. Grasshopper bends down and lifts the dog just behind his forelegs. In wry amusement he watches the skin gather in folds and wrinkles about Hellbender's neck. He looks down at his overalls, tugs playfully at the sagging folds of fur and laughs,

"Looks like the same feller made his skin that made my suit!"

He throws a leg stiffly to the side to display the baggy overalls. His brand new tennis shoes and socks flash brightly beneath the coarse and faded denim. Several people choke back chortles. Hellbender now opens his eyes and looks dolefully back at the crowd, patient in the awkward clutch in which he is toted.

"I do hope y'all don't mind me and my friends joining your parade," Grasshopper drawls on in an amiable southern whimsy.

"Back home we had to stop all the parades. Everybody wanted to play the bass drum or the glockenspiel—it was awful. Worst noise you ever heard!" A luminous wash of marching uniformed musicians swirls about Grasshopper, accompanied by seemingly distant duels of bass drums and glockenspiels.

"And we had the worst fights! Glockenspielers mad 'cause nobody could hear'em over the damned drums . . . and drummers mad because the glockenspielers wouldn't let'em solo. Except at the end of the piece, you know . . . like ba-rump! Only for a finale!"

Grasshopper twists one side of his mouth in poignant resignation," So—no more parades!"

Then he smiles impishly, "I can't wait to get back home and tell everybody how y'all solved our problem. And it's so simple, too! Just parade around without music! It's so clever. And really, it's just as much fun! By dingies, I guess what they say about you yankees is true, all right! You can invent a solution to just about any ol' problem. I don't know why we never thought of that!"

He smiles laconically, shifting Hellbender under one arm as he glances about the warehouse, apparently enthralled by banks and banks of unused fluorescent lights above.

"I can't figure out why you came in here, though! It seems like a dead end—and it was going so well, too. Police escort and everything."

"We came to fight!" a voice yells from the crowd. There is a rumble of agreement as everyone recalls their mission.

"Fight?" Grasshopper looks bewildered. "Gollee! We never thought of that either." He grins in embarrassment over such shortsightedness. "We never could figure out how to finish it off. I mean, we'd get to paradin' and since there wasn't any music to end, we'd just keep on goin' 'til everybody dropped. Sometimes it'd go on for days. Fields didn't get plowed, children didn't get schooled, papers didn't get published. Just a bunch of fools marchin' all over time . . . like we didn't have nothin' better to do! Glockenspiels and bass drums making nasty remarks about each other—sometimes even about relatives like marimbas and snares."

He pauses and looks back at the crowd, boring in on its implicit stupidity and outrage, "Now a fight would really have done it! We coulda smashed each other up real good and nothin' would have gotten done just the same. And it wouldn't have taken nearly as long!" Grasshopper pauses, then goes on, "My hat's off to you people—you're pretty damned efficient! When I think of the time we wasted doing nothing useful and I see how fast you can accomplish the same thing, why it just makes my head swim at the promise of yankee know-how."

The crowd begins to murmur as it flinches against the sarcasm rising in his voice.

"Come on. Let's get on with it," someone shouts, "Get that damned fool out of the way!"

Grasshopper sets Hellbender onto the floor, then leaps nimbly up and down as if trying to see over the heads in the front.

"Who's that calling me a fool?" he demands.

"I did!" the voice challenges.

Now Grasshopper springs onto the table to peer intently into the dimness, his hand shielding his eyes from the few lights overhead. Mr. Stevens and his two colleagues watch in astonishment as Grasshopper's sneakers leave dirty tread marks on their neatly arranged papers and file folders.

"Now who's really the greater fool?" Grasshopper snorts, "I like parades. You idiots seem partial to funerals."

"There'll be lots of funerals!" another calls.

"So much the better—there'll be less competition for this fool!" Grasshopper taunts, "I'll look pretty smart leading my parade around your funeral!"

"Get off my papers!" Mr. Stevens snarls at Grasshopper who now has begun to pace indignantly up and down the length of the table, unconcerned that his feet are scattering documents right and left.

"Oops, sorry," he apologizes and steps out of the disheveled papers, but remains on the table. "Now, people, it seems we've got us a real problem," he counsels, his accent shifting suddenly into clipped accusation, "We've got two crowds of fools with only a bigger fool to keep them apart." He glares menacingly at them all, then continues, "This has to be the silliest thing I've seen in a long time and for as long as I can remember—and I've seen some pretty ridiculous things, lately. Neither of you has anything and you're both angry just because the other is empty handed!"

Grasshopper screws his mouth up in amused consternation at the prospect before him, "Now, I'll tell you what I'm going to do. I'll give you one last chance.

"Mr. Stevens, here," he reaches down and pats the startled manager on the head, "is going to trade places with Mr. Waldo over there," he nods toward Waldo. "Each one is going to argue for the other side."

Grasshopper jumps down to stand at Steven's left shoulder. The plant manager sputters in outrage at such a preposterous idea.

"I-I will not!"

"Oh, yes you will," Grasshopper counters with the patient determination of a parent inevitably bound to have his way, "Get up and go stand over there where Mr. Waldo is." He points sternly to where Waldo stands grinning at the nonplussed steelyard manager.

"Mr. Waldo?" Grasshopper snaps his fingers like a practiced maitre d' summoning a table, "Your place is ready."

Waldo grimaces at his two companions as if to say that anything is preferable to the conclusion he has dreaded. He walks over toward the chair just vacated as Grasshopper herds the reluctant manager toward the mob of workers now startled, if still sullen, by the sudden turn of events.

"Now!" Grasshopper chirrups in satisfaction at their progress. He returns to the center where Hellbender has given up and fallen asleep. "All you have to do is pretend that you're the other man. Just put him on like you'd put on his shirt."

The two opponents glare at each other.

"I won't do it—I can't think like he does," Stevens scowls in protest, "this is crazy!"

"Do you mean to tell me that you're admitting that the man seated at that table is smarter than you?" Grasshopper screams sarcastic incredulity. He throws his hands into the air as he turns to the crowd in despair.

"Well, there's your problem folks," he bemoans the revelation as the crowd pushes forward about the embattled manager. "You've elected a man who's afraid of management. He thinks they're smarter than he is."

"I'm not afraid of him, I mean, of me. No, him!"

Stevens tries to express disdain, but finds his lips maundering in angry confusion, uncertain who he is talking about and suddenly feeling even more foolish. Waldo, seated at the table, starts laughing and relaxes a little. His ease is premature, however. Grasshopper whirls about to advance upon him, whining cynically. "Aw, come on. Don't laugh at him. That's the trouble with you management people, anyhow. You pay them a wage and you start thinking you own them—you're bigger than that, aren't you?"

Waldo reddens in sudden embarrassment. He seems to recoil from echoes of his very own words turned against him. Grasshopper returns quickly to Stevens who, glancing about at the hostile faces now surrounding him, is seeking some way to escape from a sea of discontent threatening to overwhelm him. Grasshopper seizes him by the shoulder of his suit jacket and ushers him and the two men beside him up to the table behind which sits Waldo, a ferocious scowl reflecting his ire.

"Tell him what you want," Grasshopper urges Stevens. His voice has become clipped and impatient, like that of a third grade teacher waiting for hands to be washed.

"I want—I mean," Stevens fights the confusion, "we have to close down the mill."

"NO!" Grasshopper thunders, yanking the poor man's ear, "You ninny! That's what he wants! Are you going to sell out all those people back there? Your friends and neighbors? Your fellow workers? Jesus! What a hypocrite! You Judas!"

"I'm not a hypocrite!" Stevens barks his frustration.

"Then tell him what you want, what you all want—and STOP SCREWING AROUND!" Grasshopper screams in a frenzy.

Even the security guards seem to enjoy the manager's predicament. The two men on either side of Waldo laugh nervously as Waldo, glancing surreptitiously toward Grasshopper, attempts to maintain a straight face.

"We want our jobs!" Stevens suddenly blurts out. "Just let us keep our jobs!"

Grasshopper points abruptly, cuing Waldo impatiently. The man, startled, seems to search his memory for lines he has heard so recently.

"Er, ah, no . . . No. The company's earnings are, ah, negative, and, ah, falling. My decision is final."

"Then we'll tear the place down and crack a few skulls in the process!" Stevens threatens, suddenly getting the feel of his new role.

"Like hell you will," Waldo leaps to his feet, waving a finger to threaten his adversary's face.

"I'll have the place swarming with cops—troops, if necessary!"

"All right! If that's the way you want it then that's the way you'll have it!" Stevens whirls about and shouts furiously at the mass pressed behind him. "Come on! Tear the place apart!"

His arms hover in the air as he suddenly realizes what he is urging. Hundreds of men and women looked uncertainly back at him, wondering if they should follow order from a voice they have come to despise, even if it urges them toward fulfilling their own intentions.

Stevens turns back to the table, confused, embarrassed that he has so easily been suckered into his antagonist's role. He is angry that Grasshopper now lies stretched insolently across the papers on his table, his head propped on one arm as he watches the two men, a curious twitch snatching at his mouth. For a moment total silence holds sway, then abruptly yields to Grasshopper's softly censuring laugh. "Now, wasn't that silly!"

The two adversaries glare at each other then back into a luminous image wash of raucously laughing crowds swirling about and beyond Grasshopper. Then Waldo laughs lightly. Stevens smiles. Grasshopper sits up, dangling his legs from the table.

"Just sit down and work out a way to help each other. One man can argue for the company, the other for the workers.

Take turns, it doesn't matter who does it. You're both bright fellows."

"It won't work," Stevens sighs heavily.

"There it is again. That stupid parade. Why won't it work?" Grasshopper asks wearily.

"Because the company has no assets. The stock is worthless."

"Then give it to them in exchange for their labor contract—that's worthless, too, isn't it?" Grasshopper's explanation grows irate in dealing with dullards.

"Come on, Stevens," Waldo mellows his voice, "it's worth a try. There'll be a steel shortage soon—we'll defer payment until then. We can make it somehow, maybe. If only there's some hope for the future. Each man will kick back in two-thirds of his paycheck. If we can make it six months, we'll have a chance. Just give it a try."

Stevens looks at each of his two assistants for a moment as if they are pans of a teetering balance. One of them says cautiously, "It might work if we can hold off the creditors. Given the clogged court dockets, we might be able to beat out the judgments. It probably won't work, but it might. If they defer their salaries and then buy stock out of the first receipts."

The plant manager works his lips back and forth as he considers alternatives, weighing them in dread against further months of worrying. Finally he says, "all right. Okay. We'll try it. I'll have to persuade the board, but I'll try it."

"Just try," Waldo smiles in appreciation of his taking a chance, "that's all we ask." He extends his hand and the manager takes it slowly, still dubious that any settlement has really been reached.

"Go on home!" Waldo yells to the crowd, "We'll be working all night and we need peace and quiet!"

"Harry," Stevens said to one of his assistants, "Go tell the chief that it looks like things have blown over. Thank him and tell him to send his men home. After the crowd disperses." He adds the cautionary note as he watches hundreds of people begin moving slowly back outside.

Grasshopper has grabbed up the somnolent Hellbender and walked quickly to a side door. A guard moves to stop him, then changes his mind. Grasshopper lunges through the door, his lips working furiously in a twitch as he chastises everyone in the room, "Stupid Ants! Stupid, dumb, nearsighted damned idiotic Ants!"

His companions gather to hurry out after him. They find him standing with Hellbender in his arms, looking up at the night sky, searching among the stars for an antidote to the rage working its way through him.

"Grasshopper!" Kristin calls out to him.

He spins around; a rage savages his expression as he hisses back at her, "So you finally know who I am!"

Chloe comes up to stand by Kristin. The two of them wait for a moment, then Kristin says, softly, "Come on, it's time to go."

Grasshopper purses his lips spitefully. The numbing disdain that has frozen his features now dissipates as easily as any caricature he might have wrought and then discarded. He watches Kristin suspiciously, then lifts his eyes in mock astonishment. His lips fight against a twitch rippling at his mouth. "It was a moving performance. You would have noticed a long time ago if you hadn't dragged your feet all over my material!"

Kristin suddenly draws herself up erect, disbelief rivets her gaze on him as she stares beyond his casually sardonic expression, "Because I couldn't keep my knees apart?" she murmurs, recalling a half-forgotten swipe that Dooley had tossed at her during an argument in another life.

"You said it, KayBee!"

"There!" she commands, pointing at him, "Right there!"

Grasshopper is startled by her outburst, "Where?"

"Where did you get that?" she inquires urgently.

"What?"

"What you said about me, KayBee, dragging my feet all over your material," she probes in steely resolve.

"I don't know," Grasshopper says, "I just said it."

"Peter Dooley said it! You know he did! Let him say something else—even if it's to insult me."

"You wouldn't drag your feet all over my material if you could keep your knees apart!" Grasshopper laughs sarcastically. Then suddenly his eyes narrow and his hands began to fidget nervously as he looks back and forth between Kristin and Stacy, who has moved near them both. Slowly he draws his head back and pulls his shoulder up into a martial stance. The iLink limner comes back on, bringing forth Laurel and Hardy to bracket his sides, arms clasped across his shoulders like schoolyard chums. Seeming to ignore everyone but Stacy he puckers his lips in imitation of a light kiss as he giggled. "No." Stan and Laurel look away in a huff of denial.

Kristin screams in exasperation, "Turn that off! Let go, damn you! Let go!"

"Are you all right, lady?"

They turn to see a police officer watching the exchange closely. Startled, Kristin gropes for something to say. She stares silently at the officer, frozen in realization that she is a refugee without identification and liable to be swept up and returned to the evacuation route. The possibility of being returned to the freeway has suddenly sucked the wits from her.

Rico steps forward, flashing press credentials. "No trouble, officer. We're just arguing over top credits on this story, that's all. We're all tired and need to get back to ConRelCo to get something to eat."

"Oh," the policeman relaxes, then looks about. "Where's your equipment?" he asks suspiciously.

Rico laughs heartily, "We didn't need it. Nothing happened."

The officer is not amused. "Don't get smart kid. If you're with ConRelCo, where's your equipment?"

"Sorry," Rico apologizes for his flippancy, "Actually, we already sent it back to the office up on Maypark."

The officer nods his head slightly to indicate that he knows the location of ConRelCo's field office. Again he scrutinizes Rico's press card, then grumbles. "Okay, sorry to interfere."

"It's okay, officer," Rico calls after him.

"Ant!" Grasshopper spits toward the departing officer, but the policeman neither sees or hears it.

"Don't do that, man!" Rico step forward to intercept any epithets Grasshopper might hurl after the departing lawman. "Why do you do that? He's just like you."

"Oh, no, no!" Grasshopper crows, "I am lovely elseness. Not ugly old likeness. Ants live in likeness! There are no Ants in perhaps or otherwise—it's too else!" Grasshopper intones each syllable as if each needs his full attention to flower into voice.

"What makes you so ideal?" Jason grumbles, tiring of these endless accusations against practicality.

"Ideal? Ideal?" Grasshopper scowls in disdain, "You goddamn, crummy little Ant suck." He spits at Jason as Hellbender suddenly growls a low rumbling warning.

"Now wait a minute!" Jason is taken aback.

"You wait a minute," Grasshopper snaps viciously, "I'm fed up with your hanging around and simpering about this and that. What the hell do you know about comedy? Have you ever laughed at anything? Ideal? Where the hell is it? You're only after PERFECTION!" he suddenly shouts to the stars, raging as he storms about, shaking his fist, "OH, SWEET PERFECTION! Don't you know that perfection is the enemy of the good?"

"Hey what's going on over there?" a voice yells from the line of police by the gate.

Grasshopper ignores their warnings and avoids any attempts to quieten him. His eyes flash hatred as he scrambles up onto a pile of girders, railing against legions of Antness laying siege to his proper being, screaming irate accusations down at Jason,

"MAMA! MAMA! BRING THAT BEAUTY BACK TO ME!" he howls as Hellbender joins in what seems to be a falsetto chorus at the moon.

"It's here and now! There is nothing beyond! Elseness is me! Elseness is Grasshopper. I am always new and never finished. You can't finish me off with Peter Dooley or any other Ant-sucking wannabe! Stay away from me! All of you!"

Whereupon he turns abruptly on his heel and scurries up a pile of girders and over the fence. Hellbender hurries after, squeezing through a narrow opening in the wire links below. By the time his companions have crawled up on top of the steel pile, he is out of sight. They have no idea where he may have gone.

Chloe whimpers. Jason wonders if she is accusing him of driving away her hero.

Kristin looks off into the distance, then plaintively moans, "Peter Dooley . . . please, don't give up!"

Charley is preparing to visit one of the few nightspots on the south side of Harrisburg when Al finally tracks him down and tells him about the final details of the deal worked out with ConRelCo. Under protest his comrade listens politely to Al's revised proposal, reluctantly agrees to shoot a few more segments and then as he gathers his gear, he begins to share

Al's enthusiasm. They discuss how best to weave together the threads of the narrative. Stopping by the office to drop off Mel's copy of the copyright waiver and the completed distribution contract, their ebullience suddenly evaporates—Grasshopper and his companions have abruptly disappeared! Even Rico is absent.

The two newsmen are at first paralyzed in commiseration at such an abrupt termination of the joint project. Then as they return to their rooms a sullen and disconsolate mood engulfs them as they argue bitterly for the first time in their professional association.

But gradually they work through their differing takes on the revised circumstances and their discontents abate. Forging a tacit agreement to stop glaring at each other they begin to peruse possible combinations of material already gathered. Obviously the piece as it now stands will of necessity be peripheral to Grasshopper. Everyone but him had willingly participated in the interviews and all had proven rich sources of information and color.

To their dismay, only one shot clearly portrays the little man and they curse themselves as again and again they play the abysmally short segment, watching the subject of the piece interminably holding Hellbender, scratching behind the hound's ears as he absently listens to Kristin's account of depredations out along the freeway.

A first rough cut of the essay surprises them both. More than two hours of material survive their first critical perusal of possibilities. Al thanks Providence repeatedly that Charles du Lac has been at the camera; even the most ordinary sequences seem visually alive, holding the viewer's attention in sheer power of rich imagery and skillful framing.

Their first cut ranges from Clara Abbot's frenzied exhortations against the Antichrist to Chloe's rapturous and breathlessly lucid account of the healing of Grady Abbot. They intersperse airborne views of hordes laconically trudging along the freeway with file segments gathered from ConRelCo archives. Details already known about the attempt on President Crown's life are punctuated by Chloe's account of how Grasshopper saved his life from some mysterious scarred-face assailant and end with impromptu hand-held shots of Ira Crown commandeering their steproller.

A few electronic tricks manages to extract and isolate Grasshopper's face from the brief segment they have of him, augmented by excerpts taken from old Ant and Grasshopper routines. Dim in the pale light of the motel room, faintly glowing in warm illumination of a nearby lamp, his demeanor is distinct in apparently serene contentment despite haunted, sparkling eyes. His presence seems to pervade the entire piece. The fact that he was rarely in a shot seems hardly important and even heightens the ambiguous nature of his exploits recounted in the narratives.

After a day and a half of revising and editing down to absolute essentials, an hour and twenty-seven minutes is available for Mel to preview. Their triumph is even more sublime in that it has been wrested from what at first had seemed merely an impossible conglomerate of unrelated sequences. The success they feel grows even sweeter when after a moment of silence Mel whistles approvingly and bemoans the fact that he has signed away the rights. He reminds them that their individual contracts with the company still provide that ConRelCo holds a first option on any independent productions.

Through Mel's good offices the essay is previewed and consigned to the ConRelCo central archives; a preliminary scheduling of the essay has been booked by ConRelCo's subsidiary, *Continental Broadcast Corporation*. Wendy Iverson has even called Al and told him to stop wasting his talents blood-scraping for ConRelCo and come work with the elite at CBC.

Then at the last minute the piece is bumped into a late night slot where it is likely to be ignored. Ironically The *Makewell Kisser* has been replaced in the schedule by a special news report prepared in part by Al and Charley for ConRelCo. President Crown announces the displacement of Generals Michaels and Flach from active command and details avenues of investigations being pursued. Crown then celebrates measures taken to restore integrity and order to the higher echelons of the military and within his own administration.

At midnight the two veteran newsmen open a couple of beers to watch their offspring leap into the world, but soon switch it off. Suspecting that they are the only ones in the country still up watching teleview, neither feels up to the task of cheering the other. Their moment has passed; *The Makewell Kisser* will reside forever unnoticed in the labyrinthine banks of the ConRelCo data lattice. They get very drunk and talk over old times, comparing life before catastrophe with a probable future and eventually passing out.

Next day they exchange condolences for hangovers, tell Mel they didn't think they'd be retiring after all and then turn their attention to finding and interviewing a notorious gang leader from Brooklyn who now is actively combating the federal agents from strongholds in Newark and East Orange, New Jersey.

Enough response is elicited by the late night run of *The Makewell Kisser* to persuade CBC programmers to reschedule and boost it with a dollop of preliminary publicity. All the old propaganda identifications with Grasshopper and Ant combine with suspicions unleashed in the subsequent revelations of government corruption and ineptitude. A public hunger feeds on bitterness engendered in transition to austerity and banqueted in terror, deprivation and desperation of dispossessed millions. Tribulations of the past months echo through oceanic depictions of ongoing distress. Critics and viewers alike find the video essay "reassuring, heartening, both incredible and poignant." CBC buys an option for a third showing in prime time, building a promotional campaign to link it with other documentary features.

Alfred Statler and Charles du Lac can now approach Mel and ask if he wants to make counter offers against several tempting proposals from rival news organizations. Mel is flustered and desolate in demands for explanation that have descended from above, wanting to know why ConRelCo does not own a documentary created within their own news division. He asks the two newsmen to put in a good word for him at some of the rival organizations.

Within a week nationwide furor erupts over what has become of Grasshopper. The image of him standing, holding Hellbender as he placidly gazes into the camera seems to pop up everywhere on teleview, in magazines and on posters. Wealthy invalids, seeking an end to their ailments, post rewards for information leading to an interview with Grasshopper. Some obscure religious sects hail him as a benevolent extraterrestrial or as the Returned Redeemer. Others side with Clara Abbot that he must be the Antichrist. Major religious organs decline comment and disclaim any intimations of divinity in Grasshopper's celebrity.

Look-alike contests are sponsored across the country for people with dogs who resemble Grasshopper and Hellbender. One state legislature briefly entertains a bill proposing that the grasshopper be named the state insect until someone comments upon its resemblance to the hordes of locusts that have just recently destroyed much of the local wheat harvest.

Statler/du Lac Productions generates more video essays detailing experiences of refugees, soldiers, farmers and many other individual tribulations and triumphs born of upheavals rending the land. Although well received, none ever achieves the widespread magical appeal of *The Makewell Kisser.*

Economic and social lamentations are shunted and swept aside. Novel speculations boil through what might be transpiring, manifesting most intensely among faithful believers, even as confirmed skeptics seek to debunk outrageous fraud. Whether yearning for comfort and release or hoping to expose charlatans, the mission coalesces among diverse constituencies: Find Grasshopper.

~ 21 ~

Grasshopper awakens. His eyes are not yet open, but he knows that he is awake. In a drowsy gray haze issuing from beyond his closed eyes there seem to be creatures moving idly about. He listens to occasional rustling of tall, autumn dried grass. Sporadic, complacent moans, some near, some farther away, indicate that he is not alone after all, but neither does he seem to be of any particular interest to the creatures meandering beyond his tightly closed eyelids.

He keeps them closed. Let sleeping Grasshoppers lie! He likes it there inside the vague gray space wrapped around his head. He doesn't need to see anything or anyone. It's not yet today until he opens them—it's just now. It can be now, forever, if he chooses.

His lids agree and remain closed. Only the left eye seems mildly curious about peculiar noises populating the void beyond the shuttered lids. But it isn't about to go out there and investigate alone. Vague conjectures seem acceptable for the moment.

As drowsiness glides into resolution against waking up, Grasshopper senses that a strangely meandering, shuffling kind of movement is happening inside as well. He finds he is not alone within the envelope drawn as a scrawny boundary between two perplexing arenas of vitality. An anguished nausea echoes from somewhere deep within as he struggles to locate and fend off an inner intruder.

He is slippery. He will not be forced into an open confrontation. Waiting. He's just waiting.

Grasshopper squinches his eyelids tightly together as if to pull the encircling boundary even more firmly about himself. It helps a little. His fingers quickly locate the switch to activate the iLink. Has he shut him out? Yes, a bolstering flood of iLink-enhanced awareness warmly pervades his sentience, blending local smells of poultry and dry grasses with distant vistas of cached memory stores. He hears, too, that Hellbender is bored, wanting to explore. The Ant beacon, so long vacant, still achingly polls for response . . . where did his contrary brother vanish to?

Maybe he should not have left the others. They might have known what is happening and how to stop it. They seem always certain of what is happening, though often caught by surprise. The Jason Hood, especially—so often wrong, but never in doubt! Like, where things come from. Where they are headed. Such Ant-suck.

They live such confusing lives. He can bear their mulling wiles no longer. Memories of the Ant-eye antennae probes his recollections of cooped anxiety. Leave me alone. I'm a rock, now. I don't want to be anything but a rock. A here. A now.

His eyelids painfully protest their effort in staving off the surly intruder. He relaxes them and a throbbing orange-redness fades back into familiar glowing gray nothingness. The intruder seems to have retreated. Or is he just waiting? Grasshopper grows annoyed at such presumption. A rock does not bother anyone! Why cannot a rock simply be?

Maybe the other is an Ant! His annoyance resurges. Ants! Everywhere Ants! How can he be so alone?

He listens again to the local noises rustling dry grass. Perhaps the other will go away if he knows there are others beyond. But the ruse fails . . . the opposite seems to happen. Listening to outer noises seems to evoke the intruder, to excite him. Grasshopper squinches his eyes tightly again and the movement within slowly is stilled.

Rocklike. Only rocklike. Here. Now.

Then he knows why the intruder! Ant's Sister. She evoked him. Grasshopper grins wryly. You'd just as well go away, I'm here and rocklike, she is there and then. She'll never find me among the quadrillions of rocks being still. He feels a ripple of sadness down deep and knows that he has won the moment.

Just a rock here on a ditchbank, basking beneath a sun riding its transit from cold to warm, here I am.

No. Some feminine voice speaks quietly, but firmly: No.

Who are you? Grasshopper sneers, refusing still to open his eyes. Startled at secret effrontery, Grasshopper switches off the iLink but the feminine presence persists.

Ditchbank, comes the reply. So long as you lie here, I share your every thought. All our moments are intertwined, you know.

I am a rock, Grasshopper counters, alone and apart.

No, replies the ditchbank. You are only a silly child!

I am Grasshopper!

My, you abandon rockhood so very easily! See what a silly child you are! The ditchbank laughs gently.

It's a trick! Grasshopper rails intently as once again he tries to stave off unwelcome intrusion by squinching together his eyelids in painful fastness.

How can truth be a trick? You trick yourself, you silly child. I know only what lies here and beyond. I know that you are no rock . . . only a silly child!

I am Grasshopper!

That may be a name, but you are neither rock nor grasshopper. I share moments with them as well.

I am Grasshopper!

So be it, you stubborn, silly child!

A terrible screeching erupts just beside Grasshopper's ear. Yelping in terror he leaps to his feet, eyes popped spontaneously awake as he frantically scrambles away from the red and brown flapping thing lunging toward him. A brown leghorn rooster!

He howls mercifully, relieved in gales of laughter, then flops heavily back down on the other side of the ditch. The rooster, satisfied of the confrontation's outcome, puffs out his breast, flaps his wings and crows another ear-strobing peal.

A few hens browse about the hay brown grass, clucking happily, unperturbed as well as unimpressed by the cock's lordly bluster.

"Okay," Grasshopper gives in, "so I'm not a rock."

The ditchbank does not answer; after a pause he mutters softly, under his breath, "But I am Grasshopper!"

Then he seems on the verge of some startling realization. A flash of a perspective seen from the other side. An idea clothed in raiments of memory and imagining spun through his inner eye. It dances on toes of an annoyed grievance he holds against nearsighted construction. Against silly encumbrances placed upon the cosmic rhythms of the world.

The ditchbank is telling him another secret! To the discomfiture of Hellbender's blissful repose, Grasshopper rolls over abruptly to place his ear firmly against a grassy clod of earth. His arms stretch outward, spread-eagle, to pull himself closer to the source of faint whispering.

In fierce concentration he sees and hears and smells ancient rolling, loam-rich woodland floor where Ants always have cut their wheedling furrows and carved scabrous concrete gashes into its skin.

They work in ignorance of other motives, the ditchbank whispers. Their motives intersect in the scratch of little lines webbed on paper to score asunder the blind and uncaring hum and stealth of flies, mosquitoes, sweat and disease.

"That's nonsense," Grasshopper grumbles aloud, "they do it to annoy me."

Such arrogance, the ditchbank chides in musical ease. No, they are only Ants, restless in their swarm. Ants give no thought of Grasshopper, nor of hoppening. They do not care. You recoil from them as if from death—and they fear you as a harbinger of chaos and madness.

"They are mean-spirited and evil!" Grasshopper insists. He wishes he knew how to elevate his inner voice to the sweet persistence of ditchbank's. It sounds so sophisticated!

No, Ants are merely nearsighted, merely working with what is before them. Their reasons are but pale echoes of true motives ordering the universe.

"Their true purpose," Grasshopper tries a philosophical tenor, "is frustration of the natural course of things—me, for instance! Strip away their stupid pretensions and they are empty. There is no excuse for such stupidity!"

But that's silly, the ditchbank laughs softly. Ants care nothing about you nor anything you might bring to this world. And even you cannot see so clearly as to envision the range and depth of all there is. If you could, you would already have seen that you and Ants are brothers, that there is too much diversity for just one kind of creature to carry.

"If that is so," Grasshopper says in a hoarse whisper, "then their narrow pretensions would be true and all the rest hopeless —I do not accept that!"

Not so, laughs the ditchbank, you are something that Ants can never be, just as they are something that you now despise —because you do not know them. Neither you nor they own anything but your own choices within the universe of possibilities. You each are part of it. Besides, you lived as an Ant once. But you have grown fearful of your own origins.

"That's a lie," Grasshopper draws back his head and hisses into the grassy knoll, "I was never an Ant!"

But you were. You shall be again. For now, you are Grasshopper. By chance you are caught in currents changing courses through the motives of Antdom.

"It's a lie!" Grasshopper reiterates.

Silly child! You are now what they cannot be, but you cannot long separate yourself from them. You would only die. Then someone else would have to carry your energies and do what you will not.

"Do what?" Grasshopper demands, distraught at the prospect that something Antward might be expected of him.

Be yourself—as you must be, whether you accept it or not! the ditchbank laughs musically. Everything in the universe bears toward the same course: given its circumstances, everything discovers what it might be. All is was may be.

"Aliswasmaybe? But what does that mean?" he moans haplessly, hoping that the ditchbank might yet answer several mysteries that have troubled him for as long as he can recall. How are children born in Grasshopper resilience gradually transmogrified into rigid Antness? Is Peter Dooley only an Ant? Is he the one Ant's Sister pursues in her relentless obsession about his ways.

But the ditchbank has fallen silent.

"What does it mean?" Grasshopper demands as he rises angrily on his forearms to glare down at the grass in which the packed impression of his head still can be discerned. "Answer me!"

He returns his ear to the ground, expecting a reply. Hellbender opens a mildly curious eye to watch him rail at the obdurately silent patch of ground.

"What did that mean?" Grasshopper fumes at Hellbender. The hound lifts his head, seems about to answer, then thinks better of it and starts gnawing at the fold of skin stretching from his right hind leg up onto his belly.

"Stupid dog!" Grasshopper berates his companion. Rolling over he leans back against the slope of a knoll extending up and away from the ditch. He glares sullenly along at the strip of shattered pavement rushing east to west and back again.

"Stupid ditchbank!" his surly voice and mouth twist in grim chewing over suspicion that it has all been a trick of some kind. Nothing may be trusted where Ants have tread. Maybe they have turned the earth and planted a traitor when they built these roads!

He vows to prove the ditchbank wrong. He is Grasshopper and only Grasshopper. From beginning to end he will be Grasshopper, alone. Just hoppening.

For a moment he stares in resolute silence at the contentious ditchbank, waiting for some hint of contrary voices. Hellbender's moist nose snuffling about his left ear finally gains his attention.

"What a dog," he finally sighs, patting the hound cheerily, as if to apologize for calling him stupid. Recalling events in the warehouse on the previous evening he compliments his buddy, "What a performance! Where were you in the old days?"

Grasshopper leaps to his feet, his hands grasp locks of hair on either side as he again squinches his eyelids together, fighting back against the intruder. He had said that! A reverberating echo of an almost remembered time plays just outside his grasp. The old days. What are old days?

Hellbender attends his struggle in mild curiosity. The rooster poses a greater menace in a hound's reckoning.

Grasshopper again opens his eyes carefully, a little at a time, ready at the least hint of any resurgence of that inner voice to snap them shut again. Hellbender hassles lightly, his lips and jowls draped in a good-natured smile. The hens are scattered along the ditchbank and across an overgrown yard leading back to an abandoned building that once had been a barn. They cluck and squawk in indolent perusal of thickly matted dry grass. The rooster crows again as Grasshopper sits reflectively back down on the ditchbank. His fingers slowly relax their painful pulling of his hair.

Performance. Grasshopper looks carefully at Hellbender. Repeating the word slowly empties it of significance. It becomes a dried husk to fall back into old days long gone.

Suppose the intruder is an Ant! He shudders at the thought of Ant hordes pouring out of secret places. Sadness wends about his throat and into eyes moist in realization that something has been lost. Something that can never be retrieved and held for knowing and holding because it lies across unbridgeable separation.

He looks again at Hellbender who scratches at an intruder of his own. The dried hull of the word again puckers Grasshopper's lips and slips across his tongue as he silently mouths it: per-form-ance. Repeated, it shifts slightly. Per-fume-ants. Out of the corner of his eye he sees a hen jump. He watches the chicken's head lift abruptly from a sheath of grass. In her beak she briefly holds a small grasshopper. With a quick snap of her head thrown back it suddenly is gone. Inner desolation sweeps a dry wind across his lips still maundering permuted syllables: con-sume-ants.

The rooster trumpets a shrill scream.

Eager to avoid further Antward encumbrances, Grasshopper heads south at the first available intersection they reach. A northward trek seems inadvisable. He knows that winter soon will catch him in its miserable retributions and validate all those hateful Ant maxims. A stitch in time saves nine. Possession is nine points of the law. He who hesitates is lost—but look before you leap. Hah! Grasshoppers would never get anywhere if they listened to Ants!

After walking for most of the day, during which he and Hellbender encounter no other beings taller than a cat, he realizes that somehow he still is hostage to circumstances beyond hoppy satisfaction. His stomach begins to rumble, discontent in emptiness. Hellbender, too, seems to have gained a greater urgency in his olfactory perusal of roadside items, as if expecting eventually to discover a plate of table scraps. Perplexed, Grasshopper sits and then eventually lies outstretched on the ground, awaiting some sense of what he might do to elude this last Antward concern.

An intersection of two highways divides the landscape ahead into quadrants of gold, bronze, silver and copper—all glazed in an amber haze of late afternoon. The casually paling warmth of the September sun teases the coolness of his skin and plays in a debate of daydreams across his closed eyes.

There is no hurry. A few moments. A few days. The sun will remind him of their gradual retirement southward. He snoozes lightly in drowsy reverie with Hellbender lying close alongside, his head resting on Grasshopper's chest.

Relaxing still further, he allows contentment to expand lazily in his mind until its blowsy happiness extends all the way from Lancaster back to Philadelphia, then forward to Harrisburg and beyond, where Ant signs instruct highways to travel.

Careful to include every square inch of both ditches in the nodding, slumberous blessing, he tentatively stamps in docile approval teeming legacy of unknown legions of Ants who have left so placid and auspicious a place for a Grasshopper and his companion to rest.

Then he recalls the ditchbank's admonition to be himself . . . Grasshopper?

He is disheartened by what seems an inescapable dilemma. If he does as he wishes, he will succeed only in accomplishing the ditchbank's counsel. He does not wish to follow the prescription of anyone, much less a plot of Ant-carved earth! But in this case it seems he must not do what he must do to escape. Yet the only alternative is to not be Grasshopper which surely will thrust him back into Antness! It is just too much for a poor Grasshopper's brain!

In the bleakness of his sudden melancholy even the sun seems suspect. Its treacherous benefice teases his skin into an unfamiliar boundary, a distantly warmed periphery of where he sits within its circle, brooding through ditchbank catechisms. The preoccupation smacks of timelessness. Hours course and swell about the remote contours of his skin. He sits immobilized in a vacuous aftermath of what seems to be a cosmic prank. He once was a prankster . . . he thinks. Or had it only been standup cosmic?

Then suddenly it is twilight and a tall, impeccably dressed fortyish man with ruggedly handsome features is aggressively shaking him out of the trance. The stranger's voice combines both concern and buoyant cheer. "Hey! Can you hear me? Young man. Are you all right?"

Grasshopper stares blankly for a moment up into the man's face bobbing vigorously up and down, back and forth within his field of vision. Then he stiffens involuntarily and shrugs off the man's two enormous hands. A shudder ripples chill through his body. He suddenly is desperate for warmth.

"What?" he shivers blankly. The man seems gratified to have elicited a response. Hellbender sniffs cautiously at the stranger's expensively tailored suit pants leg. He is tastefully dressed in beige silk suit and dashingly complementary taupe tie. On his feet are tan shoes elegantly cut of smartly shined Italian leather.

Grasshopper shakes his head to clear it and to fend off more shudders. It is so cold! He looks again into the man's hopefully grinning countenance and then beyond to where a vintage, well-groomed white Cadillac waits. Its idling engine emits deep-throated humming. A powerful fragrance of after-shave mingles with faint whiffs of automobile exhaust. Grasshopper sniffs anxiously at such Ant signs.

"Can I give you a ride? Are you okay? Have thieves fallen upon you?" the voice erupts again in a profoundly practiced meter of accomplished oratory.

"I suppose so," Grasshopper mutters, still confused by the onset of chill and attendant loss of daylight. His memory offers no clues. The sun seems simply to have disappeared. Its goodbye haze lingers on the western horizon. He looks again at the large person looming before him as if there might be clues to the missing afternoon.

"I guess so," he repeats dazedly.

"Thieves have fallen upon you?" The stranger looks about in apprehensive dismay, concerned that trouble may yet be lurking about. The large hands tighten cautiously into fists.

"No. No thieves."

"Oh, good. Then you must mean either that you are all right or that I can give you a ride." His booming laugh announces relief at dispelling any threat of robbery. "Either way, I'm very pleased to meet you."

He watches Grasshopper for a moment as if expecting some plausible reciprocation of amenities, but then smiles in paternal indulgence as he looks down at Hellbender. "That your dog?"

"I guess so," Grasshopper replies absently. Then realizing that his mouth keeps saying the same thing he shakes his head vigorously and adds, "His name is Hellbender."

The man booms forth another peal of laughter.

"That's a fine Christian purpose," he replies in amusement that such an imposing name has been attached to so docile seeming a creature. After a moment's reflection he announces as he extends his hand in greeting, "My name is Wally Martinsdale."

His eyes anticipate recognition but Grasshopper thinks the extended hand is meant to help him to his feet. He pulls abruptly and the two of them nearly roll over into the ditch, saved only by the inertia of the man's enormous frame. Grasshopper springs to his feet, unaware that Martinsdale's face flickers in dual pique at not being recognized as well as being nearly yanked off balance.

"I'm Grasshopper."

The man's large hands absently rub themselves as if lathering absent soap. "Is that so?" Martinsdale reflects, shrugging his broad shoulders back into tailored contours of his jacket. "That's an unusual name."

His voice has fallen suddenly into impatience and disinterest. He glances back toward the waiting Cadillac and frowns as Grasshopper follows him back toward the waiting automobile.

"There's nobody hurt in there?" Grasshopper inquires as he peers suspiciously inside the vehicle.

"No," Martinsdale says testily, "nobody is hurt that I know of."

"Good," Grasshopper replies, "I've had it with healery."

Puzzled, Martinsdale sighs and opens the door. A charitable offer has been tendered, however reluctant he may now be to honor it. Watching Hellbender leap through the back door to settle into the deeply billowing soft leather upholstery, he asks fussily, "Is he housebroken?" Martinsdale's voice takes on a lightly protesting whine.

"Housebroken?" Grasshopper repeats uncertainly. "He won't hurt anything, if that's what you mean."

Martinsdale's head vacillates between nodding yes and shaking no. "Just so he doesn't get anything dirty. I love dogs, but dirt is Satan's victory emblem, you know."

Grasshopper plops into the front seat and waits for his benefactor to find and adjust various switches for lights, heat, radio, mirror, antenna and rear window defogger. Finally the gearshift is shoved into drive and the car leaps eagerly forward.

"Going north, I hope," Martinsdale peevishly appraises Grasshopper's overalls and tennis shoes, then looks back into the mirror to reassure himself that the dog is only sleeping. Finally he sighs again and apparently resolves to make the best of whatever inconveniences may gather. He laughs drily, "Hellbender! I really like that. I'll remember that for the next Crusade—some theme like, ah, 'Hellbenders versus the Hellbent'." He chuckles at the directness of the message and savors the fullness of its sound.

"Crusade?" Grasshopper repeats. It sounds awfully Antward.

"Yes," Martinsdale confirms, "we'll be in Pittsburgh for the Christmas Crusade." He glances again at Grasshopper and then breaks into revelation of white pearly teeth as he discerns that his own celebrity has not been recognized. "I'm Wally Martinsdale. Reverend Wally Martinsdale. Of the *Brotherhood of Christian Arms*," he smiles benevolently, "at your service."

"Brotherhood of Christian Arms?" Grasshopper seems puzzled.

"Yes," the Reverend chortles, "hugs for the good; slugs for evildoers. We're Christian Soldiers all," Lustily he launches into a snatch of the hymn *Onward, Christian Soldiers!* His right fist marks the tempo as his left uncertainly holds the Cadillac within its paved course, "At your service."

"That's very good of you," Grasshopper says mechanically, wary of the suddenly effusive tenor in the Reverend's voice.

"Good of us?" Martinsdale repeats tentatively as the automobile accelerates powerfully out of a curve in the road, "Yes. I suppose it is," He seems to find the reflected comparison rewarding. "like the Good Samaritan."

"I suppose so," Grasshopper offers as the roadside speeds precariously past.

Wally Martinsdale is at best tentative in his mastery of the automobile. When the vehicle is held in the right lane, it tends to veer dangerously off onto the shoulder, causing some very unnerving moments even for someone so resilient in their concerns as Grasshopper. Driving down the left lane, however, invites disaster even on a highway so little used as the one they now travel. In compromise, the Reverend tends to straddle the center line, the Cadillac's wake oscillating ponderously between two parallel ditches. The stratagem works fairly well, although each turn in the road suggests an emphatic possibility of head-on collision. He seems to find one lane good for eating up miles but the other better for rounding curves.

He drives in benign confidence that prevails over any intimation of disaster. Surely any vehicle traveling south will have been pulled over by angels to replace a flat tire, refuel or even to relieve the driver's bladder. Any local purpose will serve to clear the way ahead of the sonorously humming Cadillac as it speeds ardently into the Good Work. Clearing the way is God's Purpose so the good Reverend drives relentlessly, if somewhat unsteadily, into its ordination.

During their progress through the night, Reverend Martinsdale gradually forgets his earlier misgivings in which he was initially tempted to leave the pair alongside the highway. He directs Grasshopper to open a large cooler in the rear and to help himself to the bounty of foods it contains. Cold cuts, boiled eggs, cheese and cucumbers. Grasshopper feeds himself and Hellbender while the Reverend's booming baritone roves through a panoramic vista painted in current concerns.

Chief among his worries is the Good Work now being launched, trying to contend with enormous refugee populations gathering across outlying reaches of the eastern seaboard. Such numbers are overwhelming relief centers that surround major centers of urban distress. He complains in hearty despair that even the combined resources of government fiat and Christian charity his and other organizations are able to muster seem to spread so thinly across a need of such vast proportions as to seem pointless. If only there could be a miracle of fishes, he sighs in sad resignation.

It breaks his heart that Christmas is only three months away and yet so many children must trek in disappointment through a season of disenchantment into a New Year that all pray will bring reprieve to a nation in tribulation.

It is sad, he mourns, that so many innocents must suffer deprivations of economic disaster. We must rally, he asserts vigorously, in the truly courageous spirit underlying the Christmas story. The Christmas Crusade in Pittsburgh will provide such a rallying cry, he muses hopefully.

Pittsburgh has been one of the few metropolitan areas to escape most of misfortunes visited upon other large cities. Some have said that the city's good fortune results from the fact that it is only concrete, glass and steel and simply will not burn. The Reverend, to the contrary, thinks it to be an indication from the Almighty that Pittsburgh has been spared for a reason. Some Divine Event, a Sign, shall transpire in Pittsburgh. He is convinced of it—and plans to lend a hand to its advent.

Their immense *Double Bubble Super Sports Arena* will be the site of a dramatic revelation to the world. It will originate a broadcast by satellite hookup to blanket the entire globe in revelation that God's benefice and blessing are still available, in demonstration that while America may have faltered its core values still are sound and virile in opening new works of prosperity and goodness. The good people of the earth have only to wait and watch for the signs, he extemporizes.

In the December Crusade, the eyes of America, of the entire planet, will gaze upon a city that has been spared. The good people of Pittsburgh have not allowed their city to sink into depths of depravity and debauchery now consuming New York and Los Angeles in retribution. Why, the Reverend admits, there may have been a little hanky panky here and there, but on the whole Pittsburghers have devoted themselves to the good works of steel manufacture, robotics and photonic computing, medicine and steak sandwich franchising. Their architectural zeal has wrought the spectacular Double-Bubble Super Sports Arena, a large dual spheroid amphitheater and sports palace intended to be more than four times the size of the old Louisiana Superdome in New Orleans. Its cavernous immensity will be the site and symbol of the Christmas Crusade. A universe of worldly pleasures may be large, the Reverend muses aloud, but the place of the Holy Spirit is ever greater!

In December, during the Christmas season, the eyes of America and the free world will focus upon one incontrovertible fact, the Reverend suddenly thunders: God does not tolerate Sodom. He does not permit Gomorrah to prosper. Babel is confused, her children are dispersed to the winds, traipsing through ignorance of their thirst for the Renewal of the Covenant. A Covenant resolved, maintained and guaranteed in renewed dedication to the principles that once wrested a great land from the wastrel grasp of naked savages, ferocious beasts and Frenchmen.

In Pittsburgh the true meaning of Christmas shall shine forth as a beacon of our founding fathers. As their beacon was founded in lights of the Old Testament prophets, so shall the new beacon shine forth revived mandates of old . . . go down into Jericho and waste their cities, destroy their children that My children may inherit their land!

The Reverend's voice trails off as he awkwardly wrestles a centrifugal force threatening to catapult the Cadillac from a hairpin turn in the road. He considers for a moment, then inquires aloud, "Is it just coincidence that Jericho rhymes with Mexico?"

Again his voice thins, paling into silent reflection as the car heads into another tight curve. It emerges into a new stretch of straightness as he grins, "So it was in the olden days. So it is now. And so it shall be."

God's plan, the Reverend assures Grasshopper, is now becoming manifest. Infinitely complex and of carefully wrought design, one of the threads of its weft had begun billions of years ago—God had commenced long before that, of course, but for all practical purposes one could figure the beginnings with the dinosaurs. "There's a lesson in that," avowed the Reverend Martinsdale. "I'm going to have that written up, too."

"Dinosaurs?" Grasshopper echoes, intrigued with the sound of the word.

"Yep. Dinosaurs. Why, don't you see? If people had been living back then . . . say a group of university professors—they would have wondered why all of those useless creatures were hanging around. They would have run studies and taken surveys and then concluded at some conference that the dinosaurs weren't coming to very much and that therefore the world showed no evidence of planning. Therefore, they would announce, there could be no Divine Plan . . . dang'd atheists!"

The Reverend thinks it over for a moment, then adds, "Your average, so-called intellectual doesn't have much faith in anything—he wants everything proven in his own terms and to his satisfaction. Just nearsighted, that's what it is," Wally Martinsdale avows.

"Nearsighted," Grasshopper perks up at the word. He has no idea what the Reverend is talking about, but he does know of nearsighted creatures. "Nearsighted. Like Ants."

"Dang'd right," the Reverend agrees vehemently, "just like a bunch of meddling ants—mindless, faithless, blind to His Purpose. They think they know everything and can invent more!"

"Dang'd right!" Grasshopper avidly chimes into disdainful sentiments. Now he appreciates the Reverend's fervor. The Cadillac edges perilously near a pavement edge, then is absently guided out of doom into a new oscillation of recovery from tragedy.

"Sinful cynics," Reverend Martinsdale accuses unseen social hordes of faithless, "but we'll show'em, won't we, son?"

"Dang'd right," Grasshopper says ardently, "They're gonna learn about Grasshopper. And dinosaurs, too, I guess."

"Dinosaurs," Wally Martinsdale draws out the sound of the word as if tantalizing someone's ear. "Your average, so called intellectual can't see that God made dinosaurs into oil so Texans could take Texas from the ignorant Indians and the idle Jerichoans . . . I mean Mexicans . . . and discover oil to make the money to build Houston and then Pittsburgh so that the Christian Brotherhood can hold the Christmas Crusade to reveal His Purpose and Intention to all the world. And they can't see evidence of a Divine Plan! Now that's nearsighted!"

"Dang'd nearsighted Ants," Grasshopper hurls ridicule toward such fatuous creatures.

"You can't hurry God," Reverend Martinsdale announces in humble awe, "He'll make his point when He's ready. It makes me proud to be a Texan! To be part of his Divine Plan. We're gonna thank God during the Christmas Crusade for having the dinosaurs make all that oil—you can bank on that!"

"And Grasshoppers, too!" Grasshopper enthuses, not wanting to be left out of so propitious a rebuke of Antsuck.

"Grasshoppers, too?" the Reverend looks blankly at him, then has his attention snatched back to the road by a sudden pounding of tires running off the edge. He heaves the vehicle back onto pavement as Grasshopper peers into the back seat to check on Hellbender. The hound has fallen fast asleep; his muzzle lies resting on his forelegs, his black nose looks like a dimple against the dim sheen of the leather seat.

"Don't get me wrong, though," says the Reverend as he sighs deeply, "we realize the times change. Evangelism has to change with it. There are some in the Brotherhood who would carry around the Ten Commandments on stone tablets, just because that's the way Moses received them in the Old Testament. That's fine, I say to them, but how you ever gonna get that message through the U. S. Mail?" He laughs merrily at the witticism and glances to fathom whether Grasshopper appreciates his stab at such straitlaced conservatism.

"I tell 'em God's Word never changes, but how we tell others about it has to—I mean you've gotta get a person's attention to save their soul. Always have. Always will.

"Used to be you could say 'hellfire and damnation' and people would shiver and quake for salvation right on the spot. My daddy used to herd'em into Heaven with nothin' more'n a Divine Promise of Eternal Life and scorching warning of the fire'n brimstone alternative.

"It's different now," Wally Martinsdale waxes nostalgic as he shifts the position of his right arm to lean into the steering wheel. The Cadillac's tires whine unsteadily at the left edge of the pavement. Slowly he coaxes them into a new course toward some distant point fleeing along the right edge.

"It's really different, now. Why, you ask about somebody being hurt. We're all hurt. Heck! If just my daddy was still alive I'd have some terrible explainin' to do!" the Reverend agonizes aloud, "I'll declare—sometimes it even seems to boil down to who has the prettiest girls in the chorus. Now don't that beat all? My poor daddy's up in heaven ragin' hellfire and brimstone at that, but what're you gonna do?" Reverend Martinsdale twists his mouth back and forth to ponder conundrums of modern evangelism.

"Get the message across," he affirms plaintively, as if instructing some unseen subordinate stranded out on the automobile hood. "That's what I always go back to . . . the message. High impact, low content—heavy on the visuals and local color. And keep 'em moving! You can't do it without visuals! That's all anyone tunes in now. And a pretty girl's smile works as well for God as it does for Mammon." His ebullience suddenly drains into distress. He glances sidelong at Grasshopper and smiles in brave resignation. "Times really change, though, don't they?"

The Cadillac approaches an access ramp leading up onto a four lane bypass. A battered, barely legible sign promises that Highway 422 is somewhere beyond. The ramp is broken and potholed, barely passable.

"This is where I turn," Reverend Martinsdale announces, "do you want to get out here, or ride on with me? Harrisburg is on up ahead that way." He nods in the direction away from his destination.

"I'll ride," Grasshopper replies, his voice betraying dread of Ant streets and buildings. "I'm never going back to a city again."

"Country boy, huh?" Reverend Martinsdale exclaims as the Cadillac turns hesitantly up the inclined ramp. He seems encouraged at Grasshopper's displeasure with urban life. "I don't blame you. I was raised in the country myself. God may have given us Houston, but first He gave us Texas to put it in and Texas is country—God's country!" His laugh booms explosively against the windshield. "My daddy had a little ranch," he reminisces, "I'd give anything to be back there right now."

"Why don't you go back, then?" Grasshopper empathizes with a note of longing that has crept into the Reverend's voice.

"Well—for one thing, it's not there anymore. An agri-corporation bought it. Tore out everything my daddy and his daddy had built with their own two hands. Now they clone beef there.

"Used to be they'd raise cattle. Cowboys would ride and nurse those scutters and hate 'em and sing to 'em and live with 'em. Now they just keep 'em in a stall, injecting hormones and vitamins into a barely living leather bag stuffed with seaweed derivative—where's the romance in that? How can you write a song about that? It's pathetic. I can't eat beef anymore—I've seen the hospitals where they make it. Turns my stomach!"

The Reverend's voice becomes taut and fidgety, as if he fears collusion with something reprehensible. Smiling gravely, he reaffirms, "Times really change, I want to tell you. Sometimes I think we just need to get back to the simple, grassroots virtues."

His voice trails off into reverie. After an interim of silence in which the Cadillac's engine has lulled Grasshopper into a drowsy meditation of his own, the Reverend suddenly inquires, "What do you do?"

"Do?" Grasshopper replies sleepily. The question smacks of precipitous obligations. "Travel with Hellbender."

"Where're you going?"

"Away, I guess."

"Away? From what?" The Reverend envisions myriad scenarios ranging from domestic troubles to scrapes with the law.

"Ants."

"Ants?" the rich baritone repeats quizzically. Then he laughs lightly, "Oh. Ants. I see . . . that's funny."

"There's nothing funny about Ants," Grasshopper assures him. "They're everywhere. They've taken over everything—especially cities!"

The Reverend glances at his passenger, probing his suddenly anxious demeanor. "I take it you're not referring to the insects that burrow down into the ground and make life miserable for country folk." As a boy the Reverend had on several occasions suffered confrontations with disturbed hives of angry fire ants.

"I'm talking about Ants that make life miserable for Grasshoppers—especially in cities run by police!"

"Are you in trouble with the law, son?" the Reverend inquires. After all, have not all good Christians been instructed to visit the downtrodden, the dispirited, the malingerers in prison and bring them into the Light?

"The law? Antsuck! They're in trouble with me!" Grasshopper declares in stolid fervor. He glowers out into the night, watching occasional vehicles approach in the distance, paired spots of light that grow larger and larger until suddenly flaring into manifest substance of a truck or even an occasional passenger car that noisily zips on by to disappear into receding darkness.

"A lily of the field," Reverend Martinsdale observes cheerily.

"A what?"

"'Consider the lilies of the field," quotes the Reverend, "they toil not, nor do they spin.' That's what you are—a lily of the field."

"I'm Grasshopper," his voice reeks disdain.

"Oh, I see. Well, I suppose it goes for grasshoppers, too."

After a moment's reflection, the Reverend says, "You really travel around as you please, huh?" He pulls his head to one side as if considering some long ago fantasy. "Well, well. That must be quite a life. I think I'd like that," he muses aloud.

"But you can't spread the Word like that anymore," the Reverend observes sorrowfully, "people just think you're a bum and won't listen to anything you say. I preached a sermon, years ago in the early days of my teleview show. It was about Jesus and how today he'd be out living among the homeless, helping people right on the spot where they need it. The ratings were abysmal. People don't care about that. They want to hear about how good it's gonna be for the middle class!

"If you want to keep people's attention you have to preach about things that hit'em where they live. My production people make me wear three piece suits all the time—and drive this Cadillac—because it's a part of the image. Impress people, they tell me, and people will listen to you. This is just my way of helping God remind people to read their Bibles and listen to a sermon once in a while, otherwise, they only watch World Class Wrestling or jangle themselves into oblivion on StimuVision crap.

"No. Times have changed. You can't be a bum for God anymore. The prophets are all dead. People just want reassurance now—they get all the warnings they want from the weather report. That's what the Brotherhood is all about, you know. Reassurance."

"Ummmh." Grasshopper grunts from his own drowsy reverie.

"I don't even like to drive," Reverend Martinsdale assures him. "and I hate these suits—they make me look like a polite gorilla who brushes his teeth. I'd much rather get about on a bicycle, wearing sneakers and blue jeans. Comfortable, you know? But then, how could I do my work?"

"Ummmh."

"Water off a duck's back! We're competing for audience share. What kind of ratings do you think a guy on a bicycle dressed in sneakers and blue jeans would get?

"No," his tone weighs the inexorable logic of the argument, "the agency's right. To get the message across we have to match them in showbiz terms. That's what the world is today. Showbiz.

"When the Bible was written, the world was just about farming and warfare—that's why all the old prophets were so filled with admonitions and parables about destruction and sowing seed. But today—it's all showbiz and glitz! Break a leg! Keep it moving—otherwise the audience switches on a game show or watches some cop chop bad guys to hamburger with a Stinson gun."

In grave resolution Reverend Martinsdale murmurs expectantly, "We're at war. War with Mammon. But now it's a Ratings War . . . and unfortunately we have to fight it out on his turf."

Then he chuckles ironically, "Sometimes I think Miles Amberson has both accounts . . . God's and Mammon's. And I don't know if the Agency is all that good at keeping them separate." He puzzles through it silently for a moment, then adds, "We're holding on by the skin of our gnashing teeth . . . God! How I hate teleview. And I refuse to go into StimuVision! How can you preach the Word when people are writhing in pleasure?"

Reverend Martinsdale rededicates the Cadillac in its pendulous lurching toward the far side of the road. "But now we finally have something to work with! Calamity!

"We'll break the back of the competition. People love Mammon in good times and fear him in the bad. Now it's God's Turn. This Pittsburgh Crusade is going to kick off a Holy War of Audience Share. This is our chance to repulse the Tide of Stimulation. We'll saturate the channels and airwaves in balm for distressed souls and the way to their redemption. It'll be like it was in my daddy's day!"

The Reverend's voice escalates into an exultant tremor as he abruptly whips the Cadillac onto an exit ramp, passing a sign indicating that Lebanon lies just ahead.

"That's why I'm going to meet with President Crown at Lebanon. He's uncovered the Serpent's Lair at the Evacuation Command Center, and now together we'll mount a crusade to wipe the blemish of Mammon from the American heart. The Christian Brotherhood of Arms and the U. S. Government are going to put up a united front! We're gonna bring this country back into united community of Christian Souls and Patriots! Woe unto all the others!"

Reverend Martinsdale does not notice that Grasshopper has fallen asleep. He lets loose a whoop that brings his dozing passenger up hard against the leather-covered dash, eyes straining and darting about in wonder for the cause of alarm.

"Hallelujah! Wondrous days lie ahead!" Reverend Martinsdale's brow furrows briefly, then he exclaims, "Hey, I don't think they'll let you in with me. You can't go to see the President."

"I don't want to, anyhow," Grasshopper assures the Reverend, "talk about Ants . . . Presid-Ant!"

"Well, the turnoff is just ahead," Martinsdale says, "where would you like to get out?"

"Anywhere," Grasshopper winces as the Cadillac bounces along the shoulder, lurching violently enough to throw Hellbender onto the floor.

"Don't worry, son," Reverend Martinsdale assures him, "this car is in the hands of God."

"But you're at the wheel," Grasshopper snaps gloomily.

The Reverend merely laughs, then seems to realize that he is missing an opportunity to further the Work. "Take some advice, son."

"What's that?"

"Come back to God. He heals all wounds. Whatever you're running from, He has the answer."

"Will He wait 'til I know the question?" Grasshopper asks impiously, now anxious to get out of the car. Glancing into the rear he sees that Hellbender has been slammed against the seatback and then onto the floor in the force of the near calamitous turn. The dog looks balefully at the back of the Reverend's head, his brow bunched above wary eyes.

A few miles farther along, the Reverend brings the car to a screeching stop just past his turn. As he backs it up he says, "Well, son. This is it. I've enjoyed our little talk."

"Ummmh," Grasshopper answers in a faintly nasty mood.

"The Lord helps those who help themselves," Reverend Martinsdale reminds him.

"Help themselves to what?" Grasshopper asks in idle sarcasm.

"Come to Him, son," the Reverend replies, mistaking wryness for sincerity. "His Grace makes all angry seas peaceful, all wounds whole and all misfortune a prelude to bliss. It makes these cars and tailored suits and all the other flashy junk tolerable . . . even necessary. Good night, son."

"Goodnight," Grasshopper mutters as he opens the back door to let Hellbender out onto the gravel shoulder. The hound seems greatly relieved to have regained solid ground.

Grasshopper slams shut the door and the Cadillac, as if relieved of great distraction, leaps into its turn and speeds off into the night. Hellbender jumps to avoid a sudden hail of gravel and now watches the disappearing tail lights from the haven of the ditch.

Grasshopper murmurs his disapproval to the dog, "Ants visiting Ants. Phooey!"

~ 22 ~

Ira Crown stares in disgruntled dismay at his visitor. Reverend Martinsdale has retreated to a couch striped in deep blue and pale silver alternating bands that sits facing the President's desk. He holds his tall and lanky frame in a gathered posture that innovatively combines sitting and squatting within an expedient disposition of limbs to imply either contrition or intimidation, as the moment dictates. His face reflects similar ambivalence, poised between apology and haughty disavowal of responsibility. His eyes have found refuge from the Chief Executive's chagrin in a reminder that the desk has only recently changed stewards. Its nameplate still proclaims Colonel Eric Cornrath.

"Reverend, I cannot believe that you actually had him. In your car. As you approached this base. And you let him get away!"

"I did not have anyone. I offered a ride to a rather strange young man and his dog. How am I to know that you have an interest in him?" the Reverend dismisses any responsibility.

"Dammit, Martinsdale! Don't you know what's going on in this country? Don't you pay attention to news? You have heard about teleview, haven't you?" Ira Crown's leaden sarcasm elicits a superior air from the chastised and increasingly annoyed evangelist.

"I don't watch news—I don't have time. I have an agency track events for me. They provide bulletins and summaries of developments important to the Brotherhood. My staff keeps me informed of matters relevant to our mission," Reverend Martinsdale explains diffidently. "And as for teleview—well, it's mostly either filthy or depressing. I don't want that trash in my mind—you are what you eat."

"Oh, never mind," Ira Crown cuts short what seems headed toward an essay on moral rectitude. He frowns down at his fingernails, as if they might have played some role in missing an opportunity to puncture the mass delusion overtaking the troubled nation. Two of his fingertips have been drumming angrily at the glass top of the desk and now are throbbing. His other hand massages them lightly as he searches among available courses. "We'll just have to find him. At least we know he's somewhere in the area."

Wally Martinsdale clears his throat, shifts into a more formidable stance and observes aloud, "Pardon me, but I'm still not clear why you think this Grasshopper person is so important. He strikes me as being a little strange, but hardly a major concern."

"That's just it," Ira Crown grimaces as he tries to mask his chagrin, "apparently you are unaware that a substantial part of this country's population thinks or hopes that he is some kind of miracle worker—that he's going to pull them out of their problems. That he can cure everything from gallstones to bankruptcy!"

"A miracle worker?" the Reverend is blithely amused.

"Yes, miracle worker. Messiah cults are springing up all over the place. Supermarket tabloids already are speculating that he is some kind of Divine Being!" Ira Crown smirks disdainfully, unaware that the remark has ignited the ensuing apprehension in the Reverend's voice.

"God?" Reverend Martinsdale gulps.

"Just watch the tabloid channels! You'll get a snoot full of Messiah returning. That the Latter Days, or whatever you people call this stuff, are upon us. And that fanatic, Clara Abbot, has somehow established a presence on teleview promoting an idea that Grasshopper is the Antichrist. And that's brought another bunch of crazies into the uproar!"

"Jesus? Latter Days? Antichrist?" Reverend Martinsdale gathers his brow in furrowed consideration of suddenly troubling pronouncements, "But that's silly—isn't it?" he asks as new possibilities rush in unruly riot through his mind.

"Of course it's silly!" President Crown snarls impatience with the cloddish pace of their exchange. He is growing in awareness that the Reverend is not altogether certain in maneuvering his own critical faculties. "Do you have any doubts in that respect?" he inquires skeptically.

"I mean, if he were, if he really were the Christ Returned, I mean, surely I would be aware of it. Don't you think? I mean, like, he was sitting right there in my car with his dog asleep in the back seat," Reverend Martinsdale worries that Divine Favor might have overlooked him. "I mean, after all, I've devoted my entire life to His Word—well, most of it—my life, I mean."

"Yes, yes," Ira Crown interrupts the rambling introspection, "of course. You're absolutely right!"

President Crown chews in futile exasperation above a stony weight he feels inserting itself into his abdomen. Small wonder that the country has fallen into such straits, he muses to himself, annoyed at evidences of incompetence and credulity revealed in the renowned Evangelist. As he watches the Reverend agonize over doubt and uncertainty abruptly filling his spiritual cup, Ira Crown realizes that he will have to lead his visitor cautiously into a course that he has gradually mapped in his mind.

"Like a thief in the night," Wally Martinsdale murmurs haplessly. His dilemma is most apparent in his vacillating blue eyes.

It occurs to Ira Crown that the Reverend bears a striking resemblance to one of his military aides, a Marine Captain who once had been featured on recruiting posters and TV spots nationwide, his craggy good looks peering across a recruiting slogan, *Semper Fi Your Life!* The same eyes—they see whatever their credo points out to them, the President speculates. In the one case, a product of military discipline. In this one, perhaps just poor focus.

For a brief moment Ira Crown despairs of releasing any of the brood he has patiently gathered from of a nest of fledgling stratagems and contending schemes. Martinsdale may simply be unable handle the role envisioned: to take the initiative and very publicly puncture this distracting bubble of speculative delusion distending and diverting the larger national agenda.

President Crown strikes a spark to ignite the Evangelist's zeal. "Actually," Ira Crown says slowly in an air of polite candor, "you and I both know that he is an impostor . . . a fraud, a charlatan."

"A fraud?" Reverend Martinsdale's eyes brighten for a moment, then sudden light breaks into fierce gleam. "An impostor, of course!" His gaze quickly foments dark pondering of forces against such stark impiety. "A false prophet!" he growls angrily.

President Crown waits patiently for this garden of complicity to take root. "But of course the people must be made aware of his duplicity!"

Wally Martinsdale frets aloud. "I can't imagine why my staff has not brought this matter to my attention. A false prophet at large in the land. A wolf among the flock!"

"You really haven't heard of the 'Makewell Kisser'? Of Grasshopper?" President Crown asks in dismay. "It's incredible. Both that it is happening and that you missed it! This nation has very real—excruciatingly real problems—and half the country is on an excursion to Never-Never Land!"

"Grasshoppers, yes, of course," Martinsdale exclaims, slightly confused at the shift in topic, "Now I do recall some items about grasshoppers being all over the country. But I thought they were talking about the little insects. You know—like a plague, a horde of locusts. What's the difference between grasshoppers and locusts, anyhow? We prayed for the crops every day, right through a bumper harvest."

"Never mind," Ira Crown gestures impatiently. He arises to walk slowly around to the front of the desk where he pauses briefly to think, his left hand toying absently with the nameplate left behind by Eric Cornrath. He picks it up and inspects it without acknowledging any import, then puts it back down as he prepares to broach his idea to the Evangelist.

"Is that Eric Cornrath, the PR agent and producer?" Martinsdale interrupts the Chief Executive's train of thought.

"What? Oh, this . . . yes, I believe that's what he does, at least until he showed up here."

"I know him very well," Martinsdale says expansively, "he worked on one of our most successful Crusades about eight years ago . . . what does he do here?"

"At the moment he's confined to the base, under house arrest."

"Arrest? What did he do?" the Reverend's voice evinces dismay.

"It's not really clear, but like a shit-shanked housefly his name seems to hover around every pit of corruption we look into."

"When I last worked with him he was civilian and I must say, the best PR producer I've ever known. Now he's a colonel?"

"It appears he was given that rank to facilitate a government public relations campaign that applied military technology to expedite media production." Ira Crown sighs, "Ironically, it was the *Ant and Grasshopper* campaign that was his brainchild. That's where Peter Dooley took on this Grasshopper persona that somehow has misfired into a lot of unholy crap.

"Later, Cornrath seems to have been assigned command of the Harrisburg evacuation route by mistake. At first he seemed implicated in the assassination attempt. Did you hear about that, by the way?" Ira Crown asks sarcastically, but does not expect an answer and continues, "But now that's looking like the work of a splinter group of military renegades calling themselves the *Morning Star Militia*. They seem to have been working with a local gang of biker thugs. The only certainty on that score, thus far, is that a Lieutenant Pomeroy, killed in crossfire with a bunch of bikers, seems to have told his squad that they were under orders to disrupt some kind of criminal transaction. It looks like a setup to attack us as we landed. We're still trying to sort out what really happened. So far, no clear leads."

"So what about Cornrath?" the Reverend insists, "He'd be the ideal executive producer for what you have in mind. For unmasking this Grasshopper fraud."

"At the moment, the only charge against Cornrath is a rather vague dereliction of duty, because he turned over base operations to FERAL personnel. But even that may not stick since it looks like the military made a mistake in assigning him to this duty in the first place. Since he is militarily incompetent, totally without experience, a tribunal may decide that, under the circumstances, deferring to FERAL *was* his duty." Ira Crown lifts his arms as if to fend off coils of unseemly red tape.

"Anyhow, back to Grasshopper," Crown continues, "we have to discredit him. Puncture the bubble of notoriety whipped up about him. I think we can work together on this. I don't care what it costs or involves. Even if we have to bend the Constitution into a pretzel—he must be debunked. This administration may be the last, unless . . . ." He decides against pursing a thought and stresses, "Know one thing. He is a trickster. A clever lunatic. And he undermines loyal impulses that keeps us on track. And without that, control is impossible. Without control . . . well, you can imagine."

President Crown holds up his hands in a sign of resigned futility. Wally Martinsdale inspects the two outstretched palms as if they represent a puzzle he must solve. "What can I do?"

President Crown let his arms fall to his sides, then brings them in front and folded across his chest.

"Your deal in Pittsburgh. In November and December."

"The Christmas Crusade?"

"Yes. That's it . . . we expect to find him by then. I've put every available investigator on it. We'll pull him in on a tax charge, or something."

"I want Eric Cornrath working with me on this," the Reverend urges, "according to your own account he knows about this Grasshopper character better than anyone else. That makes him the only real choice for executive producer," Reverend Martinsdale insists, "And he'll need a budget for any extras needed beyond the Crusade itself."

"All right," Ira Crown reluctantly agrees, "But I want Marty Roasdahl in the Psychiatric section of Walter Reade to take a look at him. It may seem silly, but I don't want to take any chances."

A gleam of wariness suddenly appears in the Reverend's eyes. He laughs as he recognizes something he has been looking for in the Chief Executive. For a moment he simply shifts from his pose of dejection into one of resolute intention as he closely inspects lines of consternation now wrinkling the President's face. He rises to his feet as he announces in cool confidence, verging toward impudence, "Mr. President, Ira Crown. Chief skeptic. You're afraid it might really be Him, aren't you?"

"Of course not," Ira Crown snaps vehemently, "but I do want to know what we're dealing with! I want to ensure that he is not some sociopath with enough demented talent to parlay confusion and disaffected chaos into a full scale revolution!"

Wally Martinsdale gazes unconvinced back into Ira Crown's demeanor. His own eyes are still large and rounded from a nascent fear that It might really be happening. What if He returned and the Nation's Minister did not know Him? Dread of such divine snub grates against his ministerial faculties, old tangles of disowned doubts resurface to unravel in gloomy prospect that he might somehow be found unworthy.

Then he summons from the zeal of lifelong dedication energy to expel these confusing doubts, to summon threads of cautious inference into a gathering knot of conviction. Such ties always have bound his life and girded its fruitfulness for anointed renewal. He clears his throat and says in the calmly deliberate prospect of what must be, "You are right. Such chicanery, it's hideous. He must be a fraud. I'll do whatever lies in my power to reveal his true nature."

"Good," Ira Crown smiles in satisfaction, "now. If we can just find him in time. We've tracked down his friends. It's just a matter of time until they blaze the path to where he is hiding."

"Friends?" Martinsdale is irked with himself that he has thought the word 'disciples'.

"Yes," Ira Crown explains, "they're doing odd jobs for ConRelCo's field teams covering the brouhaha here at Lebanon Command Post. We know where they are. Now it's just a matter of keeping them under surveillance."

"So you've even found your Judas," the Reverend fails in an attempt at humor.

"No. No Judas. Just some people that were traveling with him when I ran across them."

"You ran across them?" Wally Martinsdale grabs at the startling disclosure. "Do you know these people? This Grasshopper?"

"Well, yes, after a fashion," Ira Crown reddens, immediately regretting the slip, "It was just after the ambush near here. I ran across them while escaping. They . . . well, they're definitely not typical. They were running around naked and insisting this Grasshopper—his real name is Peter Dooley—had just healed one of them, Grady, a local farmer who was shot in an altercation."

He breaks off the reminiscence and shrugs sheepishly as he adds, "And, yes, you're quite right. I've kicked myself a thousand times for letting them go on their own way. But it was just before that teleview documentary stirred up all this nonsense. Damn!"

Wally Martinsdale has walked to the window beside the desk and now stands gazing through his reflection into its darkened pane and out toward dots of light punctuating the night-blanked horizon. After a long pause he breaks the silence. "I suppose you are right. But how is a poor mortal to know?"

"What do you mean?" President Crown asks warily. He senses a resurgence of reluctance in the Evangelist's resolve.

"It's just that skepticism is such a two-edged sword."

"What are you talking about?"

Wally Martinsdale turns to gaze hesitantly back at the politician, then seems to cast out this residue of doubt as he sighs, "Oh, either way, I suppose it doesn't really matter. We all are blind to His Purpose."

"Ummh." Ira Crown grunts ignominiously. He decides that the interview has just concluded.

~ 23 ~

Jason is chagrined that his new colleagues among the technical crews of ConRelCo do not acknowledge his previous experience at WNYX in New York. He is regarded as a novice. Worse, his association with Grasshopper quickly becomes target of sarcastic taunts: Hop to it, now, Bugboy. Kiss the noisy cable and make it well? Some clever wag even drops a grasshopper into his coffee when he goes to get some sugar.

Somehow Al and Charley garner acclaim for producing a video sensation while Jason is seen merely as callow accomplice of a charlatan. Still worse for his flagging esteem, Chloe is adopted as everyone's kid sister while Stacy and Kristin soon gather a protective squad of male admirers. Even Grady, who often comes by to hover about Kristin, is regarded as some kind of notable for his close brush with death.

They all busy themselves in their new roles. Chloe manages field equipment inventory and checks personnel assignments. Stacy—well, Jason has never been clear just what Stacy does—her duties seem to change constantly. He thinks one day she has been assigned to driving visitors about in the steproller. Another day she is reading copy into an uplink segment. And toward the last she is even on screen as a reporter. Mel obviously finds her a visual asset on camera, cynically remarking to Al that teeth and eyes are the hallmarks of a born teleview anchor.

Kristin, operating from previous media experience and a natural eye for gathering disparate threads into a sturdy fabric of visual presentation, seems to have found an enduring role as field producer and general production maven: she expertly resolves disputes, especially between Al and Mel. Compared with shepherding Banner and Dooley mediating between the two contending newsmen is a piece of cake, she confides to Jason.

Chloe tries to salve Jason's ego and soothe his impatience. Every evening they talk over the day's events, she listening more than speaking. But it is her listening that lets him endure. Somehow her eyes gradually banish any sting of invective and coax reflection that perhaps he has at least in part elicited abuse from coworkers. Gradually he accepts that he had rather puffed himself up, basking in Grasshopper's reflected celebrity, even as he belittled its import. It seems that his own callow self-esteem has been met squarely by coworker disdain.

It is a difficult lesson, but finally he accepts that it merits contemplation. At Chloe's urging, he begins to appreciate the importance of patient forbearance and quiet acceptance. As he learns to relax, defer and diffuse his own indignant response to teasing and tricks inflicted by his work mates, gradually they cease. The puddle gradually clears as his fretfulness ceases churning. Again Chloe has revealed her canny appreciation of social interaction. And yet again she has proven a ken of Jason's psyche far clearer than his own sense of himself. At times such as these she seems wisdom itself.

Nevertheless, Jason is the only technical crewman with a stake in the documentary's royalty pool. As promised, Al and Charley have given the four of Grasshopper's companions a piece of the production credit on *The Makewell Kisser*. Even as teasing subsides, it becomes apparent that his production credit holds him apart from general camaraderie, that his colleagues do not expect him to hang around much longer. He grows increasingly isolated from all but the Grasshopper contingent.

Sam Horowitz appears one day, saying that he and the two agents with him have been sent by President Crown to check on the well-being of Grasshopper and his companions and to follow up on some leads regarding their experiences prior to the assassination attempt. Kristin is disheartened and Stacy indignant to hear that thus far no hard evidence has implicated either Colonel Eric Cornrath or Eddie Minnow.

Jason is suspicious of a cover-up and whitewash, especially when questioning turns to the assassination attempt and whether there might be any connection with the Morning Star Militia. In earlier conversations with Hector and Grady, Jason has gleaned an impression of the notorious organization as a loose confederation of protective associations, especially targeting criminal enterprises encroaching on the peace and integrity of local communities.

Hector has scornfully dismissed likelihood that any Morning Star clusters might be involved in any sort of insurrection or political hit. Morning Star Militia, he assures Jason, was originally created by a group of former soldiers in West Virginia communities in resistance to extralegal pressures exerted by coal mining operations determined to level mountains and destroy communities so they could more easily extract deeply buried energy resources. In latter years, as formal police effectiveness declined in the face of budget shortfalls and pressures of outbound hordes of urban evacuees, the Morning Star Militia rapidly blossomed across Pennsylvania and Virginia. Hector had known their leader, Eli Whitney Penny, years before when he worked as a coal miner. "Eli Whitney Penny wouldn't assassinate a President. He'd tell him, face to face, to back off—and dare any devil to differ."

Before the agents leave Sam casually inquires about Grasshopper's whereabouts. Jason curtly replies that they have not seen him since he trotted off back in Lancaster. Until Sam raises the question it really has not occurred to Jason to look for Grasshopper. He has no idea where to begin searching and no way of knowing whether he still is even nearby.

In the wake of the agents' visits it dawns on Jason that Grasshopper is a major news story and ConRelCo, a major conduit of information about every significant event transpiring on the globe. Perhaps Grasshopper might be located via the ConRelCo cybernet! He begins scanning raw reports for any intimation that Grasshopper has surfaced and suggests to his friends that in ConRelCo's data fields they may run across hints regarding Grasshopper's whereabouts.

In his spare time Jason learns to probe and query the ConRelCo data engines that link thousands of host data lattices via satellite and opcable within a global cybernet. He soon recognizes that the sheer mass of data requires automated digital processing if he is ever to gather any useful information regarding Grasshopper's whereabouts. He gradually incorporates elements of the AiPAL he has been developing to help automatically scan and categorize every news feed and field report, wedding each to a burgeoning tree of data elements gleaned from the operating system image he previously copied from Grasshopper's malfunctioning iLink.

Watching data flows from anywhere on the globe, the AiPAL evaluates informational relevance of every item in the

ConRelCo cybernet in relation to the search for their elusive quarry. As he delves farther and farther into the cyber-sorcery of AI agent synthesis Jason adds new traits, drawn from open-source as well as proprietary code leased from the ConRelCo in-house store of linguistic processing routines.

He is especially pleased with his own unique solution to the problem of how a computing device might emulate intuitive conjecture, an essential undergirding of linguistic prowess that guides even the simplest instance of human utterance and reading of inferential meaning.

For years he has toyed with a digital vector notation based on a rectilinear spiral matrix, the *SPOT System*. Its derivative *Lattice Articulated Vector Analysis* using *String-Indexed Lattice Keys (LAVA/SILK)* makes possible rapid analysis and comparison of configuration-expressive vector data. He originally stumbled across the technique in archives of an early computer graphics journal where it had been presented as a way of specifying spatial configuration. Jason learned that its author had also played around with linguistic applications of the methods using a *Concept Deployment Matrix* that maps categories of a modern thesaurus onto the SPOT spiral matrix. The result is a simple but powerful means of digitally encoding inferential meaningfulness.

Now the LAVA/SILK inferential engine proves to be a crucial feature Jason adds to the AiPAL that enables it to navigate effectively through torrents of data in search of patterns similar to Jason's verbalized items of interest.

Working with the augmented and empowered AiPAL Jason is able now to simply tell it, in oral descriptions, the kinds of information he wants to it consider, give a few examples that quickly set its SILK keys for efficient combing of LAVA data sets to extract the few crucial nuggets that hopefully will guide them toward Grasshopper.

To keep the AiPAL from being absorbed back into the ConRelCo stable of data agents and thereby to keep secret its mission, since management probably would not approve of his intentions, Jason assigns the now flourishing data agent a share of his own royalties from *The Makewell Kisser* to pay ongoing license and storage fees. A curious twinge of satisfaction infuses the knowledge that for so long as ConRelCo maintains *The Makewell Kisser* video documentary in its archives, his AiPAL will have its own root origin, sustenance, and independence within the ConRelCo cybernet. In a perverse nod toward aspirations of immortality he dubs the AiPAL: *JasonHood.*

Crude at first, gradually the AiPAL is elaborated into a veritable data-cruiser, able to plunge through oceans of information, to scan latticed horizons for any semblance or mention of Grasshopper or Peter Dooley or anything that faintly smacks of him.

There are many false reports, ranging from an ostensible stint in Las Vegas as a lounge comic to flaky assertions that he has been reincarnated as the Dalai Lama. With practice the AiPAL's conjecture engines become ever more adept in ranking important clues highly and demoting frivolous ones down into insignificance.

Curiously, a news conference held by Wally Martinsdale following his meeting with President Crown is indicated among several routine news stories. Jason calls up the raw segment and indexes it to the reference linked to Grasshopper.

Reverend Martinsdale crows and marvels at wondrous works afoot, especially in the new accord of cooperation between the Brotherhood of Christian Arms and the U. S. Government to sustain and relocate urban evacuees. In response to a reporter's question about a Second Coming the Reverend scowls and launches into a diatribe about the charlatan Grasshopper, who is known personally to him! In the same segment Jason notes that the reporter has also identified Clara Abbot as leader of a group called the *Sisterhood In Christ's Arms*.

Jason plays and replays the segment mentioning Grasshopper. Kristin and Chloe concur with his logic. If he has gained his information from Ira Crown or any other source, would Reverend Martinsdale say that Grasshopper is "known personally to him"? Jason doubts it. Which means that he must have encountered him recently! Probably on the day of his meeting with Crown. And it was on the next day that Sam had appeared to inquire about their "welfare".

Without knowing precisely why, Jason now is convinced that Grasshopper must be nearby. And if he is nearby, then there are only a few places he might be. They all agree on one certainty: Grasshopper will not be found anyplace that is at all Antlike, which rules out cities, towns and military or evacuee encampments. Most likely he will be in the most secluded spot he can find, some solitary abandoned shelter in the midst of uninhabited forest. While many towns and suburbs have fallen empty over the past several decades, they are persuaded that Grasshopper would find even such town ruins far too saturated with Antward offenses. Further, his addiction to being online via the iLink will cause him to gravitate toward some secluded place that is within range of a maxnet signal.

Jason calls down satellite shots of their locale into an image processing console. While providing seemingly inexhaustible scanning and focus, the amount of information available means that one needs to know where to look. He gives the AiPAL new routines to filter image data and process various strata of frequency data. Chloe doesn't like looking at the generically mechanical visage of the AiPAL, so Jason scans in his own face and provides a language processor that can sample and speak in his own voice. This final flourish at last sparks her interest in how AiPAL: Jason Hood has been spun out of mere cyber-gossamer!

Setting the imaging routines to mask out densely populated areas, he begins first with an infrared scan of all the remaining sparsely populated regions within range of maxnet access points. Just as well begin with image data closest to his own optical understanding! Working mostly on hunch and what he can learn of the Reverend's movements just before his meeting with Ira Crown, he coaxes the AiPAL into finally developing two locales he thinks merit searching. One is a wooded area near the intersection of the Lebanon Command Post road and highway 72; the other is outside Lancaster, in the direction of Lebanon. Both locales are well within maxnet coverage and give evidence of habitation by a single human being and a medium sized animal. Other clues urge him into conjecture that he is close to finding Grasshopper and Hellbender.

The women are amazed that he has so quickly become convinced that Grasshopper is not far away. They are dubious that Jason could possibly locate him without once leaving the lattice console.

"Okay, let's see if the AiPAL can hack into his iLink," Jason shrugs off their skepticism.

He opens a code channel, tweaks a few bits and says, "We'll have to break through the encryption to get into his iLink. If it succeeds then we know he's there." After a lengthy interim of cryptological negotiation the AiPAL, already equipped with some of the finest black hat decryption technology available, decodes an iLink challenge and successfully merges with Grasshoppers iLink. A burst of communications traffic blossoms into a display window on Jason's console; its vista opens into a virtual rendering of Grasshopper's movements, surroundings and sounds accompanied by a hound. The ladies are convinced.

They decide to walk to the location, since a steproller can be checked out only in pursuit of a story. Not wanting to invite scrutiny of their mission, they head north toward Lebanon as if they are simply going on a hike. Everyone is too busy to notice, anyhow. They set out for a walk they estimate at less than two hours, beginning shortly after sunrise.

Within two miles the landscape becomes wilderness. Rural communities that once dotted the region have long since been abandoned, their sites falling back into natural proclivities of weed, scrub and tree. Along the road that leads northward they trek into ever-deepening forest of hardwood trees. For most of the way they are relatively quiet, perhaps for not wanting to admit that, even if successful, they really do not know what to do next.

After little more than a couple hours of walking past ever fewer old ramshackle buildings and ancient remnants of farm equipment, Chloe suddenly stops and listens. In the distance a hound is baying. "Hellbender!" Chloe squeals. Her skepticism has evaporated immediately upon hearing the distant howl.

Jason, too, hears the dog, but does not care to chase after just any canine noise. "How do you know it's Hellbender?" he asks.

"It is. He has that little trill at the end. Like that." She squiggles an uplifted forefinger to mark the air as in the distance the wail rises again in a quavering pitch and then gurgles a little falsetto note at the end.

Now Kristin, too, is convinced. She leaps the ditch and runs off into the woods, pausing only to look back in impatient excitement calling to Stacy, Chloe and Jason. "Come on! It is him!"

~ 24 ~

Yet again, Jason finds running through woods exasperating. Littered and uneven ground trips him up while bushes and brambles reach out to snag his clothing when he is lucky and slash his skin when he is not. Somehow the three women evade most such obstacles and quickly outdistance him, dashing on ahead without waiting for him to catch up.

Finally, out of breath, he stumbles out of a brush thicket to see them with Hellbender over near the edge of a clearing. All are peering up into the gold-tinged crown of an immense oak.

"Hellbender!" Jason exclaims astonishment that they have actually found him. Grasshopper must be nearby!

The dog glances back over his shoulder as he approaches, wags his tail in subdued recognition and then rests on his haunches to again look expectantly up into the crown of the tree.

"What's up there, boy? A squirrel?" Jason follows Hellbender's gaze up into the massive limbs of the old oak. Perched on a branch, leaning sideway to hold his ear against the trunk as if following a conversation in the next room, is Grasshopper. He does not move, but merely looks down at the newcomers as if they have been with him all day.

"Grasshopper!" the newcomers chorus.

He says nothing, but raises a finger to indicate that he will be with them in just a moment. Eventually he removes his ear from the trunk, tilts his head to one side as if considering something puzzling and then says to the groundlings, "Hi."

"What are you doing up there?" Stacy asks.

"Waiting for you, I suppose," Grasshopper says after a moment's reflection.

"Well, I guess I'll take that as an invitation," Stacy laughs and begins pulling herself up to where Grasshopper waits sitting on a sturdy bough, his back supported by another. The couple kiss as though she has just returned from a day at the office. Jason sees Chloe give a little shrug of acceptance as she, too, begins the upward climb. Kristin offers her shoulder in support of Chloe's legs to give her a boost up.

Hellbender frets at being left on the ground.

"Hush!" Grasshopper scolds the hound, "I can't help it if you don't climb trees. You don't hear me complain about being left out of all the smell gossip you're always nosing about. So just be get over it!"

Hellbender seems to abruptly lose interest in attaining a tree perch. He saunters in widening circles about the trunk and eventually disappears into distant brush, apparently finding pursuits more amenable to houndly talents. Jason chuckles that Grasshopper's tone reminds him the way his father and mother used to bicker.

"He sounds different . . . more coherent," Jason observes in a low voice to Kristin, "something's changed."

"He does seem less agitated," Kristin agrees. She looks up to where the tree-borne trio have amiably gathered in what appears to be wordless chatter. "They're conversing without talking!" she exclaims.

Jason looks up. "I want to check something. Hold on." He pulls his euPhone from a pocket to tap and stroke its touchscreen several times. After a moment he says to Kristin, extending the display for her inspection, "Jeez! It looks like my AiPAL modules have been downloaded as apps into his iLink OS."

"But how'd that happen?"

"I guess when I hacked into the iLink to show you ladies that I'd really found him." Jason speculatively winces, then after a flurry of finger strokes on the euPhone brings up an animated display of graphic signs imposed on a gridded map. "And look at this. See that triad of icons? It shows both Chloe and Stacy online, as well as him. That big question mark is a polling signal. I think it's looking to link with your brother's iLink, you know, as Ant. And see those rolling lines of text? That's LAN traffic with his iLink.

"They're all interacting with it. They don't need to actually make sounds, but their lips move, I guess they're lip-echoing, kind of like reading silently . . . I noticed them doing it back at the Abbots. It just seemed really weird then. But now I see what's happening." Jason's eyes widen in muted excitement," This is really incredible!"

"It's still weird," Kristin furrows her brow, trying to mesh what she is seeing with what she knows of the nanotech. "Oh, God," she exclaims, "the replicators and nbots—they're both infected. But how could that happen? It was all supposedly customized for Pete's DNA. How about you and me? Are we?"

Jason checks again and shakes his head slowly, "No. Just them. When I check their signal traffic, it's as saturated as his. Maybe the nbots have adapted to them."

"There was some discussion about auto-adaptive versus custom versions of the technology," Kristin muses, "Cornrath said they decided to go with custom, even though it would take longer. Hell, he may have lied about that, too. He could very well have been trying out a generic version for some military or industrial use."

"How did Stacy and Chloe get them?" Jason dreads the answer.

"Peter and Stacy have become friends with benefits, if you know what I mean. That's what broke up her relationship with Jim Meerling," Kristin raises a mildly regretful eyebrow, "and from the story I heard it sounds like Chloe was in the thick of Grady's healing. She may have picked it up from all the blood splashing around. Maybe physical contact spreads it from person to person, you know, exchange of body fluids during kissing and the like." Kristin's eyebrows rise in realization," And, of course, that means Grady is infected, as well!"

"Chloe also was swept up into their . . . mania, or whatever," Jason muses over how Chloe had plunged along with Grady and Grasshopper into seizures of nakedly leaping enthusiasm. But he prefers not to mention their bouts of kissing during those exuberant gyrations. "Is the nanotech what healed Grady?" Jason searches for some rational explanation."

"Maybe. Probably. Yes, I think so . . . the replicators work off DNA and ontogenetic sequences encoded into cellular structures. That's how it generates the biotic interface to the iLink. Suppose that as a side-effect they also rebuild and repair cells wherever they encounter damage. Or maybe they hyper-accelerate the body's natural repair mechanisms somehow. You know, kind of like in psoriasis where an errant immune system puts out messages to repair damage, even when there is none."

"What should we do?"

"What can we do?" Kristin frowns. "We need to get everybody back to the lab. They can terminate the replicators by radio signal."

"Do you think they're in danger?" Jason frets aloud as darker weeds sprout unexpectedly in his garden of tech fantasy. Chloe and Stacy are now interacting with Grasshopper through the AiPAL enhanced iLink across maxnet. What unexpected prospects of new physical affliction or psychic peril do such transformative powers bring into play?

"Apparently Peter has lived with it just fine. As long as they don't get too intoxicated or high on drugs, maybe they'll do okay until we can get them help. On the positive side, it seems that if they get hurt somehow they'll heal really quickly. Who needs doctors?" she smiles ruefully.

"Should we tell them?" Jason asks, "Or would it just worry them too much?"

"Maybe wait 'til we talk to Stacy and Chloe alone. I don't want to set Peter off again. He can be so touchy."

"And take off on another Ant tirade," Jason agrees.

"What are you two doing down there?" Stacy asks from her tree station.

"Come on down," Jason urges, "we have something important to tell you"

"You come on up," Grasshopper says curtly, "it's no farther up than it is down."

Perhaps it is the days of teasing he has endured as a result of notoriety. Or perhaps it is simply the unheralded triumph of his succeeding in finding the object of their search. In any event, Jason wants to let Grasshopper know that he is the dominant story of every tabloid and gossip media vehicle between the Atlantic and Pacific. He takes secret satisfaction at the displeasure Grasshopper certainly will find in knowing that so many worrisome Ants are searching for him. He decides to defer his discovery about Chloe and Stacy for the moment and answers, "Okay, if you insist."

Chloe's hand reaches down to help Kristin as Jason pushes her up onto the bottommost branch. He sighs and follows up after them, not recalling anywhere near as much effort in his boyhood scampering among tree limbs.

Chloe reaches up to pat Grasshopper's foot affectionately. "The Grasshopper band is back together," she sighs happily.

Kristin looks at Dooley, worried, "We have to hide," she says urgently, "everybody's looking for you!"

"Hide?" Grasshopper gestures toward screening crowns of the surrounding trees, "I think we're pretty well hidden." He laughs at the pointlessness of such concerns. "Who do you think will look in this tree out of all the trees in the world?"

"We found you," Jason rebuts, gloating in his own detective work.

"Maybe, maybe not," Grasshopper dismisses the claim in a drowsy tone, "you were just walking along the road. I saw you way down there." He points through a break in the clusters of leaves that permit a glimpse of the distant road. "Now which one of you said, 'I know where there's a tree that Grasshopper's sitting in?'"

"Well, the whole country is looking for you," snaps Jason, annoyed that his lackadaisical whimsy has not mellowed, "Like it or not you're the most famous guy in America! The hero or scoundrel of the moment, depending on who's talking."

Grasshopper shrugs in pity usually reserved for unfortunates. He refuses to concern himself with implications of Antwardness.

"I'm so glad we're all together again," Chloe enthuses, trying to quell any bickering as she attempts to move closer to the branch where Grasshopper sits. Holding onto a small branch, she hugs his legs with her free arm.

"And in such a magnificent tree, too," Grasshopper pats the trunk affectionately. He wraps an arm about Stacy, "And now I'm in the thrall of ecStacy."

Stacy smiles pleasantly, lightly tickling his ear with her index finger, "Peek-a-boo."

"Machu Picchu."

"What are we going to do?" Kristin whispers to Jason. Her face is gathered in worries how they might coax him down into more dependable sanctuary.

"Kristin, Kristin," Grasshopper chides her gently," Pay attention. Really pay attention. You'll see that no matter what happens, there's no cause for alarm."

"But there is. You don't understand," she protests, "if we can find you, so can they. And there are real problems we need to talk about."

"Pshaw!" Grasshopper puffs, "Just look around. See how many trees there are! And you know how Ants do things—first they'll alphabetize the list. Then number everything so each has its own little squiggly place. Alder, ash, birch, boxwood, catalpa, china berry, cottonwood, dogwood. We're way down the list at 'oak'. And even after they get to oaks, think how many oaks there must be before you'd ever get to this one. It's silly to worry them anytime today. I doubt they'll even make it to catalpa."

"But we can't stay here!" Kristin despairs of dealing with mangled logic.

"Well, I don't know why not," Grasshopper snorts impatiently, "but even if we can't, there are plenty of other trees even farther down on the list. How about walnut? That'll take forever! Really," he assures her, "they are very nearsighted. And forgetful. That's why they have to be so methodical. They get lost without maps and lists, always going about squinting at things, trying to figure out where they are."

"He does have a point, in his own inimitable way," Jason tries to soothe Kristin's upset, "if we're careful, we have a good chance of avoiding them. The way things are screwed up, the chaos and confusion—that's all on our side. We just have to decide where we want to go."

"Southwest Virginia?" Chloe reminds him of their original destination.

"Why Virginia?" Stacy asks.

"Chloe grew up there," Jason explains, "she has relatives all over the place that can help us. Maybe we could just disappear into the hills."

"But what about ConRelCo?" Kristin suddenly realizes that she may have to choose between her concern for Peter's well-being and her own unexpected hopes for a resuscitated career.

"Al and Charley would love to get their hands on him again. And Mel!" Jason exclaims, "Mel would sell him by the pound just to reinstate himself with the corporate bigwigs in Rio."

"He's right," Chloe nods to Kristin. "ConRelCo is out. We need to get someplace back in the mountains. Then maybe we can all live in peace."

"But we have to cure him," Kristin protests. "Surely there must be some therapy."

"Cure me?" Grasshopper spits out the words, "Of what?"

"The nanotech. And your amnesia," Kristin says sadly.

"Amnesia, ambrosia, Ant tease'ya and throws'ya," Grasshopper lampoons such trifling concerns.

"Well, you could be helped!" Kristin argues, her eyes imploring Jason for support, "I want you back . . . I want Peter Dooley!"

"I don't want any help," Grasshopper scowls, "and I don't want to go anywhere. And if Peter is so important, then you be

Peter, Peter, pumpkin eater had a life but couldn't keep it—phooey!"

"But how are you going to live out here?" Kristin argues hopelessly.

"I have a place, a very good place!" he insists.

"It wouldn't be a tree, by any chance, would it?" Jason groans, already weary of trying to coax him down, much less lead him to some mountain aerie.

"Oh, heavens, no!" Grasshopper chirrups merrily, obviously amused at the idea of living in a tree, "it's an old Ant house."

"And you're living in it?" Jason is surprised he finds any building accommodating.

"Of course. It's very peaceful."

"Isn't it a little Antward?" Jason is perplexed.

"Oh, a ditchbank told me all about that. And I must admit, Ants do make things that can be very convenient. But that doesn't mean I have to agree or become like them, just because they build houses I like."

"A ditchbank?"

"Yes . . . she is very wise . . . knows all about moving and carrying stuff." His mouth twists into a philosophical pucker.

"Do trees talk to you, too!" Jason growls, recalling his earlier position sitting with his ear planted firmly against the tree trunk, apparently listening.

"Oh, yes," Grasshopper is unperturbed by the irony in which the question is posed. "Trees are not very talkative. But when they do get going, they can tell you an awful lot. Whispering down the ages on very old breezes. That's what we were doing when I saw you walking out along the road."

"Talking to this tree?" Chloe, of course, is immediately intrigued.

"No. Listening," Grasshopper carefully emphasizes the distinction.

"What did you hear?" Chloe squeals in delight.

"You'd be amazed!"

"I bet," Jason grumbles to Kristin, who rolls her eyes in dismal appraisal of any hopes of ever pointing Grasshopper back toward familiar mental norms.

"It's next to impossible to say in words," Grasshopper frowns like a haggard explorer recounting remarkable vistas, "But you can feel it and smell it if you let your whole body be like a tree. That way you begin to get a sense . . . the closest word I know is 'confidence'. Confidence that every way is ultimately just one viewpoint, part of one way and every world is just some peculiar time, part of one larger world. Ideas swimming in iLink can help think it through: kairos hoppens wherever there's dynamic fractal expression of a whole in each part of its moments. 'As above, so below,' says Hermes Trismegistus. I say, 'One world, one way.'"

"Oh, grok it with Gurdjieff!" Chloe and Stacy exclaim in unison, then alternate utterances during Grasshopper's pauses, "kairos, ancient Greek for the right or opportune moment," Chloe trills, followed by Stacy, " . . . a period of undetermined time in which something special happens." and back to Chloe, "Chronos is sequential and quantitative; Kairos, fractal and qualitative."

"Events in chrono-time," Grasshopper brings the finger tips of his two hands together, then reaches out in similar gestures to lightly touch finger tips with his two adherents simultaneously, "form relationships that shift and interact as patterns in kairos-time. Patterns emerge and fade, sometimes resonating schematically to generate rudimentary opportunity."

"Accidental direction," laugh Stacy and Chloe, bouncing shoulders to shake imaginary pompoms, "R—O—A—D."

"You two sound like you're channeling *Wikipedia* with *Revenge of the Cheerleaders*," Jason knits his brows as he looks askance at the two women clinging on to Grasshopper.

"We are, kind of," Chloe and Stacy giggle together, "Among so many other ready vibes. Grok Subud?"

"The whole universe, everything, is alive, opening, expanding, every point rushing away from every other point all throughout the opening of ongoing creation," Grasshopper nods approvingly toward Stacy while Chloe's dancing fingers flutter like butterflies to suggest expansive opening as she warbles, "Points drawing lines that sweep planes through nested volumes that interact in events within larger events of interbraiding synchronies of opportunity that vest archetypes of probability . . . and so it goes, on ineluctably to the Divine. *Aliswasmaybe.*"

"There's maxnet out here?" Jason interrupts their convened soliloquy in blank astonishment. He pulls his euPhone from a pocket, taps it several times and then exults, "Ha! I didn't notice before. There is . . . we're within range of a skypod port over that way."

"One world, one way?" Chloe ignores Jason's discovery to echo Grasshopper's earlier surmise into compact epigram.

"O, wow," Stacy nods.

"Fractals?" Jason gives delayed reaction to the word heard earlier, even more surprised that Grasshopper has any familiarity with mathematics, "What about fractals?"

"Oh, that everything," the convocal trio jointly explain, "Aliswasmaybe," stresses Stacy, "is in a single place of multiple dimensions expanding outward and upward through diverse modalities and implications, drawing lines across planes to move through volumes that persist as events enduring through reminiscence into meaningfulness. Aliswasmaybe: the origin and author of all being and meaning," explains Grasshopper. "Ants mostly live in *chronos* while Grasshoppers hoppen in *kairos*," Chloe helps the triadic chorus along. "Yet all live in Aliswasmaybe," Stacy beams.

Grasshopper rubs his hand appreciatively over rough bark. “And this tree knows all about being and upping, too.”

“Upping?” Jason is taken aback by their effortless interweaving of individual snippets to vocalize rounds in a musically progressing dialog of evocation.

“Yes,” Stacy interjects, “getting up. Growing up. Lifting up. Being up . . . upping.”—”It used to be way down there,” Grasshopper points to the ground where the tree’s huge roots descend into leaf strewn earth.—”It’s such a wondrous branching,” Chloe agrees as she pats the enormous trunk in awe of its solidity, “every child branch echoing and extending its parent.”—”I used to spend a lot of time in trees when I was little,” laughs Stacy, ‘just sitting and listening, swaying in the wind, there’s a feeling that’s just . . . enthralling.”

“Upping,” Grasshopper smiles at her, then pats Chloe’s head affectionately, “Ants forget to do that. Ants don’t up. If they did, they wouldn’t be such nasty little turd counters”—”They don’t up because they don’t pay attention to where they are,” agrees Chloe, “Only where they’re going to or coming from or why they can’t stay because they don’t have chronos-time”—”You have to put down roots and really be somewhere before you can expand outward and upward,” Stacy sighs— “hoppen,” surmises Grasshopper sadly.

“O, wow,” Chloe lightly raptures their joint reveries, “One world, one way.”

“You’re too much,” Jason snorts, “Down to up, up to down. If I keep listening to this, I may actually begin to understand it. And that worries me. O, wow,” he teases Chloe.

“This banter is all very opaque and mildly pleasant,” Kristin interjects impatiently, “but it isn’t very comfortable up here and I’m getting chilly in this wind. Do you think we could get down now?”

“If you want down, then get down,” Grasshopper is serene, “Hellbender will be glad for your company.”

Kristin looks about at her companions sprawled awkwardly on their respective branches. She disapproves Chloe's credulity and Jason's reluctance to persuade Grasshopper to face unpleasant facts. The four all know that he is in for some very difficult times if he does not manage to avoid legions of news teams and celebrity-hounds chasing after him.

"One thing is certain," she proclaims bitterly, "this tree can't tell you how to deal with the cranks that are chasing after you." She shoots an accusing glance toward Chloe as the chief apologist for his abdication of responsibility. A hint of animus wells up from her failure to penetrate their maddening disavowal of any disquieting cares. As she steps down to a lower limb tears flood her eyes to cloud her vision. She miscalculates the distance to the next branch down and loses her footing. Kristin screams and gropes awkwardly for a handhold, misses and then drops through buffeting branches to fall heavily onto a broad tree root. She gasps and shrieks at stabbing pain tearing through her left arm just above the wrist. The others scramble down out of the oak to help.

"It's broken," she sobs pathetically in a mix of rage and frustration. "I know it's broken!"

Grasshopper climbs down, looks casually at the arm she clasps protectively to her breast and then squints back up into the tree crown. After a moment he squats down to whisper into Kristin's ear.

"I won't! Dammit, I will not do it! That's insane. You're insane!" she winces in pain, hissing through clenched teeth her rejection of some upstart suggestion.

Grasshopper looks back up into the tree. He seems truly disappointed by her angry disavowal. He shrugs as if signaling to someone still up on a branch. Then he rocks forward and whispers again into her ear.

Kristin vigorously shakes her head in obstinate refusal, but the movement inflicts new pains upon the injured arm and finally she gasps and reluctantly acquiesces, "Oh, all right!" she glares up at Grasshopper as if looking for some hint that he might be mocking her. He smiles encouragement and nods for her to proceed. Kristin inhales deeply and then says in tones clipped short by firmly clenched teeth, "I'm sorry. I didn't mean it . . . it's a very nice tree."

"You have to mean it." Grasshopper lifts an eyebrow as if she should know better.

She readjusts her torso to a more upright position, gingerly holding her left arm as she moves to lie back into the embrace of the tree roots. For a moment she gazes back at her companions, looking deeply into their expressions of concern as if pondering her fate within ongoing lunacy. She takes a deep breath, sighs sadly and shakes her head slowly back and forth in resignation. Fresh tears wet the corners of her eyes as she says slowly, "I just don't know anymore. I just don't know."

Kristin looks blankly into Grasshopper's probing gaze and then finally says, as if admitting something she has been holding back for a long time, "I shouldn't have doubted. I should pay more attention to everything around. I've been a very silly Ant and I'm sorry."

Grasshopper beams his approval, standing upright as if everything has been restored to its natural order. Chloe, Stacy and Jason exchange gazes, still waiting expectantly for something to happen. They look tentatively from Grasshopper to Kristin and then back to Grasshopper, who now merely leans against the tree.

Kristin looks dolefully toward the ground as she absently rubs the injured arm still clasped firmly to her side. Then suddenly she leans forward and probes tentatively where the root had struck her arm. "Look!" she exclaims, "I can wiggle my fingers!"

She tentatively extends the injured limb and moves her fingers in sporadic display that seems to work an invisible marionette. "It was broken! I'm sure it was! It had to be—didn't it?"

She stares in wide-eyed amazement, as if seeking Jason's critical opinion. He shrugs ambiguously.

"What difference does it make?" Grasshopper asks quietly.

Kristin ignores his question and looks up into the tree, then at Stacy, Chloe and Jason. She seems to be considering all their narratives of his exploits. Finally her gaze returns to his face and she gasps in conjecture, "You really can! You are! Oh, God, you do!"

Unable to further articulate her astonishment, she stares off in saucer-eyed bewilderment. Standing up, she carefully tests the arm and retreats back into more skeptical appraisal, "It must have only been twisted."

Grasshopper howls in merriment. "Rearsight and nearsight," he cackles, then says in the most expansive mood Jason has ever found in him, "Come on. I'll show you the little house I found."

~ 25 ~

Following Grasshopper a short distance away from the road into a small clearing, his friends approach a two story brick building, still in fairly good shape. Several sheets of weathered and peeling laminate panels have been bolted where once there had been a large window. Over the years the panels have bleached and warped in vicissitudes of rain and sun until their fastenings worked loose. They now express each gust of breeze in a grinding scrape and thump. Behind the building is a rusted remnant of metal shed, its wooden roof members thrust awkwardly up at each end, seemingly headed toward exhausted collapse onto the ground.

"An automobile garage . . . most of these went out of business when they started taxing petro fuels so much." Jason explains to no one in particular.

"Somebody lived here, though," Chloe adds as she points to window boxes overgrown with weeds hanging at the ledges of several of the upstairs windows.

"Mechanics have to live somewhere, I guess," Jason says, "but this is a strange place for a garage. It's a long way from the highway."

"There's a coal pile," Kristin announces, pointing to the far end of the dilapidated shed, "let's build a fire. It's getting cold."

"Yeah. And maybe we can eat," Jason agrees, noting that they still have provisions they brought along

"That's no problem," Grasshopper declares. He motions for them to follow him around to the back of the shed. Suddenly, in the brush, a rooster crows and they see several red hens.

"Chickens!" Chloe shouts, "we can have chicken!"

"Not chicken," Grasshopper corrects, "eggs!"

"Oh," Chloe thinks for a moment, then jests, "don't eat the chicken that lays the poached egg, huh?"

"Look, I found one," Stacy announces as she plucks a white shelled egg from the brush."

"It's just like Easter!" Chloe shouts and scampers about looking for more.

Within half an hour they have collected a dozen eggs or so. Chloe carries them carefully in a fold of her jacket as Grasshopper leads them into the upper room of the abandoned garage.

At the top of creaky, but serviceable stairs are three tiny rooms comprising a kitchen, toilet and bedroom. A table and several laminate crates furnish the outer room. In the corner where they now stand regarding the premises is an old mattress shoved against the wall. Hellbender trots over and claims it as his own by circling casually, lying down and promptly falling asleep.

"I can't say much for the maid service," Kristin winces.

"Oh, we can clean it up," Jason says, "it's better than I'd hoped."

"Remember that you said 'we'," Chloe lifts an eyebrow in recollection of his housekeeping tendencies.

They tacitly agree that they are not going back to ConRel-Co, at least not today. Then they set about cleaning the living quarters, dragging out assorted junk, sweeping, then hauling in various items found downstairs to be recruited into new roles as furniture or utensils.

By twilight the stove works, the rooms have all been swept out, the mattress dragged outside and beaten until only moderate puffs of dust exhale from its interior and the hand pump in the kitchen primed and worked until its water loses the color of rust and fetid odor of stagnation. An old piece of transmission housing has been scrubbed clean with sand and water and now becomes an awkwardly managed frying pan. Eggs are scrambled into rubbery but satisfying consistency and after a comically impromptu, but greatly relished meal, they settle back on the mattress with Hellbender and watch darkness gradually invade the room, illumined by the opened stove firebox door and a couple of strategically placed candles found in a box of junk.

"Boy, it sure could be a lot worse," Jason remarks. The wind is picking up outside, vigorously bumping and scraping the laminate. "I wish I had thought to tie that panel down, though."

"Do it tomorrow," Chloe urges him to ignore the noises.

"We can't stay here forever," Kristin says, "what are we going to do?"

"Going to do, going to do," Grasshopper mocks her gently. Then he chants,

> *The ivory-flecked bubble-horse machine*
> *Was never checked nor tamed.*

It rides demented through dreams and drives of nations,
Breeding in the rear of unattended service stations,
Feeding on the wounds and moans
And severed sighs of maimed mechanics.
Who inquires of its children?

"It's going to be rough getting over into the mountains," Jason ignores the recitation.

"Trees have no desire to go anyplace else," Grasshopper chides him, "in fact, I don't think they can even conceive of such a notion. So they stay where they are, tasting seasons winding around their trunks. Their whole world is like a fat doughnut, with them at the center. A doughnut of earth and air glazed in stripes of day and night."

"Is this philosophy or poetry?" Jason asks sarcastically.

"It's tree-thinking," Grasshopper sniffs.

"Do we have time for this?" Kristin breaks in impatiently.

"Trees don't use clocks," Grasshopper seems to think she has asked a question of him. "For a tree a year is a spiraling loop that winds up the trunk and then back down from its leaves into its roots. An old tree is a big, concentrically involved coil of years looping up and down, up and down in a twisted skein."

"I like that," Chloe muses aloud, enthralled by mystifying images forming in her head.

"They aren't concerned about being different or like anything else. Although they do respect apartness. They can't worry because they never move. They don't need to move because wherever they might be it's just 'being there'."

"What happens if someone moves them?" Jason toys with the idea of catching him without an answer. No luck.

"Oh," Grasshopper whistles gravely, "that's terrifying for a young tree that doesn't have many years looped in its trunk.

But for an older tree, it isn't too stressful. It's like a bad case of hiccups in time."

"I thought you said they didn't understand time," Jason lunges at the apparent inconsistency.

"I said they don't understand clocks," Grasshopper sneers impatiently, "Clocks are not the same thing as time. Change."

"What's the difference," Kristin cannot resist asking. Her voice sounds puzzled at reckoning time apart from a clock.

"The same difference as between a flag and a breeze," Grasshopper says lightly, "in time-streams, moments follow moments, clock tics. In time-places, moments find moments, opportunities open.

"A hiccup in time," Chloe smiles happily, "I like that."

"Yes," Grasshopper says. It would sit very still and erect for a while. Like this." He abruptly hiccups as his eyes flash open in the glare of light from the stove as he gazes sternly about, inspecting his imaginary new station. Everyone laughs, even Kristin, as he relaxes and says, "See? First it is and then—blip—it is again. Only the place is replaced, changed—new is apart from used to be. He feels different and yet the same. Like when you're daydreaming and suddenly a whole pack of time disappears and you can't remember anything about it except a ditchbank's voice. If somebody moved you then, it would be like a hiccup in time."

"Oh, I understand," Chloe remarks excitedly.

"You would," Jason mumbles.

"You're just a near-sighted, forgetful Ant," she punches him in the side with her elbow.

"It makes more sense if you lose your Ant, I suppose," Jason says sardonically.

"Yes!" Grasshopper chimes in, "That's what the ditchbank said! You forget your Ant!"

"The ditchbank soliloquy again," Jason groans, "I guess I've wasted my entire life studying. I should have been out talking to ditchbanks and trees."

Grasshopper looks blankly at him for a moment, then shrugs that he confuses the two. "Ditchbanks are very different from trees," he admonishes.

"Of course," Jason growls.

Again Chloe punches him in the ribs with her elbow, this time more forcefully. "Well, they are," she says in defense of Grasshopper's premise.

"I know, I know," Jason chuckles, "God, I can't believe how ridiculous this sounds." The elbow hits again. "All right, all right. I'll shut up!"

"Good," Chloe murmurs.

"He doesn't know because ditchbanks don't talk to him," Grasshopper says to Chloe.

"Trees, either," Jason reminds him.

"Jason, please!"

"All right, all right."

"Anything can talk to you," Grasshopper says slowly. "If a little of everything is in you and a little of you is in everything, then all you have to do is listen. Pay attention. Imagine.

"Anyhow. Unlike trees, ditchbanks know absolutely nothing about being still. They're always running somewhere—like roads, except ditchbanks are more informal and friendly. Anything a ditchbank says, however, has to be accepted as something coming out of a ditch."

"What do you mean?" Chloe asks.

"Well, they always come from someplace and go to someplace else and that's all they understand and sometimes even that gets clogged and confused. Coming, going, carrying, filling, emptying and all the things that can happen like that. A tree and a ditchbank cannot converse. Not only do they not understand the same things, they don't even know about each other!"

Grasshopper revels in his realization that ditchbanks and trees don't even conceive of each other. "The ditchbank said that I was once an Ant and that I would be again," Grasshopper admits diffidently. "I was furious! I couldn't be one of those stupid, frivolous, hungry little beasts. Those little thieving managers always piling up things they collect. Greedy, possessive, snarling little," Grasshopper pulls back from an unpleasant thought. "Well, anyhow, that all became very obvious after a while . . . like when Chloe thought I could help Grady."

"But you did!" Chloe insists, her voice is enraptured at the recollection.

"No," Grasshopper shakes his head from side to side with firm conviction. "Of that I'm certain. I did not heal Grady."

"Well, at last—thank you!" Jason exclaims. It is his turn to jab Chloe's ribs.

"But what happened, then?" Chloe's eyes shine in bewilderment.

"If a tree and a ditchbank suddenly taste each other, then marvelous things can happen," Grasshopper explains quietly.

"I don't understand," Chloe moans in anguished confusion of her own recollections, "How? What about the little boy on the freeway?"

"He was tired . . . and very afraid. And his mother didn't understand his fear and help him learn from it. Everyone was telling her that he needed a doctor. She too became frightened. And then they were going to be driven away, out into the wild. And so the little boy became even more afraid and finally he just left—went to sleep and wouldn't wake up. I could . . . almost taste it . . . I didn't like it. So I got his attention. And he knew I wasn't afraid and so he woke up. The ditchbank and the tree . . . see?"

Grasshopper spreads his hands wide apart, pleased with the clarity of his explanation.

"But what does a tree and a ditchbank have to do with . . . ?" Kristin asks suddenly.

"Jump, Kristin. Jump!" Grasshopper says quietly. "A grasshopper jumps. If you have to ask Antsuck questions you'll never know how a tree grows out of its ditchbank," he growls in exasperation at her laggard pace in following his exposition. "The tree world lifts up whatever is delivered to its roots without knowing how it got there. And for the ditchbank, the tree is just an interruption, maybe even an obstacle to whatever it carries."

Grasshopper looks sad, as if Kristin is deaf as well as a little dense. "When something happens like the ditch carrying something bad for the tree or the tree closing over the ditch then they both are in trouble. Neither really knows about the other but still they are closer than spit on a tongue. The ditchbank is a sideways world. The tree is an up and down world. Trees and ditchbanks are just ways of being. But you have to jump to see it."

"But it doesn't make any sense," Kristin wails.

"Kristin!" Jason advises, "Don't try to make sense of delusional chatter!"

"Jason, stop it!" Stacy and Chloe join the fray.

"There he is!" Grasshopper points a finger at Jason.

"What? Who?"

"Your Ant! The one that's all clogged up!"

"What are you talking about? It's me, Jason."

Grasshopper drops his hand back to his lap and shifts his shoulders. Finally he looks at Jason and says amiably, "This is no way to get acquainted."

"Just don't go pushing people around," Jason warns him.

"I don't think you'll ever know who I am, Jason. You are a tree, stubborn and unmovable. But until the day you die and fall back into that ditchbank, you won't know that things sometimes happen that don't want to be explained. You won't know upping.

"You say that your world is made of this and that. For you it acts in very specific ways. And you simply ignore the rest sweeping past you, unawares, in the ditchbank out of which you have grown. I'm telling you," he says hoarsely, suddenly grabbing Jason's arm, "every particle—even the tiniest little morsel of you—everything in the universe is a little tree sitting by a ditchbank. And every person—even the most clogged up stupid Ant—is a little tree sitting by a ditchbank. And the whole world is a tree sitting by a ditchbank. You can talk to the sky or shout down holes, but you'll never know what the earth delivers to you until you awake and pay attention! That is what is clogging your world and that is what will kill you."

"Yeah, yeah," Jason is bored by preaching.

"But you are a healer," Chloe picks up the old thread.

Grasshopper looked at her sadly, as if sorry to disappoint her.

"I've lost my Ant. Somehow that affects people. But they soothe themselves when they pay attention."

He looks down at his two open hands lying palm upraised on his thighs. The fingers of the left hand wiggle slightly, then those on the right. After a moment of repeating the movements as if testing the outer extremities he says quietly, his voice distant and hollow, "It's all so confusing—when everything is a center."

His voice trails off into silent meditation. It seems for a moment as if he has left them all there alone to ponder their own plights. Kristin and Jason exchange glances of mutual resolve.

"Jason and I think we know what's happened. And it affects you and Chloe, too, Stacy," Kristin says evenly.

"What?" the two women chorus; Grasshopper seems disinterested.

"Peter's nanotech, the nbots and replicators. You both have them. Jason and I think that's what healed Grady."

"What? But how about your wrist, when you fell from the tree?" Chloe is reluctant to embrace new conjectures.

"I suppose I didn't really break my wrist. I just thought I had." Kristin arises to throw some more coal into the ruby glowing hotly in the belly of the stove. Hellbender suddenly lifts his head and utters a low, ominous rumbling. "It's the panels on the window downstairs," she suggests.

"Hellbender wouldn't growl at that," Jason says, "those panels have been making noise all afternoon and evening."

The wary rumble in Hellbender's throat grows more challenging as he arises to stalk cautiously toward the door, his nose sampling the air with quick, tentative wriggles.

Kristin whispers nervously, "There's someone out there!"

Jason jumps up and runs to the window near the door. Peering out he can easily make out figures of men moving slowly toward the building. "There are people out there! Lots of them!" He hisses.

Jason and Kristin jump up, looking for some means of defending themselves as footsteps begin ascending the outside staircase. Hellbender barks furiously at an approaching presence. Grasshopper, flanked on either side by Stacy and Chloe, sits impassively, looking at his hands and flexing the fingers.

Outside a bullhorn announces, "Dooley! Peter Dooley . . . aka Grasshopper! This is the FBI. Come on out!"

"The FBI! How did they find us?" Jason at first is mystified, then realizes what must have happened, "Sam. He planted something on us. They've been tracking us!"

The door bursts open and intense flashlights blaze into the room. Hellbender backs away from the door, snarling and barking in outrage.

"Nobody move! FBI!"

"We're not armed!" Jason shouts, stepping back, his hands over his head. "Don't shoot!"

"Okay! Just take it easy. Everybody."

"Sam!" Chloe exclaims as she recognizes his voice.

"Sam? Sam Horowitz? Is that you?" Jason demands squinting into a blinding light.

"Yeah, that's right," Sam's voice confirms gruffly, "say, aren't there any lights in here?"

"No. Just the fire in the stove and a couple of candles," Jason says, somehow suspecting that lack of lighting has deprived them of some civil rights.

"Which one's Dooley?" another voice asks.

"Him," Sam's flashlight swings to spotlight Grasshopper still sitting cross-legged, watching his wiggling fingers and ignoring the sudden brouhaha.

"What's he doing?" the voice asks.

"Who knows? He's loopy."

"You bugged us!" Jason accuses, "that's illegal!"

"Tell it to the judge, kid."

"But how?" Chloe asks, puzzled that she has not noticed anything attached.

"It's biotronic, dropped in your food," the other agent explains proudly, "it gets stuck in your appendix."

"Shut up, willya?" Sam snarls.

"But what if she gets sick?" Jason demands.

"It won't hurt her," Sam says gruffly, "we use them all the time. They're designed to dissolve and be absorbed by the body."

"Come on, Dooley," the other agent walks toward Grasshopper who now looks up, "you're under arrest."

"A rest?"

"He's a material witness in the attempted assassination of Ira Crown and the murder of a number of Federal officials."

"That's no reason to arrest him," Jason protests.

"Stow it, kid or you can come with us, too, for interfering with federal officers."

"Jason. Please don't!" Chloe begs. Hellbender still is barking vehemently.

"Somebody quiet that dog before he gets put down!"

"Hellbender! Quiet!" Grasshopper orders sternly as he gets up, adding in resignation. To the officers he says softly, "I'm coming."

Two more officers enter, take hold of Grasshopper and snap on wrist restraints to hold his arms behind him. "Take care of Hellbender," Grasshopper pauses to lean down and awkwardly rub the dog's head with his manacled hands. "Goodbye, fella. You're a great guy." Then he calls to his companions as he is led out the door. "Maybe you can get him back to Hector."

"Put Hellbender in there," Jason says to Chloe, pointing to the other room.

"Come on, Hellbender," Chloe sniffles, "this is just too much for any of us. Be good now." Leading by his collar she ushers the reluctant hound into the adjacent bedroom and closes the door.

"Be careful with him," Sam cautions two agents escorting Grasshopper from the room. To Jason he says, reluctantly, "I can arrange for you to be taken somewhere, if you want, after we finish our debriefings here."

"Debriefings?"

"Outside is a skypod with field equipment. We need to ask you some questions and have you make depositions before you can go." Sam motions four men to move everyone outside to where a skypod has set down to anchor itself by the road.

Debriefing turns out to be far less involved than Jason expects. He, Chloe, Kristin and Stacy are all led inside a small cubicle enclosed by a metal cage. There, after a few preliminary questions that seem more like delaying tactics than useful inquiries, a staffer enters and says, "All clear."

"Okay," Sam sighs heavily, "let's wrap this up and get back."

"Wait. That's all? What kind of debriefing is that?" None of the proceedings makes any sense to Jason.

"Did I say 'debriefing'," Sam gives him a blunt stare, "maybe 'decommissioning' would be more accurate."

"Decommissioning?" Chloe asks, "What's that?"

"That's what they call the termination phase," Kristin says, "they switched off the nanotech, the replicators, by encoded radio signal"

"Something like that," Sam nods, "I don't pretend to understand it. I just follow orders. But it turns out not everyone had them. Only Chloe and Stacy. The techs had to screen you all inside this unit just to be sure. We already processed the Abbots and everybody back on the highway who traveled with him."

"A Faraday cage," Jason exclaims. Suddenly he understands the elaborate precautions taken to isolate them all from ambient electromagnetic interference: to ensure a thorough reading and purging of any vagrant nanotech.

"Why didn't you treat Dooley?" Kristin asks, dreading that she already knows to be the answer.

"We're ordered to bring him back as is for further study prior to decommissioning."

"Is Eric Cornrath still involved in this?" Kristin demands angrily. "Has Crown made some kind of deal with that devil?"

"I have no idea," Sam answers, his expression blank and disinterested, "I just follow orders."

"Will someone please explain what you're talking about." Chloe pleads. Kristin quickly explains the nanotechnology that let Ant and Grasshopper interact within a virtual media environment. She recounts hearing from Eric Cornrath how rogue replicators had fatally amplified her brother's drug high and speculates that somehow the nanobiotic agents had spread from Peter to Grady, as well as to Chloe and Stacy.

"Then that's how Grady was healed? Those replicators repairing damage done by the bullet?" Chloe is deflated by sudden demystification.

Sam says to Jason, "President Crown has ordered me to assist you and your friends in any reasonable manner."

"I guess we could use a lift back to ConRelCo," Jason sighs in resignation; to walk back tomorrow would seem so dreary in defeat.

"Can do," Sam answers impassively. "Okay, let's wrap this up," he barks to two agents who lead the entourage out of the skypod cabin and back into the evening air where they are escorted to road vehicles for departure. The skypod releases its earth anchors to float lazily upward, assisted by faintly whirring electric fans.

As they forlornly wave goodbye toward Grasshopper's lackadaisical smile beaming at them from the rear of a sedan, Jason remarks, "Like he says, you never know what may pop up from a ditchbank."

~ 26 ~

"Eric Cornrath! You old Lucifer! Come in, come in!"

"Wally Martinsdale!" Cornrath exclaims heartily. The two men crow like old war buddies. "What are you doing out here, Reverend?" Cornrath enters his own former office and shoves his hand into the Reverend's big paw. Eddie ensconces himself on a nearby sofa as the two men gravitate toward two chairs.

"Doing the Lord's Work, Eric. Just doing the Lord's Work."

"One of the things I've always admired about you, Wally," teases Cornrath, "is that you seem to be able to find the Lord in just about anything you want to take on."

"Now, you old rascal. Don't go trying to get me in trouble with The Boss," Martinsdale booms effusive appreciation of Cornrath's joshing, "judge not."

"I know that sometimes the Lord's Work seems to pop up in the Devil's playground," Cornrath laughs heartily, "didn't you get your start preaching to sinners on Bourbon Street, down in New Orleans? Way back when? But then, I suppose even you have to grab market share where you can!" Cornrath chuckles effusively, then glances over to the sofa, "Eddie, you know the Reverend, don't you?"

"Yeah. Sure. We've never met, but I've seen him in the news. Hi, Doc." He rises, reaches across the large coffee table between their two chairs and shakes the Reverend's hand.

"Eddie was regional director of FERAL. Now he's working with us at Amberson Agency."

"Pleased to meet you, Eddie," Martinsdale blathers effusively as he settles into his seat, eyeing the two men expectantly. "What're you boys up to these days? And where have you been keeping yourself, Eric? I tried to get in touch with you when we first thought of this Crusade—Miles said you'd been called up from the Reserves. I didn't know you were an old soldier."

"Well," Cornrath replies in tones that in other circumstances might have told of great campaigns, "now it's time to 'fade away', as they say." He glances over to wink at Eddie. "Now that the worst seems to be over the Army has . . . ah . . . released us. We're considering a few opportunities we've come across during our duties here."

"Great!" the Reverend booms, "That's just great! Maybe I can talk you into handling our Christmas Crusade in Pittsburgh." The Reverend hesitates for a moment, then says with a degree of impatience, "Miles put a guy by the name of Wilkes in charge of Crusade publicity and organization."

"Fred Wilkes? Jeez," Cornrath winces, "Miles must really have been shorthanded."

"His ideas sounded okay at first, but somehow they never evolved into anything effective," Cornrath complains. "There is a hunger—a real hunger—in this nation for a return to simpler, more basic pioneer values," Reverend Martinsdale thumps his seat arm to emphasize the paucity of their current producer's concern for fundamentals. "Virtue and goodness. Straightforward ideas to be embroidered and hung over the fireplace beneath the trusty old flintlock—you know what I mean?"

"Be sure you're right, then go ahead," Cornrath echoes a familiar litany of plucky yet chary wisdom, "if at first you don't succeed, try, try again. But look before you leap."

"Exactly!" the Reverend nods in gratification, "But Wilkes doesn't seem to know what I'm talking about. He keeps pushing glitter and production numbers. It's all he knows how to do."

Cornrath laughs scornfully, "You're right about that—that is all he knows. Let me guess. He wants a full chorus line with girls in coonskin caps and not much else, right?"

"Exactly!" Martinsdale thunders, "that's precisely what he suggested. And it just won't fly! It won't fit the theme of our Crusade. He's going to wash the whole thing down—he'll draw people, sure. I mean we'll get a certain amount of attention after thirty-five cheerleader types parade across the platform shaking their bazooms to wake up the boys. But then what am I supposed to say? I mean, I'd just be embarrassed. If my daddy saw something like that—Lordy, I'd be explaining 'til doomsday. Heck! I'd be explaining ON doomsday!"

He chuckles heartily. Cornrath smiles sympathetically at the Reverend's plight. "Fred always pursues the line that you can't sell a person what they already have plenty of," Cornrath ponders the promotional aspects. "He's stuck, though, in trying to sell stoic endurance. It just won't take on people who are out of work and looking for something better."

"But that's just what we must get across," the Reverend stoutly insists. "Stoic endurance. Pioneer virtue. Bite the bullet. The President and I agree on that."

"Crown?" Cornrath tries to mute an inflected note of concern, "What about Ira Crown?"

"Oh," Martinsdale gloats in the wake of the most formidable alliance he has ever gained and shifts his position to more comfortably face the two men. Eddie darts his eyes toward Cornrath, eyebrows lifting apprehensively. "The President is in such a fix with the evacuation of the northeastern seaboard. Then this Grasshopper thing. It has him stymied. He can't seem to get the kind of media coverage he needs to get the country back to work—he says its because they're out chasing Grasshopper. But they've found him now and want you to come back onboard for the final study and decommissioning. Then Crown will have him checked out by psychiatrists over at Walter Reade. I'm supposed to arrange a multinational media event during the Christmas Crusade to reveal to the world what a charlatan and impostor this Grasshopper really is."

"Grasshopper?" Cornrath's interest is piqued, "You mean they found Dooley? We need to get him back to our lab before he goes to Walter Reade. Tell Crown that's imperative. Even more important than solving his PR problem. We have to get him back into the lab to evaluate what's happened with military technology before any academic or medical doctors get to him. If he'll do that, then maybe we can work something out . . . otherwise, count me out."

"That'll work," Martinsdale nods vigorously, "sounds like we're converging toward a common purpose. I'm sure he'll agree."

"Good," Cornrath raises his eyebrows toward Eddie, who slowly nods that he'll be okay with the developing accord.

Martinsdale muses, "Crown chewed me out—up one side and down the other. Just because I'd never paid any attention to this Grasshopper. But I guess I'm not the only one."

"Don't feel bad, Reverend," Cornrath shrugs, we've all been preoccupied with bigger concerns. He flicks a knowing glance back toward Eddie, then continues, "I created Grasshopper. I made him the sensation he is but lately I've been out of touch. But we can bring him back in line. You might say he's already rigged for the occasion. We just have to get him into the lab before any academics screw with the rigging."

"I'll try," Martinsdale sighs as if a pervasive fog is lifting. "Maybe it's not quite the problem Ira Crown thinks."

"Oh, we can make a deal," Cornrath calculates rapidly. "There is a slight, ah, discrepancy, though." He exchanges winks with Eddie. "You see, Eddie and I left the Army under what might best be called dubious circumstances. Not that we did anything wrong, ourselves, mind you. But the Seaboard Evacuation had a lot of problems—you can imagine. I got reassigned here by mistake and when Crown started taking the fallout and looked for somebody to chop before he got chopped, I happened to be the convenient scapegoat. Do you follow?"

The Reverend nods, his eyes rounding in awe of the myriad legions of hell contending for control of the nation. "I'm not concerned about any of that, Eric," Reverend Martinsdale assures him. "I know you and I know that you are the best promotional talent in the business. I have to worry about what Wilkes is doing to the Christmas Crusade. He just doesn't understand the Christian message like you do, Eric. We're spreading the Good News, not opening a pleasure palace."

"That," chortles Cornrath, "is always a problem—getting your kind of message across—over the competition. Communal singing and magnificent spectacle; that's what it's all about."

"You can do it," Martinsdale beams in confident entreaty, "I've seen you do it before. Take it over, Eric. Take it over and we'll work out an accommodation with Ira Crown."

"Do you think you can do that?" Cornrath inquires casually. He raises an eyebrow in response to Eddie's searching glance.

"Ha!" Martinsdale snorts, still smarting over the dressing down he had received from the Chief Executive, "He's barely holding things together as it is. Right now he can't afford for this Pittsburgh Crusade and the Grasshopper finale to fall through. He needs you as much as I do—actually, he needs you even more!

"You're the best, Eric. You know how to talk to the ordinary man and woman and to do it with a touch of class. That's what will put this across. I know it. Crown will know it. And if he doesn't go along, well, he just might lose the whole show!"

Eddie chuckles contentedly at this abrupt turn in their fortunes. Cornrath settles back into plushly upholstered comfort and smiles through savvy eyes,

"Reverend. I think we can do business. Just get me back in the lab and bring Grasshopper to me."

~ 27 ~

Fred deWitt watches the deep blue Sterling police cruiser slide to a halt by a sign declaring, *HOSPITAL VEHICLES ONLY*. He surmises that the FBI can park anyplace they want, any time. A warm, ebullient awareness of the importance of these men spans out to include himself. After all, he has been ordained by Dr. Roasdahl to shepherd Peter Dooley through transition from federal to psychiatric custody. Fred inhales deeply and pushes perfunctorily through double glass doors into the aura of his new mission.

Four men are in the automobile. Fred watches intently, wondering if he should smile a welcome or mirror the profound gravity of their enterprise. After all, he has heard that Ira Crown himself has asked Dr. Roasdahl to drop everything and tend to this case. One of the four men emerges from the front passenger seat, closes the door carefully as his eyes scan about, searching for indications that the area has not been cleared as requested.

"Are you deWett?" the man flashes a badge in a brown leatherette case.

"DeWitt," Fred corrects anxiously. His feet part slightly to match the man's calmly professional poise.

"Do you people have everything ready to accept custody?" the agent inquires closely. Fred quickly reviews a mental list of items and procedures prepared and rechecked for two days. Every detail is as required; he nods. The agent seems to be waiting for more emphatic verification.

Fred stammers awkwardly, "Oh, yeah. Sure. Bring him in —what's he like?"

If only he had gotten a closer look at the agent's ID card. He doesn't know his name, but the agent knows his. Fred dislikes such asymmetrical social exchanges. It leaves him unprepared, vulnerable to the unexpected. Now that he has found his voice, its thinness irks him and he suspects that the agent's smile is really just a veneer over a barely civil smirk. Fred adds an appendix to the question, "What's your name again?"

"It's still Horowitz. Sam Horowitz. Secret Service."

"But I thought this was an FBI operation," Fred howls. Someone has misled him.

"We all try to cooperate," Sam chuckles casually, "mostly with the paperwork and surveillance. This is all at the President's request."

"The President?" Fred crows. It's true! The significance of the moment has been restored. "What's he like?"

"Ira Crown?" Sam seems baffled by the question.

"No, Dooley—Grasshopper."

"Oh. Well, he's finally home," Sam jokes as he looks at a sign identifying the psychiatric wing of the hospital. " He's a looney. Not much of a problem though. At least he hasn't tried anything with us."

"I guess not," Fred replies in awe of Dooley's escort under the banner of Grasshopper's notoriety, "he'd have to be crazy."

Both find themselves laughing politely, without really knowing why. Sam turns to gesture with his hand and two agent escorts emerge from the cruiser. The driver remains at his post, fiddling with a source of some pop music. Fred is intrigued to note that Federal agents listen to ordinary channels and ordinary music. It makes the world seem more accommodating, just realizing a thing like that.

He waits for his new ward to emerge from the car. Then as the two agents approach, it dawns upon him that his charge is one of them, a curious looking little man in baggy overalls and a red plaid hunting shirt. His shaggy, unkempt silver hair and beard make the top of him seem rather disheveled and horrifically unimportant—transient. Vagrant, even. Tennis shoes and bright yellow socks punctuate the other end of his slight frame with a sense of festive celebration. Despite his being led by wrist restraints, he walks in a jaunty gait and seems to be having a great time. Overall the impression is in ludicrous contrast with the neat businesslike presence of his escorts, dressed in suits, and with Fred's own spotless ocher uniform. He fears there is some mistake.

"This is him?" the question soars into an embarrassed falsetto. He had envisioned someone of more saintly or perhaps even demonic demeanor, preferably with beatific or spooky eyes or at least a large, dominating frame like prophets in the movies. "He looks like a weird jerk!" Fred whimpers.

Sam chuckles, "This is him—Grasshopper. This is what teleview has promoted to pull this country out of hell. Isn't that something?" The agent shakes his head in wonder.

Frowning, Fred inspects the little man more closely, straining to detect any hint of extraordinary powers or perhaps some mystic force field. The object of his scrutiny fails to conform to any of the various images he has concocted out of tabloid and television renderings. He feels slightly betrayed and suddenly reduced to ordinary straits. "Okay. If you say so. But I don't know what all the fuss is about!"

They enter a short corridor terminating in another set of double doors. As they move along successive hallways and up an elevator, Fred points out to the agents all the security features of the ward. He emphasizes that they routinely accommodate some very dangerous people. His expression indicates that he does not anticipate Grasshopper to be much of a challenge.

"Nobody has ever escaped," Fred enthuses proudly, "not for as long as I've been here. If one gets out of hand—blam!—that's it man! The whole place locks down like a vault and we zap'em with tranquilizer gas."

"This should hold him, all right," Sam agrees, "but like I said. He talks crazy, but he's really a low key kind of guy. We could have brought him in on a bicycle."

Fred glances ruefully at Grasshopper who seems genuinely unconcerned that he is the topic of their conversation. His demeanor reminds Fred of a child at an adult party, reticent and subdued, but taking in an unfolding spectacle. "Mr. Dooley will be a model patient, I'm sure," Fred remarks peremptorily as they move through a final set of double doors and arrive at a desk manned by two orderlies playing cards.

"Let me have the paperwork on Dooley," Fred postures stiffly. Without glancing in his direction one of the card players picks up a clipboard and tosses it clattering onto the desktop. Fred clears his throat. "That's all right, boys. I can handle this. Just keep on with your game."

Neither orderly seems to notice him. Once or twice they glance toward the agents and Grasshopper. Only the cards sustain their interest.

Fred nods in acknowledgment of an imagined response and hands the clipboard to Sam. His right hand produces a pen.

"Just sign each copy on that line."

"Okay," Sam squiggles the pen with a tired flourish, produces a document from the inside breast pocket of his jacket and adds, "If you'll return the favor, we'll be on our way."

The paper transaction complete, Sam tosses a mock salute and says, "Okay, Mr. deWitt. Nice doing business with you."

The other agent speaks up as he removes the restraints from Grasshopper's wrists, "What's the quickest way out of here?"

Fred points in the direction from which they came, "Just go back—follow the red line. We'll monitor you from here and open the access doors." He nods with his head toward a battery of video monitors, each depicting a different set of double doors. Occasionally a bell sounds faintly and one of the orderlies reaches mechanically over to press the appropriate button. Somewhere in the building a person moves from one corridor to another.

"Okay, Harry," Sam says, "We're through for the day."

"Fine with me. Is this typical of the fluff duty you guys in Treasury get?" the other agent laughs. "Over at Justice we work for a living."

"Ah, get off it," Sam growls as they passed back into the exit corridor.

Fred watches the agents depart, expecting them to turn and wave goodbye. They do not. Chagrined, he moves in the opposite direction, then pauses, turns to his charge and motions impatiently, "C'mon. I'll show you your room."

Grasshopper follows silently after as they walk down a barren hall to the end where Fred throws open a door and says in condescending monotone, "This will be your room. You'll wait here until Dr. Roasdahl can see you. I'm afraid you'll have to empty your pockets."

Grasshopper shrugs as Fred methodically searches through the overall and shirt pockets and then frisks him police style. He discovers, and places one by one on the small table by the bed, a smooth round pebble, a brownish red and bedraggled chicken feather, a broken dog collar, an oak leaf and a piece of crumpled, water-stained magazine cover bearing Ira Crown's likeness.

"Why do you carry this junk around?" Fred asks in disdain.

"Ditchbanks," Grasshopper replies simply.

Fred shakes his head tiredly to and fro, "Crazies—that's all I ever get."

He watches Grasshopper sit idly on the edge of the bed and look back up at him with eyes that remind Fred of his cocker spaniel, Westy. Nothing in the face suggests either violence or miracles. Not even mischief. To Fred's thirsty eyes there is only an ordinary vagrant dirtying bed linens at the taxpayers' expense.

"Did you have a good trip?" he asks disinterestedly. The calmly appraising gaze directed at Fred begins to make him feel awkward and alien.

"Pleasant enough," Grasshopper answers simply.

Fred awaits the usual protestations or inquiries of people committed to his care, typically against their will. Grasshopper, however, says nothing; he merely sits and watches the young orderly fidget. To escape the pressure of his charge's gaze, Fred moves abruptly to the window.

"If you lean way around to the right and look over the ledge there you can see the entrance," Fred announces. "A lot of important people come here, you know."

Grasshopper remains silent. Fred's eye catches sight of the sculpted eagle poised over the hospital entrance and suddenly it seems especially significant. Words well up to announce themselves as he fights against a curious sense of self-abasement.

"Boy, that's a great eagle! I think I'd be an artist if I wasn't planning to be a doctor. I like helping people—that's important, you know. Helping. Of course artists are important, too. Like that eagle—he looks like he's gonna take off any minute now!"

The flood of words halts abruptly. Fred grins sheepishly as he realizes they haven't made much sense. He turns to find Grasshopper leaning back against the headboard, his torso stretched out on the bed as his eyes rove in idle inspection of the room. His tennis shoes already have shed dried flakes of mud onto the crisply clean blanket. Fred tries to force himself into an easy posture. He leans his skeletal frame against the chest of drawers and pushes his thick glasses up onto the bridge of his nose.

Still grinning stiffly, he tries to salvage the discrepancy between what he had said and what he had meant to say.

"Like I told Wanda—she's my girl. At least she says she is when she's out with me." He grins slyly at the implication. "George—George Parsons—he works on the sixth floor. He says Wanda likes anything in pants and oughta be in sports so she could get paid for chasing balls."

He slaps the bureau top, laughing loudly. Grasshopper laughs, too. Fred wonders if the patient has really laughed or simply mimicked him.

"That's pretty good, huh? George is something else! Wanda hates his guts because he's always trying to trap her in the supply room or the exercise club. Anyhow, like I said to her. Everybody's got problems and unless people try to help out they just get worse and pretty soon all the really important stuff that has to get done gets left to hang around getting undone." Fred loses track of the sentence, but then recovers.

"Like Dr. Roasdahl. He's gonna be working on . . . er . . . with you. They say President Crown personally asked him to drop everything else and work just with you," Fred's tone implies that Grasshopper should be impressed. The orderly looks again at the wrinkled fragment of magazine cover with Ira Crown's face gazing back.

"Hell, he should have dropped everything a couple of years ago. At least that's what Wanda says. She really got mad when Dr. Roasdahl's wife left him and moved in with some other guy. She says Dr. Roasdahl deserves someone more understanding of the pressures on the poor guy. He's always helping people and finally they say his wife just got tired of him never being home and left just after their son ran off somewhere. Now their daughter is staying with his mother—Dr. Roasdahl's. She drinks a lot and won't let the kid go out on dates because she's afraid of muggers, rapists and germs."

Fred sucks in his breath to underscore the ponderous weights bearing down upon his mentor. "I mean, Dr. Roasdahl is a great guy just trying to help out the sickies of the world and, I tell you—it's just one thing after another.

"But like I tell Wanda, everybody's got problems, you know. And we all gotta help each other."

An inspiring thought slowly works its way into Fred's concerns. Without fully understanding how it might work, he suddenly announces, "Like you could maybe help Dr. Roasdahl, maybe."

While Fred rambles, Grasshopper's eyes have slowly closed. Now he opens them slightly, one more than the other so that he resembles a recumbent silver hooded owl. He glares suspiciously at the orderly still leaning haphazardly against the chest of drawers. "Help him do what?"

"Get some rest," Fred suddenly completes both diagnosis and prescription. He is buoyed in the idea that he and Wanda might yet help Dr. Roasdahl. It's inspiring!

"How?"

"Simple. Just make it easy on him. He's already dropped everything else. You're all he has to concern himself with right now. If you aren't too big a problem, you could be a vacation for the poor guy. He sure needs one!"

Grasshopper seems amused. "I'm a vacation?" His expression suggests that he has perhaps underestimated Fred's insightfulness.

"Sure. We can work it with the tests and interviews and all. You make it easy on him and he takes it easy and relaxes without knowing it. Relaxation, that's what Wanda says he needs most."

"I'm a vacation," Grasshopper muses aloud. He likes the taste of the novel thought.

"Yeah!" Fred urges enthusiastically, suddenly feeling such kinship with his new ward that he no longer resents the flaked mud on the blanket. "Just go along with him. Once he thinks you're responding to therapy he'll relax and maybe loosen up again. Like his old self. He's a really great guy. Wanda just loves him. She says he's the kind of guy you can't do enough for."

"I don't plan to go anywhere yet," Grasshopper advises, still musing the ditchbank prospectus of vacationing.

Fred takes the ambiguity as tantamount to agreement. He presses forward in exploration of Dr. Roasdahl's vacation itinerary.

"He likes to start off with Rorschach—the ink blots. He'll show you a bunch of weird pictures and ask you what you see in them."

Fred straightens up to make an emphatic point. "Now, for God's sake—don't see anything in them! Just play dumb. He's such a great guy that he'll probably help you along by telling you what they mean."

"Hmmmh," Grasshopper murmurs idly.

Fred continues happily devising the respite so desperately needed by Dr. Roasdahl and begins to pace about the room, crossing back and forth near the foot of the bed. Grasshopper follows with his eyes, tracking ochre-traced trajectories bumping the bed with every turn.

"Then he'll probably do dreams. He's not too keen on them, but he says they sometimes help to fill in details that might otherwise go unnoticed."

"I don't dream," Grasshopper interrupts, closing his eyes again to make a quick internal check.

"Great!" Fred proclaims, "That's fantastic! He'll read you some out of his books and ask if you like any of them. Now this is important. Still play dumb and let him pick out the interesting ones. It'll be the ones he already knows and so already understands. He won't have to sit up late at night and puzzle over them. He can get to bed early and get some sleep!"

Suddenly Fred wheels, almost falling onto the bed. He regains his balance and says ominously, "Uh-oh."

"What?" Grasshopper's eyes blink open. He shifts to a more comfortable position, fluffing up both pillows to support his head.

"The word association test—I don't see how we can get around that one. You'll just have to come up with some words."

"Oh," Grasshopper replies disinterestedly. He folds his hands patiently across his chest and again closes his eyes as Fred ponders the new dilemma. "Why don't I just leave before we get to that one?" he offers.

"Oh, no!" Fred considers the possibility but quickly dismisses it. "He'd feel like he's responsible. He'd go back to work and his vacation would be cut short and he'd have the same work load and a pile of guilt and frustration on top of that. I mean, he'd have to explain it all to the President! That's no way to end a vacation!"

"I'll just say anything, then," Grasshopper suggests helpfully.

"NO! No, no, no," Fred gestures emphatically. "That's what you're supposed to do! He'll be sitting up late at night again, figuring it out and looking all hollow-eyed the next day. And then he'll go into depression and start taking those little blue pills. And then, oh, God—he might try it again . . . no, no, no. That just won't do! Not at all!"

Fred recoils at the threat of demons returning to torment his mentor. He pluckily tries to approach the problem along a new avenue. "Let's see. He's a stickler for detail, especially in word association."

Sighing despondently, Fred searches for a route around the quagmire of word association.

"Why don't we just skip that one?" Grasshopper proposes. "If it's such a problem. I don't mind."

"No. That's just it. He won't skip it," Fred moans. "I know him. We'll just have to take our chances on that unless . . . I've got it!" he beams. Giggling with delight, Fred hops through a few steps of a would-be jig. Grasshopper is impressed. "I've got! I've got it!"

Grasshopper waits for more terpsichorean invention.

"I'll copy off the clues and answers from his crossword puzzle. He does them everyday at lunch. I'll get an old one and copy the clue words. You'll have the answers. You won't even have to memorize them. You can just read them off the list because he never looks at the patient during word association. He says it interferes with the mental processes. You can have the list in your pocket and pull it out whenever you need it. It'll work! It'll work! Wait 'til I tell Wanda!" he gloats, contemplating in advance her admiring favors.

"Hmmmh," Grasshopper reflects briefly, more impressed by his enthusiasm than by its occasion.

"It'll be downhill from then on. After the tests, when he's got an idea of the problem, he just talks to you. If you need it, he'll help you with the symptoms, too. He's such a great guy—you just don't know!"

"Symptoms?" Grasshopper seems suddenly concerned for Antward accusations. "What kind of symptoms?"

"Oh, headaches, depression, constipation, dizziness, hiccups, hyperventilation, insomnia—you name it, he can cure it. That's his specialty, chemotherapy. 'Talk is just to classify and confirm the problem,' he says. 'Give me bullets every time and keep the mumbo jumbo!'" Fred makes a pretend pistol with his forefinger and thumb. He blasts away at some fanciful syndrome.

"Or feelings—that's his favorite. Like he'll say, 'Maybe you have such and such feelings' or 'Perhaps you often think of' and so on. Just agree with him and you'll probably get from diagnosis to therapy and then to closure in a week or so!"

Fred exults in the prospect of a week or two of rest and relaxation. Dr. Roasdahl will be happy. Wanda will be grateful. He will regain his own throne in her eyes.

"It'll do wonders for him. And the rapid progress with your case will cheer him up. I think he's afraid he's losing his touch. Wanda says he worries about getting old, but won't admit it. He used to be recognized as the best in his field in the old days . . . that was before my time."

"Therapy?" Grasshopper responds in abruptly evident hostility toward the Antward prospect. "The ditchbank told me to be careful of Ants—especially the ones that want to get into your head!"

"Now, for God's sake, be careful!" Fred demands anxiously. He moans at the eruption of unexpected symptoms in his ward. "You don't say anything like that to him! Not ever! You'll have him up all night worrying!"

"Sorry," Grasshopper shrugs, "but in this Ant Palace even Grasshoppers have to be careful."

"Now look," Fred frets in exasperation. He sits down on the foot of the bed; his hand grasps Grasshopper's right tennis shoe as he implores, "There's nothing to worry about. He always recommends the same things—more sex, less work and do your own thing—Follow your bliss, he likes to say. But if you start talking about ditches and ants you're gonna get both of you hung up in diagnosis and this thing could go on forever! Then you'll never get out of here!"

"Don't bet on it!" Grasshopper retorts firmly, "When it's time to go—I go!

"Look," Fred cajoles coolly, "trust me. It'll be easier for everybody—you included. Just do it my way. Even if you escape, they'll just find you again. You're big stuff. You can't hide from the FBI. This way you can help Dr. Roasdahl and yourself, too. And I will greatly appreciate it."

Grasshopper looks across his tennis shoes at the hopeful entreaty in Fred's face.

"I don't make promises," he avows slowly, "they just clog your roots. But I will stay here until it's time to go. And until then I'll try to do what you say." Grasshopper pauses, then appends a fiercely cautionary note, "But he'd better not try to push me back into an Ant! Not an Ant!"

Fred murmurs in hapless frustration at his ward's implacable hostility toward some spectral insect.

"Just don't say anything and he probably won't even think of it," Fred urges hopefully.

"Okay," Grasshopper nods, "Okay. I won't mention Ants if he doesn't"

Fred smiles gratefully as he makes a mental note to check the Rorschach and all the crossword puzzles for any hint of insects, especially ants. And ditchbanks, too. His eyes settle upon the articles he has removed from Dooley's pockets. For a moment the enterprise seems doomed to failure from the outset. Then he smiles amiably for lack of a better approach and reminds the patient, "Remember, now. If you don't mention it, he probably won't even think of ants."

Grasshopper snorts, "Hmmph."

"Roofing adhesive."
"Tar."
"Omen."
"Bode."
"Santa."
"Monica."
"Shocked white."
"Ashen."
"Jackie's ex."
"Ari."
"Cupid."
"Eros."
"Hippodromes."
"Arenas."
"Caftanlike."
"Loose."
"Textbook section."
"Tableofcontents."
"Pointed."
"Aimed."
"Encore."
"Again."
"Frontward."
"Fore."
"Marriageable."

"Nubile."
"You've done this before."
"Molt."
"No—I mean this test."
"Dire."
"Wait, I'm talking now."
"Claire."
"This is not a list."
"Oct."
"This is not a list!"
"Indecent."
"Please, stop."
"Happen."
"No, don't happen. Stop."
"Snow."
"Dooley!"
"Rung."
"Wrong?"
"Right."
"Did you say 'wrong'?"
"Right."
"Why did you say 'wrong'?"
"Dooley."
"Dooley is 'wrong'?"
"Right."

"Just look at the picture, Mr. Dooley. Tell me what you see."

"Grasshopper."

"Tell me about the grasshopper."

"Well, he is being chased around by a bunch of Ants. They want to take him back to their little underground burrows."

"Why would they want to do that?"

"They are very nearsighted. Because Grasshoppers jump around, Ants think they are magical. They want the magic."

"That's very interesting. Tell me about the magic."

"Magic is magic. It only happens once. You can't tell about it because words happen a lot. Things that happen a lot can't touch things that happen only once."

"You mean that the magic is unique?"

"Uniqueness! It hops through Grasshopper. Or Grasshopper hops into it and makes it different each time. Ants don't like that."

"Why not?"

"Because it confuses them, I suppose. And then they get angry. But really it's the magic that they want all along, but won't admit it because it would spoil all their greedy little games. The magic won't let you own it."

"So what do you think is the reason for that?"

"For what?"

"For their wanting the . . . ah . . . magic, but not admitting it."

"I suppose the same reason that your mother won't let your daughter out of the house. They're afraid of muggers and rapists and germs."

"My mother?"

"Yes. Your mother."

"But how did you know that?"

"Another Ant told me."

"Another Ant?"

"Yes."

"You think I'm an Ant?"

"I know you're an Ant."

"How's that? I mean, what are the signs of Antness?"

"The nervous laugh. The way you bend over when you walk, because your stomach is knotted up. The way you look when you don't understand something—like it's my fault. All the new books over there that have never been opened. Neat piles of papers that others write and send to you for others to read."

"That's very interesting. You're speaking metaphorically."

"No, I'm speaking truthfully."

"I didn't mean to imply you weren't. The greatest truths are often couched in metaphor."

"Ants. Ants! You just have to carry something bigger than you. Then, when you can't, you break it up into little pieces. You don't care if it gets destroyed. Just chop it up with your little teeth . . . own it, get it back into the hole."

"Mandibles. Ants have mandibles."

"Teeth! Teeth! Just tear and chew the world into bits and pieces for the little Queenie. Never care what it might really be like. Just wad the pieces together . . . and pretend you've done something. The Taj Mahal! The Sistine Ceiling! All the little tourist Ants trooping through, pissing and spitting and acting like they own a piece of the heart."

"That displeases you?"

"You're damn' tootin', it does! Ants, all of you. You won't leave me alone! Kristin wants a consort. Chloe wants a healer. Jason seeks the Grail. Clara wrestles with her Antichrist while Martinsdale just eggs her on. And Cornrath conspires to own to world and everyone in it while Crown only cares to sit at its top. And you! You're going to make me into a sideshow."

"What do you mean?"

"I know why you're asking me all of these questions."

"And why is that?"

"Because you want me to be crazy."

"I want you to be crazy?"

"Of course. Cornrath has already taken away my nbTwin. The ditchbank was right; here I am, just another Ant. But I'm still not Dooley, I'm this . . . before Dooley. And I'll stay here as long as I can. It's the only magic I have left. But you think that after you've chewed through it all and made it into a syndrome, gotten me back in my armor, then I won't be a threat to your Presid-Ant, or to Clara Abbot and those Ants. Then you can sell all the pieces to people who want to get in on the action. I know what you're up to!"

"What am I up to?"

"You're the Medicine Ant! You think sickies are helped by you. But all you do is milk their magic, being careful not to get any on you. The Ants in here are the one's closest to the tit of wonder. You milk it, you spread it around and sometimes you manage to lure the defenseless back into some neat and predictable carapace. It's a damned shame! All these Ants on the verge of liberation and you pull them back into that madness!"

"What madness?"

"Look at the teleview, Doc! Read the history of the world! Ant swarms mauling and torturing Ant swarms. And have you ever seen them clean the cadaver of a Grasshopper? Or an entire city? It's hideous!"

"Tell me about the magic."

"Figure it out for yourself, Doc! I am Grasshopper! Nowhere in the whole universe is there another me! I am unique, indivisible and possessed of hydromatic awareness."

"Hydromatic awareness?"

"Yes . . . hydromatic awareness. It shifts forward and backward, leaps and bounds with the infinite grace and smoothness of quantumless energy in imaginary time. It sees everything in everything and needs no dreams to restore a unity carved and dismembered by impious little Ant teeth. It draws no conclusions and so makes no little Ant turds! It is the ever flowing renewal of all that is. It awakens every morning. That in itself is a miracle—reborn into all the other miracles of a single day!"

"But . . . you seem to be talking about . . . ."

"Stop! Get your goddamned little teeth off of me! Don't you dare try to give me or mine another name. I won't stand for it! Do you hear? I won't! I am Grasshopper. I have no mother or father. I am a child who's everybody's orphan. I cause nothing and I come from nothing. I am the elseness of wonder that flows through all moments. Aliswasmaybe. I will not be pinned down or hemmed in. I grow from happenstance to grease channels of chance, promote the improbable. I do not create miracles—I am the miracle!

"Now . . . good doctor . . . how do you propose to cure that?"

~ 28 ~

Thanksgiving Day launches the Christmas Crusade in a spectacular parade through downtown Pittsburgh, inviting the eyes of the world along a route through the city to the spectacular though lately dormant *Double Bubble Super Sports Arena*.

The Arena had been built in a burst of regional civic bravura glorying in a last gasp of urban stimulus revenues. In this marvelous paean to civic pride, in its fully opened flex-configuration, from high up in his aerie seat a patron could have watched, simultaneously, sixty-four basketball games or seven-hundred and sixty-eight boxing matches taking place across the expansive floor. Because the unaided eye does not detect much useful information from so great a distance, a complex system of sensory aids, both visual screens and acoustic speakers enable one to relax in comfort beneath a panoply of virtual competitions, opportunely interrupted at commercial intervals to allow a quick dash to one of myriad refreshment stands ringing the arena corridors.

It was too big.

Of course, no one ever even tried to find sixty-four basketball teams willing to participate in a "simultaneous season" as one wag called it. How could the cheers and jeers of one contest be kept apart from others in such an immense, upside-down sea of spectating? Further, simple logistics of carting so many participants and spectators to a single location and then moving them out again quickly proved prohibitive. Rarely did the arena exploit any but the most conservatively dwindled configurations.

Austerity had finally doomed such immensity to abandonment. Electrical, gas and water demands were far above allotments for entertainment facilities, even competitive commercial sports. Police forces were downsized below any ability to monitor and control ordinary traffic, much less large event crowds. In its latter days of operation, insufficient power for air conditioning induced a greenhouse effect wherein occupants suffered a tropical amalgam of sweat and locker room stench. The Pittsburgh Chamber of Commerce finally capitulated, lamenting that inadequate utility services had forced the closure of the ninth (or tenth?) wonder of the world.

A sarcastic observation in the online editorial page of the Hobart, Oklahoma *Sentinel* once noted that Hobart has just as much emptiness as Pittsburgh but the astute citizens of Hobart have not thought it worth underwriting its enclosure. Such irony in outlying regions typically has little impact within Pittsburgh; no one seemed to notice or care when the gargantuan arena closed. People just grew tired of wrestling with this renewal-osaur.

Now suddenly it springs back into life. Exceptions to austerity restrictions permit feeding whatever power it needs. The

Brotherhood has leased it for the entire holiday season and fronted money to resuscitate and refurbish the vagrant facility. Soon it is evident that tourism will be booming in Pittsburgh and that the true meaning of Christmas is not lost.

Thanksgiving awakens to jubilation arising across the entire metro area. Rousing from a sullen slumber of soggy sales and sagging profits, surly energies of recession, stagflation and unemployment are transmuted overnight into enthusiastic anticipation of revival, spiritual and, especially, material progress. Billboards, bus cards, media specials, as well as a round of appearances by the Reverend Wally Martinsdale, focus concerns upon Pittsburgh's role in unraveling the tangled fortunes of the entire nation.

Somehow a rumor is incited in the renewed fervor of the community and leaked into the national and international press: Grasshopper is Coming! Immense signs are erected on buildings to display a maxim rendered in needlepoint, suspended above a trusty flintlock: *HOPE * VIGILANCE * HARD WORK*.

Local media organs of the Brotherhood are rife with intimations that the *Double Bubble Super Sports Arena* is not unlike the ancient dinosaur. Citing the facility's lampooning sobriquet, renewal-osaur, it argued: had not Infinite Wisdom once created immense beasts of seemingly useless lives, only to be interred in deep mortuaries of eventually precious petroleum?

Now, too, enthused the Brotherhood, the Double Bubble would be transmuted via mystery of rebirth into a new Ark of the Covenant. Pittsburgh would become, by dint of blind yet energetic enterprise, the vital center and source of reawakening to Christian Truth. With that Truth God would sweep the nation and all the world clean of the debris of Mammon's Lust and Confusion. The hapless children of the world would be consecrated to His Purpose, mystically infusing their poignant desperation and purposelessness, like those hapless reptiles of yore, into basic and irrevocable trust in His Will and Implacable Mercy. As foretold in *Revelation*, that old Serpent, that old Dinosaur, would be consigned to the pit in shackles of Adamantine Divine Will. Its energies, miraculously transformed, pumped up and transmuted into Divine Power and Utility—In Grid We Thrust!

Meanwhile work crews labor continuously to apply new paint, work and replace decrepit fixtures and control systems, hang banners and bunting. A week before Thanksgiving, in a giddy, celebratory ceremony the two immense globes of the double domes flare with light as Reverend Martinsdale and Mayor Connally join hands to pull the electrical main switch that reconnects the Arena and the regional power grid. It seems for miles around to all who view the moment that the city has bared her glowing breasts to the approaching fall of night.

A relentless barrage of scintillating innuendo electrifies local public consciousness and galvanizes attention from distant regions. Money is no problem in Pittsburgh. Austerity simply evaporates. In the heady flourish of new prosperity throngs of pilgrims journey to the city anointed by Reverend Martinsdale as the New Jerusalem.

Television crews and news personnel arrive from all over the world. The aged, infirm, despondent and dying respond to swarming rumors that Grasshopper will reveal himself during the Crusade. Insurance companies, obdurately cynical of mortality, take out advertisements across the nation proclaiming that no medical policy covers trips to Pittsburgh for any reason whatsoever. Real estate promoters, farmers selling produce from wagons and carts, prostitutes and pimps, entertainers and show business hopefuls, accountants, engineers, spouses and singles converge upon the city by rail, truck, air charter and foot. At least two of every conceivable kind of person and animal descend upon Pittsburgh for healing, salvation and at least one more crack at the good life.

"God is in Pittsburgh today," the Reverend Martinsdale proclaims to the world from the forum of a network talk show. "Tomorrow," he promises confidently, "He will be in your city. And if not tomorrow, then surely the next day. Friends, you have only to believe and trust in His Mercy. He'll come to you . . . but please, don't come to Pittsburgh in person. Stay home and watch the entire Crusade on teleview. The city is unable to adequately house and protect any more pilgrims. Please, for your sake and that of the Crusade—do not come to Pittsburgh!"

Still they come.

"What about Grasshopper, Reverend?" a reporter asks on another occasion as Wally Martinsdale is accosted in transit between his hotel and the local Brotherhood offices. "Is it true that he has been found? That he will appear with you during the Crusade?"

Reverend Martinsdale smiles enigmatically, "Rumors usually are aromas of some sort of fact. I can tell you that no one who has truly believed the Good News has ever been disappointed." The smile shifts subtly to broad reassurance.

"Then Grasshopper is indeed here with you?" the reporter persists as Reverend Martinsdale pushes his way through a rapidly consolidating crowd of pedestrians. He turns briefly back toward the cameras; his teeth flash a bravura message of detached compassion. "Believe. Have faith and God will tender His Own Mercy and His Own Plan!"

"But Grasshopper?"

"Is a child of God—as are we all," the Reverend abruptly interrupts. "And we can no more predict His Purpose than a jackrabbit can hop backwards!"

A program schedule for the Crusade is printed in the Brotherhood's Crusade Herald. While there is no overt mention of

Grasshopper's advent, there is a listing of titles for all sermons. The topic for the eleventh of December is *Healers, Prophets and Folly*. By the listing is an asterisk indicating that seats in the middle and front of section A, near the platform, will be reserved for all those with a valid doctor's certificate authenticating an incurable malady. Knowledgeable people begin referring to the eleventh of December as Grasshopper Day.

The Thanksgiving Day Parade is a two mile procession: turkeys, Steeler footballs, inflated dinosaurs with friendly faces, model oil derricks covered in aluminum foil celebrating the first oil wells drilled just north of Pittsburgh at Titusville in 1859, costumed Indian Fighters and defiant Anglo Fighters, Mayflower Pilgrims, teleview celebrities, beauty queens, Disney characters and finally a cotton-bearded and tiredly jovial Santa Claus who sporadically casts handfuls of brightly colored plastic grasshoppers to clamoring hordes lining the parade thoroughfare. On the chest of each grasshopper figure is stamped the numeral eleven. It is instantly read by the crowd, expressed in exuberant cheers as the realization sinks in, that December 11 is indeed Grasshopper Day.

Chloe, Kristin, Stacy and Jason have managed to gain assignments with the ConRelCo crews dispatched to cover the Crusade. Convinced that Grasshopper is indeed somewhere in the area, they have become increasingly apprehensive of his fate and determined to extricate him from its fruition. A crewman back at the Lancaster office has shown Jason how to extend the capabilities of his AiPAL to search even more powerfully among data exchanged along utility oplines and satellite channels. It is very illegal and he knows he could be whipsawed for it, but does it anyhow.

They are astounded to find in the AiPAL's cache copies of excerpted video records from Walter Reade Hospital depicting tests and interviews with Grasshopper. The AiPAL has intercepted a transmission from a Doctor Roasdahl to a colleague at the University of California at Berkeley. As with many academics, he has not bothered to scramble the contents.

There is evidence in it of serious usurpation of constitutional authority, not to mention unwarranted intrusions upon Peter Dooley's civil rights and various other Presidential "high crimes and misdemeanors". Jason is tempted to show the material to Mel. Kristin reminds him that he has illegally obtained the evidence. She convinces him that since it cannot be used in broadcast nor in any court proceedings, disclosure would only divert them from our immediate purpose. They should, she argues, let others discover and deal with evidence of conspiracy between Ira Crown and Wally Martinsdale's Brotherhood of Christian Arms. We cannot afford to be diverted by anything, she insists.

In days following Thanksgiving, hints and intimations of Grasshopper's imminence keep the city alive with rumors of sightings. The four frazzle their nerves and energies attempting to track down every lead. Grasshopper posters and T-shirts are hawked at sites of ostensible miracles where dubious witnesses recount his sudden arrival to redress some anomaly of nature or man. A chain of stands opens up throughout the city to dispense Bibles, incense, guitar strings, plastic crosses and the hottest souvenir of all—a plastic bobble-head Grasshopper doll complete in wardrobe of overalls, red plaid shirt, yellow socks and tennis shoes. Accessories are available at extra cost and include a replica hound that fits snugly into the doll's movable arms. Attachable angel wings fabricated from flameproof synthetic fiber and a series of plaintiff dolls afflicted with various ailments and snap-in plastic cures are available to petition the healer doll.

Hysterics dance sporadically in streets and on sidewalks in seizures of jabbering visions and frenetic glossolalia, celebrating Grasshopper's ever more stupendous legend. Shaggy minstrels play a variety of musical instruments in alleyways and porticoes to extol his deeds in unevenly rendered and impromptu borrowings and conversions of the tales of Jesse James, Pretty Boy Floyd and Big John.

The Pittsburgh Police Department attempts preparation for every contingency. Extra officers are borrowed from other metropolitan areas, sworn in and given last minute seminars in crowd control by veterans of New Orleans's Mardi Gras patrol squads. The hosts of citizens and supplicants now swarming streets in the day and descending nightly upon the Double Bubble to hear Reverend Martinsdale are too enthralled in anticipation of marvelous wonders to become embroiled in collective discontent. Nevertheless, city officials nervously attend each day as December 11 approaches, aware that the collective experience now moves in its own primal momentum and cadence. It has grown far beyond their meek powers to contain. Thus far, pandemic enthusiasm has seemed to ward off incipient disaster; nevertheless, many officials send their families out of the city in sheer panic at possible scenarios now being discussed daily. Helicopters perch day and night atop the Municipal Building.

Nightly, Reverend Martinsdale reassures avid crowds within and outside the Arena. He speaks to teleview audiences across the nation and throughout the world. He tells them that they are not alone in their hour of trial. He lathers them in hope even as he regales them for complicity in wickedness. On one night he berates their moral, economic and political collusions; the next finds him consoling their mangled souls.

All have sinned and brought the nation into its tribulation! God has not intervened, the Reverend asserts, because He created man free to choose between the opposed paths of Good and Evil.

Reverend Martinsdale exhorts all to spiritual aspiration, citing a long catalog of evils for which everyone in earshot will be held accountable. Very soon, he thunders into a subtle gasp of fear edging the taut atmosphere, an accounting shall transpire —lies, deceit, lust, greed, vanity and all manner of sinful delusion shall burst forth in torrents of iniquity from blasphemous dams ruptured upon the Final Impact of God's Truth.

On the next night he salves the very wounds salted on the previous. Be not afraid. God is merciful. His Grace is our rescue. Reverend Martinsdale emphasizes the latent goodness of man in God's Love and gently rails at them for allowing their baser natures to dissuade them from intrinsic, yet unrealized, goodness. He hints that the weaker among God's Children must be forced in law to accept the dictates of Parental Love. He suggests that government must become the Word of God crystallized on earth for the benefit and edification of carnal man, aiding good people everywhere in threading intricate mysteries of God's Purpose. Money, power, knowledge . . . all will mislead them, he admonishes.

Consider the Dinosaur! Pittsburgh could not have existed except as heir to the bounty God provided in that ponderous beast. For oil pulled from Pennsylvania's earth ignited and drove the industrial forces that created Pittsburgh and the rest of America's prosperity. And only God could have foreseen the connection! People must learn to distinguish between true succor and false seduction. They must embrace their authentic representatives in their vote and renounce and cast out those who would mislead them in fakeries of welfare and semblances of healing.

By early December a few media critics complain about the tenor and direction of Reverend Martinsdale's nightly sermons. Political commentators begin probing implications of his messages. They point out that the logical extrapolation of the Reverend's moral imperatives would arrive at tyrannical oligarchy manifest in a theocratic re-fusion of church and state. To rule by revelation, as one veteran expresses it, is indistinguishable by any test from rule by whim.

Bales of letters, many suspiciously identical, protest ConRelCo's editorial news services. One letter Jason recalls hearing about complains that the word "okracy" (the okracy?) was not in his dictionary, but that he doesn't like the sound of it. Sampled opinions and statistical analysis makes apparent that protest campaigns are being generated and orchestrated, probably from regional offices of the Brotherhood of Christian Arms. The general opinion expressed in them all is that the weak and morally inferior are unfit to govern themselves. Similar examples in logic and ostensible syllogisms are repeated across a spectrum of handwritten, printed, typed, word-processed and faxed sheets of orchestrated discontent elicited by any criticism of Reverend Martinsdale's ministry.

The Reverend does not himself respond directly to any of the scattered critiques of his sermons. He does not need to. Outcries and indictments of "intellectual buffoons" and "pointy headed professors", disdain of "media magnates" and "puffed up political pundits" and choruses of scathing refutations do the job for him. Secular alarmists become truly alarmed.

On December 10, the eve of Grasshopper Day, Reverend Martinsdale announces that President Ira Crown will, on the next evening, address the congregation by remote hookup following an event of momentous importance. For those troubled by the thrust of recent events, implications now seemed unmistakable. A few members of Congress, exiled from office upheavals in Washington, D.C. attempt to launch an investigation into a now evident cabal between Ira Crown's administration and the Brotherhood of Christian Arms. Others consult with their constituents and meet a monolithic reluctance to question the process of "pulling the country back together". Technical difficulties plague feverish efforts by uneasy Representatives and Senators to form ad hoc committees by means of electronic caucus. A quick probe by the General Accounting Office of records of the Brotherhood is derailed by investigators working at cross-purposes and a stolid lack of concern within the deeper levels of an increasingly feckless bureaucracy. Arcing over it all is a national consensus that what is transpiring in Pittsburgh is an event of sacral significance.

The Eleventh dawns into a crisp and frigid December day. A clear, crackling blue sky seems to detach the city from the rest of the world. Anticipation has built toward an implied, yet mysterious revelation that will revive a weary and strife-ridden world. Streets and parks are again packed with armies of the curious and the faithful as well as not a few cynics who have painfully learned to keep their opinions to themselves.

Patrolling police officers remark upon a strangely contemplative mood as they stroll among the watching and waiting throngs. An invisible strain seems to draw through every face. Even solitary prophets and seers wrapped in admonishing placards and handing out cheaply printed tracts are pacified in a ubiquitous and taut awareness that today is the day.

His friends have given up trying to find Grasshopper. The only bright moment in several weeks of searching, on the streets as well as within traffic of the ConRelCo data lattices, is when Chloe and Jason run into the Abbot brothers on a sidewalk outside the Arena. Both Hector and Grady are in a disheartened yet determined search for Clara.

"You wouldn't believe the change that came over that woman," Hector complains loudly, "she can't let go of the notion that Grasshopper is the Antichrist. Now she's taken to visions and talking to angels."

"The angels told her to get us out. She kicked us out of the house!" Grady says sadly, "Then she got her Ladies Circle church friends together and started holding all kinds of meetings. A local cable company even gave her a regular time to preach. And of course she had to come to this revival! It's terrible when angels take over—nobody cooks!"

"It's awful," Hector moans in agreement, "just me and Grady and Hellbender roaming around, waiting for her to come to her senses. I don't know what we're gonna do!"

They invite the two of them and Hellbender to come back and stay in the rooms they privately have dubbed "Grasshopper Central". Hector and Grady have been living out on the streets with thousands of others, subsisting daily on handouts and food disbursed from the Brotherhood, an alliance of local churches, the Red Cross, and even the Buddhist America Mission.

Their hope to rescue Grasshopper benefits mightily from the solid common sense of the two brothers. Hector and Grady both, perhaps because of earlier military experience, prove to be adept at planning how they might yet rescue Grasshopper from the debacle apparently planned for him. Hector's idea is one of simple desperation—they will create some sort of disturbance and whisk him out in the confusion. Hector, Grady and Jason will each wear extra articles of clothing for Grasshopper to don at the first opportunity. Kristin and Chloe will carry gauze and bandages with which to wrap his face. They will hover near the section reserved for medical supplicants and at the first opportunity grab him.

Grady and Hector manage to acquire a few firecrackers, salvaged from previous July Fourth celebrations. In such a crowd, at the right moment, they reason that a few firecrackers will divert security and create sufficient confusion to allow them to rescue Grasshopper from his captors. It seems all chance and all very desperate. Their worst fear is that Grasshopper might be so drugged or restrained that he will be unable to take advantage of whatever diversion they might create.

By seven o'clock on the Eleventh they are in position in the area assigned to ConRelCo personnel, watching thousands of the faithful thread along aisles to their assigned stations beneath undulating contours of the intersecting domes.

"I've never seen anything like this!" Hector marvels as he gasps in awe at the distant interlocking laceworks of exposed steel girders supporting the canopy.

"It's spectacular!" Chloe exhales slowly, "So magnificent."

Jason is enthralled by huge video screens placed strategically at intervals within the canyon space to allow every viewer, even those high above, in distant aeries, to witness what the naked eye cannot easily perceive down on the arena floor. "It's gotta be the world's biggest living room," he laughs derisively.

"What do you mean?" Kristin asks as she scans about, evaluating exit routes.

"It's just a seat to watch a gigantic teleview screen," he explains, pointing to the nearest billboard size monitor. She smiles ruefully, not in the mood for joking.

"How are we going to get Peter out of here?" Kristin asks of no one in particular.

"I don't know," Grady reassures her, "but we'll do it. Don't worry." He pats her knee affectionately, where a few firecrackers have been taped to her lower leg, beneath the pants. They all are carrying them similarly hidden. "In this place these things are gonna sound like the wrath of God!" He laughs at an unintended joke. Again Kristin merely twists her lips in the direction of a smile.

The far tiers look like a canyon wall textured with closely packed, colored bumps that, in the eye's sweep across the arena's expanse, became rowed patches of wavering people and finally emerge into the pajama and bandaged-swathed ranks of infirm gathered in Section A. Only their press credentials have allowed them to penetrate so close to the stage. All seats near the podium are assigned to those either in need of healing or assured of their own ordination into the elect. The *Brotherhood of Christian Arms* has subsidized waves of faithful immigrants from other states, bringing them in and ringing Reverend Martinsdale's dais with their ardently supportive voices.

The dais is a platform erected at the end of a long runway. It extends from a richly draped and festooned entrance over which a choir loft is poised astride rows of Corinthian columns. Crimson velvet swathes the air behind a millennially voiced choir and falls in sumptuous cascades to the arena floor from heavenly domes that focus and enrich the powerful coursing of the Doxology.

Far above it all, running along tracks and rails set into the overarching domes are video cameras and microphones controlled from a booth dangling like a distant eagle's nest at the intersection of the two hemispheres. The engineers are mere dots in the window of the remote cubicle.

"How do they get up there?" Chloe asks in breathless wonder at the vast distances engaged in the architecture of the arena.

"I suppose they come down from the roof somehow. They must control the video and sound systems," Jason speculates.

"Oh, Jason, look," Chloe suddenly giggles and points toward the immense video screens suddenly hawking popcorn, hotdogs and ice cream. Somehow the space seems less intimidating in the thought of munching popcorn. Then it stops; apparently in the distant cubicle someone has inserted the wrong oprock. The screens flicker briefly and then leisurely browse through scenes of the Holy Land and Stations of the Cross.

Hector points down toward Section A where bandage-plumed and wasted bodies sit anxiously in wheelchairs or lie on litters and gurneys propped up to provide a view of the platform before them. Hospital attendants and nurses seem almost to outnumber the assembled injured, ailing and hypochondriacal.

Moans and wails of misery hover above the mass of white and pastel colors punctuated occasionally by brightly hued prints cheering up this bathrobe or that pair of pajamas. Crutch fights break out from bickering over position as each supplicant plots the shortest route by which they might reach Grasshopper first. In anticipation of such a mad rush, burly attendants garbed in white with armbands marked by the duplex crosses of the Brotherhood of Christian Arms stand in solid array between Section A and the platform below. Not until now has Jason noted that the two crosses, depicted in perspective, are positioned to cleverly hint at the letter M.

"There's our diversion," Hector mutters, "I bet we can set them off!"

"It looks like they're gonna set themselves off," Grady agrees as he looks toward the loudly impatient and fidgeting infirm.

They quickly discuss several alternatives for beating the anticipated rush toward the platform and finally settle upon a course of action should a spontaneous disturbance occurs before they instigate their own.

Then the richly embroidered voices of the choir suddenly surge into anthem as the lights begin to dim. A spotlight brings all eyes to the entrance at the far end of the runway.

Unseen, high above them, microphones position themselves by computer analyses of acoustically live coordinates at which desired sounds may most efficiently be plucked from the incipient cacophony below. A ConRelCo engineer has told Jason about the Arena's incredibly sophisticated sound system that essentially controls itself. Technicians need only to follow along in their script, checking dials and gauges in the unlikely event that intervention becomes necessary.

Now the voices of the choir rise to overwhelm and silence the roar and excited buzz of spectators. Lights above the platform shift slowly through a spectrum of colors as hush falls like twilight across the assembly and they all wait breathlessly for the evening service to commence.

Reverend Martinsdale appears suddenly, striding through a wash of colored spotlights that lead his vigorously athletic figure down the runway to the platform as the full chorus surges into thunderous exaltation: *Hosanna! Hosanna! Hosanna in the Highest!*

He stands gazing through rapturous pleasure into the dimly shadowed canyon of the faithful as the choir behind flares in brilliant illumination and shifts suddenly into the anthem that has become his trademark, *A Mighty Fortress is Our God!*

His upraised arms lead the congregation through splendid harmonies of Martin Luther's hymn as his eyes and brows flare in benign punctuation of its cadences. As the final massive chords of recorded pipe organ accompaniment fade and the illumined choir fades back into semi-light, Reverend Martinsdale reaches forth into the pause of absolute stillness. His upraised palms lead the assembly in the benediction of the *Twenty-third Psalm*. His magnificently coifed head upturned and eyes closed, his powerfully built frame ablaze in a multihued incandescence of spotlights, he then recites The Lord's Prayer, pauses, thanks the members of the audience for their attention and commitment and then locates the distant dot of red light identifying which camera to address as he gravely intones, "Bothers . . . Sisters . . . Friends in Christ . . . these are terrible times.

"'We have called upon God,' many say, 'but God has turned away.'"

He pauses to let the image of an indignant Almighty hover briefly in the minds of millions.

"We have called upon our government in its armed camp of Lebanon, fled from the wrath that has consumed the Capitol and most of Gotham Seaboard. We have beseeched our leaders and officials to salve our wounds and admonish fears that rage in the night of this great land—our cries have gone unanswered."

Again he pauses to let the multitude consider its plight.

"So say many. And these same misguided unfortunates, bereft of loved ones lost in tribulation, uprooted from the care and conscience of their homes, in pain and suffering call forth to any who will listen. And they cry, 'Who has heard us?'"

Reverend Martinsdale turns his head slightly sideways as if listening for a faint echo.

"'There is a man,' they say, 'a wondrous man who can heal our wounds. Cleanse our souls. Wash our eyes in the balm of renewed vision.'"

The Reverend gazes stolidly out into the gray haze masking the congregation. His finely tuned ears catch a subtle gasp of anticipation. He flashes a smile of ivory and then drops his voice suddenly to a bare murmur. Far above the dais, directional microphones whir along their tracks, jockeying for best position as the murmured frequencies are plucked to flow and ebb in oceanic intimacy, washing across the assembly and among millions of teleview witnesses around the globe.

"There was in olden times," Reverend Martinsdale announces in softly measured decibels, "a man who proclaimed himself to be God. Simon was his name. Simon Magus.

"Read your Bible," he challenges confidently, "it's there: *Acts*, chapter eight, verses nine, ten and eleven."

He closes his eyes and quotes from memory.

"'But there was a certain man, called Simon, which beforetime in the same city used sorcery, and bewitched the people of Samaria, giving out that himself was some great one:

"':To whom they all gave heed, from the least to the greatest, saying, This man is the great power of God.

"':And to him they had regard, because that of long time he had bewitched them with sorceries.'" Reverend Martinsdale opens his eyes, leans against the clear plastic podium and waits briefly before intoning in a still, gentle voice,

"Bewitched. The people were bewitched by his guile. They were misled by fakery and illusion. They believed in Simon Magus as they never had believed in God and the fury of their misplaced faith caused them to see many wondrous things.

"And they told their friends. And they told their neighbors of the things they had seen and heard and the fame of this . . . this Simon Magus grew and grew and spread throughout the land and then into the world until many were ready to believe that he was indeed God!"

Reverend Martinsdale sips water from a glass on the podium. He replaces the vessel and toys with its rim reflectively.

"But Simon Magus was a man. Merely a man. An ordinary man filled with guile and deceit—and hate!

"Hate? Did I say hate?" the Reverend suddenly bellows.

"Yes!" thunders his own answer. "Hate. For Simon Magus must have hated these poor people, these poor Samaritans, to mislead them so. His heart must have been filled with the treacherous bile of Satan's own hatred of God!

"And so he fooled them! He made them believe that he was God!

"But," Reverend Martinsdale's voice again falls to a tone of gentle reassurance as he chuckles lightly, "God will not be mocked. He sent the apostle Philip. And he sent the great teacher Paul. In those days Paul and Philip were his number one and number two troubleshooters."

Reverend Martinsdale smiles knowingly into the gray haze,

"You know what troubleshooters are. They go into a situation and try to head off problems. They find out what the devil's going on. And Paul and Philip found very quickly that the Devil—Satan, God's Adversary since the beginning of time—Satan had a great deal going on!

"Think for a moment," he urges, "do you recall the headlines of today's news and yesterday's and the day before that, and on and on and on? Wouldn't you say that even now, especially now, Satan has a great deal going on?"

Reverend Martinsdale sees the dot of a yellow cue light, prompting him to turn toward the left as a new camera comes on.

"But what about Simon Magus? What happened to him? It seems that as is the way of Satan, old Simon Magus was of no further use and so Satan just let him go. In the presence of God's agents, the apostles Paul and Philip, Old Satan just decided to close up shop and hand Simon Magus his pink slip. In the presence of God's Anointed, Simon Magus was powerless!"

The Reverend smoothes back his already faultlessly groomed hair.

"Simon Magus destroyed himself by desperately trying to prove that he was God. He jumped from a high wall, trying to fly. And in the presence of God's Apostles he fell to a perfectly ordinary run-of-the-mill death. He had finally come to believe his own lies and so was destroyed by them!"

Reverend Martinsdale turns slowly to take in the entire panorama of two hundred and eighty degrees of blinding, incandescent haze behind which he can hear a growing, pulsing, gnawing anxiety. He allows the tenor to climb briefly until it feels just right. Then he jabs a finger out toward the distant red light and sweeps it through an intimidating arc about him,

"Remember Simon Magus in what you shall witness tonight," he admonishes. "His death proved the folly of his believers. The folly to which all of mankind is prone—the folly of misplaced faith!"

Reverend Martinsdale casually sips again from the glass as an angry buzz of consternation rustles through the immense occupation.

"Has Simon Magus returned? Or someone just like him? Is Satan up to his old tricks again?" he laughs contemptuously.

"Are we fooled? We who are washed in the Blood of the Lamb? We who sing Hosanna to God in the Highest? We who for centuries have challenged dark tribulations in the diamond assurance that we are saved?

"Surely Simon Magus is not our folly?

"Then why is he here? Would he deliver our souls to his Dark Master? Will he deliver to that Old Serpent the unchallenged domain of the entire world? Are we to sit idly by and watch the Antichrist gather in a lost flock delivered by this Judas?

"Well, are we?" he screams in sudden, furious pain.

"NO!" choruses the unseen thousand in the choir.

"NO!" scream the unseen thousands populating the darkness-veiled congregation.

"Who is he?" demands the Reverend in a raging bellow, "Who is this Judas?"

Loudspeakers foam his righteous fury out into the cavernous gulf separating him from the faithful. There is a frozen moment of incredulous disorientation. This is very much different from what has been anticipated. The energies of avid faith and expectation, steadily fueled in a diet of media speculation for weeks, now hovers uncertainly and then crystallizes in a taut precipitate of disappointment as again the Reverend's amplified voice demands,

"Who is he? What is his name?"

Someone in the darkened distance screams in agony of savaged hope transmuting into rage, "Grasshopper!"

"Who?" thunders the Reverend in dissatisfaction at the feebleness of the reply.

"Grasshopper!" More voices pick up the cry.

"Who?"

It becomes a rising, tumultuous chant opposed by the thrumming strains of the chorus behind him droning the melody of *A Mighty Fortress is Our God!*

"Grasshopper!"

"Who?" screams Reverend Martinsdale as he lurches and bobs with the assured cadences and gyrations of a veteran cheerleader.

"Grasshopper!"

Microphones scurry out of the way as the two domed cavities reverberate, pulsing booming echoes of four hundred thousand chanting, screaming, ecstatically frenzied voices. Engineers in the control booth watch in apprehensive gratitude as automatic monitors shut down for the duration of the torrential outburst. The booth itself shudders powerfully in the violence of the dual domes resonating in the throbbing frequencies of the multitudes roaring below.

Chloe grasps Jason's shoulder as she gnaws at her finger. Desperation recoils and mounts with each powerful intonation of Grasshopper's name. Kristin shudders and watches through panic-widened eyes where tears well and blur this field of dreadful vision. Grady and Jason exchange timorous glances as the sounds and fury mount. Only the imperturbable Hector seems undismayed; he chews idly on a toothpick as he shouts reassurance to us all, "Hell, this ain't nothin' but a big tent meeting' I've seen this before—it ain't nothin' he can't handle!"

Hector leans over to a couple of old men in wheelchairs who seemed confounded by the unexpected turn of events.

"You boys better get ready. He's probably gonna be out in a minute. He can heal you. Healed my brother Grady, there. And Grady was near to death. Show'em your scar, Grady," Hector grins complacently as Grady turns and pulls open his shirt to reveal the redly knitted flesh where once had been a savage wound.

"That, my friends, is all that's left of what a nine millimeter done to him at pointblank range. Grasshopper cured him—I saw it!" Hector avows sincerely. "She saw it," he points to

Chloe who nods assiduously. "He saw it," Hector points to Jason who reluctantly also nods on cue.

The two old men blink in puzzlement at Hector, then blink again as they gazed judiciously at Grady's bared and obviously repaired chest. A brief debate of exchanged glances finally decide them. They will not take a chance and be left out of some sudden flurry of healing. Flicking the switches of their motorized chairs they head down the specially constructed ramp descending to the arena floor in an attempt to gain a head start on the competition. Other invalids, seeing the two old gentlemen racing in high gear down the incline begin clamoring for their attendants and nurses to help them down. Some impatiently begin moving under their own power as the congregation continues its thundering chant, still building toward some yet unrealized and ponderous climax.

"Grasshopper! Grasshopper! Grasshopper!"

A spotlight flares its beam onto the entrance at the head of the runway. From behind two columns, apparently shoved out by hands that quickly pull the curtain closed behind him steps a ludicrously costumed Grasshopper. An outrageously padded and awkwardly askew variant of the costume in which he had gained celebrity now renders him pathetically absurd. Stuffed limbs of the hideous garb drag ponderously behind as he moves tentatively down the runway toward Reverend Martinsdale. Gigantic video screens depict a close-up of a tragically wandering figure, only his frail face gazing out of the monstrous costume is recognizable to his friends.

"My God," wails Kristin, "what have they done to him?"

His sudden appearance excites the crowd to a new plateau of frenzy as they suddenly are reminded of his origins in the televised skits of Ant and Grasshopper. Trickles of individually sensed embarrassment at having been duped join quickly into a torrential outpouring that clamors for retribution.

"How will we get to him?" Chloe pleads into Jason's ear.

"I don't know," he yells above the clamoring tumult, "hang on a minute and see what happens."

Reverend Martinsdale turns to watch Grasshopper shuffle ridiculously toward him in oversized clown shoes. The approaching face is obviously perplexed and baffled by the noise and blinding lights. The Reverend holds up his hands and, amazingly, the noise of the crowd suddenly subsides into an excited brushing of whispers and muffled exclamations. The automatic sound system switches back on and feedback from one of the microphones not quite in position causes a momentary squawk that is quickly damped. The piercing sound seems to further quell the apprehensive congregation, as if rebuked into silence. Only a few coughs punctuate the darkness surrounding the dais and remind everyone of the vastness of the enclosed space now seemingly centered in the platform beyond.

"Ah, yes," Reverend Martinsdale exclaims, as if just now noticing the approaching figure, "This seems to be our very own Simon Magus—Grasshopper!"

"Come here, Grasshopper," the Reverend challenges, "and in the presence of God's Elect dare to show us that you are not merely a man! Show us that you are not just Peter Dooley! Show us a miracle!"

Grasshopper's face on the video screens blinks uncomprehendingly. He seems frightened and disoriented and on the verge of tears. His eyes are gathered in disoriented consternation. Kristin and Stacy moan in helpless anger.

Suddenly the section of the dais where Grasshopper stands detaches itself from the remainder of the carpeted island and begins rising slowly into the air atop a rigid column. Grasshopper loses his balance at the sudden displacement upward and falls heavily onto the cushioning pad of a stuffed leg attached to his costume. He catches himself on both extended arms and waits until the rising shaft halts ten feet above the Reverend's head. He clambers clumsily to his feet, trying to avoid being tripped up by the splayed panoply of mock insect legs.

"Jason, what are they going to do to him?" Chloe shrieks a whisper. The cited image of Simon Magus plunging to his death is vivid in all their imaginations.

"Damn!" Hector gasps hoarsely, "Come on, we'd better get closer."

They depart their seats and begin working their way back around the line of arm banded attendants whose awareness, along with that of almost everyone else, is focused on the video screens and the distant spot of light where Wally Martinsdale exhorts Grasshopper to demonstrate his vaunted powers.

"Perform a miracle!" the Reverend taunts him, "Show us that you can fly . . . or heal the sick—an entire section awaits your mercies." He gestures toward Section A where attendants are now beginning to have to exert themselves to hold back the press of people poised to rush toward the raised dais.

"Go ahead," Reverend Martinsdale challenges in confident sarcasm, "at least tell us about yourself. Who are you?"

On the screens above, a close-up of Grasshopper's face shows his eyes blink and scrunch up slowly as he tries to help them adjust to the glare of the lights. Microphones whir along their tracks in search of the acoustic center of his space.

Grasshopper squints into the dark and then abruptly straightens up as best he can under the awkward costume and stifles a yawn.

The Reverend seems startled by his noncommittal gesture and again commands, "Tell us! The people are waiting!"

As the congregation buzzes in impatience and suspensefully mounting discontent, Grasshopper kicks off the gargantuan shoes that cramp and foil his feet. He manages to unzip the side of the painted and stuffed jacket that forms the upper section of the costume to pull it off. It falls stiffly over the edge of the dais where the Reverend stands below, still taunting, "Come on, healer. We're waiting for a miracle!"

Grasshopper looks about uncertainly and listens to the buzzing din about him. He seems enthralled by his image on a distant video screen. Like a cat first encountering a mirror, he moves his head slowly from side to side as if enchanted that the distant visage responds. Then he raises a hand toward the audience, as if in greeting. His lips begin to move in an address that broils in energies of Shakespearean soliloquy. But no sound issues; nothing can be heard across the fuming noises of dejection tossing among the audience.

In the control booth high above, in the roof line recesses, the chief engineer glances at his monitor to confirm that indeed Grasshopper's lips are moving. He peers closely at the garishly painted face and notes that the lips seem indeed forming words. The audio meters registers no signal.

"He's talking . . . where's the damned sound?"

A quick check of the instrument panel shows no malfunctions. His program sheet is cued in red at this point; he must not lose any portion of this.

"Bring two more mikes on line—three and six. Number two may be stuck in a dead space."

Mikes three and six spin obediently into position above, homing in on the distant scene, still to no effect. The chief engineer fumes,

"Bring them all on line—override gain control and get me a signal."

His assistant protests, "That's pretty dicey!"

"I know, I know," the chief waves his program sheet with its border emphatically marked in red in the mutinous assistant's face, "either you do it or you sign off on the responsibility now!" The assistant shrugs and brings eight mikes on line simultaneously, all aimed at the same source in the acoustically live contour generated by the two domes. He flicks open fail-safe housing covering the manual override switch and begins slowly escalating signal amplification.

"Why can't we hear him?" the chief agonizes, watching Grasshopper now seeming truly earnest in addressing the restless crowd beyond. Looking out the control booth he watches the auxiliary microphones scoot quickly into their designated locations, their collared shafts guided by the control computer toward the acoustic center of Grasshopper's zone.

"Take it higher," he commands as perspiration breaks across his forehead. "Dear God, please don't let it come on all of a sudden."

The power meter needles quiver and green channel LED's bounce lightly, edging gradually toward a red band. Then the chief's courage breaks and he shouts,

"That's enough! I don't care! Bring it back down . . . fifty kilowatts in this place . . . nossir! Not me!"

The assistant reaches for the control just as the maundering face on the video screen gathers itself, set its tongue into its lips, puffs its cheeks and emits exultantly flatulent tremors of raspberried indignation, "@#$*@#$!!!!!"

Pseudo-fart escalates logarithmically through the amplifiers and literally explodes from surrounding speakers across the arena and back into the focusing contours of the two resonant domes. It bounds back and forth into a shuddering, whistling wind that is fed from a screaming panic of thousands below. It builds into an oscillating torrent of a million echoes doubled and redoubled in a fraction of a second into the seeming Wrath of God Unleashed.

The entire edifice rumbles in quaking throes of pendulous resonance as in the ominously shuddering control booth audio governors are frantically switched on and the control system manages to generate compensating damping frequencies that cause a gathering acoustic storm to suddenly stop. Overheated circuitry automatically shuts down in electronic self-protection as far below the Reverend Martinsdale screams futile imprecations at a stunned congregation across the hungry and panicked silence. The microphones have now all gone dead.

Reverend Martinsdale realizes that he and Grasshopper share an island from which no sound can issue across so vast a space. Still ablaze in the illumination of spotlights he frantically waves his arms up at the control booth in an attempt to convey that the sound is off and he wants it back on.

The chorus mistakes the gesture for its cue and bursts forth into the exaltation of the *Hallelujah Chorus* of Handel's *Messiah*, the number next scheduled in the program. Following the same program, and taking their cue from the chorus, the stage manager obediently lowers the raised shaft isolating Peter Dooley, who catches his balance and rides it down as the Reverend seems to rise up and meet him. But Reverend Martinsdale has turned and is now waving at the choir in a desperate attempt to hush them up.

Realizing that something has obviously gone wrong and taking it as a sign of Grasshopper's still heady powers, the entire population of Section A pours down the ramps and toward the platform for their share of the healing. A line of attendants lean against the surging wave of diseased and dying bodies, miscalculate the force needed to restrain the ailing from a sure cure, hold momentarily and then break as a tangled mass of wheelchairs, crutches, respirators and stretchers hurtle toward restoration.

Cameras zoom in on the melee, broadcasting their enthusiastic jubilation in mute images throughout the world. *Official Medical Certificates* flutter across the platform as the embattled Reverend screams furiously at the unheeding invalids, kicking them back, pushing and prodding them with a crutch he has snatched up to wield against the stampeding horde.

In the confusion house lights come back on and the remaining sections unleash their occupants in swarms across the stadium floor. Peter Dooley looks out, gasps and flees back along the runway to disappear among the columns beneath the still singing choir.

As the audio circuitry cools down, microphones switch back on and one quickly focuses on the flailing and snarling Reverend.

Hurrying around behind the stage, his friends manage to grab Dooley and swath him in bandages and the extra garments they have brought. In the pandemonium they are unchallenged as they escort him out the back entrance.

Looking back toward the video screens, Jason sees and can hear Reverend Martinsdale shrieking at massed supplicants besieging the dais, "Get back! Get off! Go back to your seats! You're ruining everything! Get back, you goddamned bastard! Go on! Do as I say! He can't heal you, you stupid son-of-a-bitch! Don't you bite me! Get off! Get off!"

Cameras zoom in for close-ups as the Reverend beats at one little old lady with the big end of a metal crutch. The angered crowd now heaves forward from all sectors of the arena to aid her, engulfing the beleaguered Reverend. In desperation he wraps himself for protection in the discarded Grasshopper costume, its padded shell fends off blows as he staggers clumsily backward for sanctuary beneath the choir stall. Private security guards and a phalanx of Brotherhood Deacons pour out of the side doors beneath the choir to head off the mob chasing the Reverend down the runway.

With every available guard summoned to assist in holding back sudden chaos, the small refugee band has little difficulty in leading Dooley out into the mezzanine where they pause for breath. Inside the amphitheater, guards and Deacons fight madly to turn the crowd, now grown into a tumultuous herd rampaging blindly and gradually force it to circle faster and faster about the arena floor until sheer exhaustion quells the riot. Cameras and microphones pan back and forth above the melee, transmitting its unseemly chaos to a global audience.

Finally someone flips a switch to cue the finale, a prerecorded message by President Ira Crown, speaking from the Lebanon Command Post. His face gazes down benignly from video screens upon the still slowly circling herd of men, women and children. His visage seems to ebb back from the injured and dying trampled at the periphery of the now empty dais. As the congregation slows and begins to disperse into fragmented and individual realizations of horror and disgust at the carnage surrounding the platform, across the multiple gigantic screens overhead President Crown smiles magnanimously and distantly from an uncluttered desk flanked by the flags of the nation and of his office.

"My fellow Americans and citizens of the World! I sincerely hope that this evening has not proven too disappointing. In times of distress and anxiety we often choose to believe that which is most remote from common sense.

"Tonight we have witnessed the reaffirmation of one of the oldest and wisest of proverbs—By their works shall ye know them.

"We must continually guard against our own individual credulity. In mankind's past so many peoples and great nations have been led by false belief into oblivions of their own making.

"Our problems are not easy ones. There are no ready solutions. But if we have learned anything from the events of this evening it is that we must accept our individual lots. Accept and confront them in the solid resolution that can never fail to heal even the sorest of wounds. Wounds inflicted by a world only too ready to aid our personal and collective downfall.

"We shall have failed only when we look to others for our defense and salvation. That has been the lesson of the American experience!"

The immense overhead screens all flash white as someone in the control booth recognizes that the Presidential message is not scheduled until after the next choir number. But the choir has dissolved and joined the trickle of people out of the arena exits, past lines of ambulances loading casualties. Up in the control booth, it becomes apparent that the program schedule is now meaningless. The two technicians begin the shutdown sequence, quaffing screens into numb gray, shepherding microphones back into their hutches, dousing lights above the dais, illuminating distant passages of the exits.

The gigantic electric sign above the scoreboard still shines brightly as house lights begin to dim:

HOPE*VIGILANCE*HARDWORK

~ 29 ~

Outside, in the west parking lot, among trailers and recreational vehicles convened for temporary Crusade housing, hordes of stunned, confused and angry faithful mill about, buzzing snatches of prayerful thanksgiving for deliverance voiced amid hysterical ponderings of meanings and portents.

Hector and Grady are lost in scrambling stampedes incited by strategically aimed firecrackers, popping like pistol shots to quickly open a panicked channel of escape. Peter Dooley, standing alone and disregarded in the aftermath of mayhem onstage, is grabbed by his friends and escorted from a storage area behind the choir loft. Hastily swathed in bandages, camouflaged in extra articles of clothing brought to veil his notoriety he is fitfully ushered out through the boiling milieu by Stacy, Kristin, Chloe and Jason moving as a protective phalanx around him.

As they make their way through the crowd, though, toward the distant city street, Dooley pulls the stifling bandages from his face.

"Not yet," Kristin yells at him, "we have to get out of here."

But it's too late. "It's him! Grasshopper! Antichrist!" nearby voices spark the crowd's attention. Shrill outcries hurl would-be stones, "Beelzebub! Get him . . . back into the abyss!" Clawing furies open chanted incantations of purgative tribal rage, "Grasshopper! Grasshopper! Grasshopper!" as they lunge in vindictive riot toward their disavowed bogeyman.

"Where are the police?" Jason yells, scanning desperately about the area for any sign of uniformed personnel or staff, "Somebody help us! Wait! Wait! Stop!"

Prodding elbows and hands knock asunder their protective huddle, reach out in concert to seize and lift the spare-bodied silver-haired quarry who merely stares blank bewilderment back into their fury. Pummeling hands pluck him up and away from his shielding friends to raise him overhead, skittling along like a perplexed platter passed across a ravenously tossing sea of clutching hands. His lean frame awkwardly bounces hand to hand up and over the gathered crowd. A cry of bloody inspiration echoes across the mob, "Over the edge! Over the edge!" Anonymous troupes of flailing arms roust spiteful fantasies to toy and tease delicious terrors of atavistic musing, then fling him over the raised concrete ramp wall, onto pavement forty feet below.

"Peter!" Chloe, Stacy and Kristin all scream in horror, helplessly flailing against the nearer drudges of this collectively roiling beast. They are shoved back, knocked ignominiously onto the pavement.

Finally exhausted, triumphant in having expelled some vicarious token of ancient discontents back into mythic darkness, fury slowly subsides. Mobbing dementia dissolves and resolves to disperse back into sullen clusters of forlorn, shaken and lowly bipeds that skulk away from a shabby detritus and into anxious disavowal of so dreadful a moment, their heads wavering bemused in awe and occasion of misbegotten wonder.

Eventually the four friends fight their way to pavement below where Dooley's motionless body, limp and bleeding, sprawls sickeningly askew at the base of a ramp support column. Beyond, the pressing crowd marches antwardly on, oblivious to their lament. Stacy wails a hideous moan as she falls, then scuttles on hands and knees, to raise, gather and lovingly cradle Peter Dooley's head and shoulders in her lap. Kristin stoops to wrap her arms about them both as Chloe slumps weeping into Jason's arms.

Eventually a young woman with an official badge and radio comes upon the distraught tableau and calls for assistance. Another official, a young man in his twenties arrives shortly after. In their dialogue and ensuing radio traffic Kristin overhears that police are occupied with a murder scene up in arena offices. A Crusade official, Eric Cornrath, has been fatally stabbed in a men's room. Police are looking for a thickly built man with a badly disfigured lip and scarred left cheek seen leaving the area just before Cornrath's body was discovered.

~ 30 ~

Jason and Chloe have hiked over, in part to wish farewell to Hector and Grady before resuming their trek southward toward Chloe's childhood home, but also to follow a hunch bearing upon further development of Jason's AiPAL.

In the front yard, among ancient oaks they find Hellbender who gleefully welcomes the pair. Jason looks at his euPhone and nods expectantly to Chloe, "Yep. Just as I thought. They overlooked him. He's still active."

"Do you really want to do this?" Chloe asks, looking searchingly into his eyes. They have talked about this prospect and the great unknowns it opens for both of them. Recalling blissful grace of the mutually hoppening ken she found in Grasshopper's band, Chloe is more confident of Jason's conjecture than he.

"Yes, I do," Jason nods, then leans down to nuzzle the hound affectionately. "We have the Grasshopper iLink OS merged with the AiPAL. That should let us swim and drink the maxnet, as he did. And the two of us resonate together, through it all." Jason looks back up at Chloe, to smile bravely, "Who knows? Let's see what 'hoppens' . . . if it doesn't work for us or we decide we don't like it, we contact the FBI. They'll come switch us off. Until then, at least, we won't have to worry as much about getting hurt on the way to your folks' place."

Jason and Hellbender rub noses, then noisily slurp tongues in avid commingling of revived canine friendship.

Intense heat kindles slowly along and through Jason's tongue, penetrating backward across his throat, expanding down his esophagus and on into his belly where it explodes like an elixir of hot whiskey to course throughout his body, all the way to tingling extremities of fingers and toes. The AiPAL server running on his euPhone acknowledges exchanged signatures and configuration data to constellate his nbTwin from sparkling myriads of nanobiotic radio strobes. Jason breathes deeply and gasps, "O, wow! Upping."

Chloe smiles coyly when at last he asks her, "From me or Hellbender?"

"From you, of course, silly." She wraps him in embracing arms to kiss deeply and lovingly, then steps back, eyes closed in ecstasy as she murmurs, "One world, one way . . . O, wow!"

The End

LaVergne, TN USA
21 December 2009
167677LV00004B/1/P

9 780578 043197